Plasmonic Optical Fiber Biosensors

Christophe Caucheteur
Médéric Loyez

ARTECH HOUSE
BOSTON | LONDON
artechhouse.com

Library of Congress Cataloging-in-Publication Data
A catalog record for this book is available from the U.S. Library of Congress

British Library Cataloguing in Publication Data
A catalog record for this book is available from the British Library.

ISBN 13: 978-1-63081-971-2

Cover design by Mark Bergeron

685 Canton St.
Norwood, MA

10 9 8 7 6 5 4 3 2 1

Contents

Foreword

Research and development in the physical sciences and engineering have changed in fundamental ways over the last few decades. Barriers are crumbling between disciplines due to the recognition that societal needs for progress in human interactions and safety, including individual and community health as well as the protection of the biosphere, require and benefit from highly integrated solutions from many disciplines. This book, coauthored by an electrical engineer and a biochemist, is a perfect example of such integration of knowledge. Christophe Caucheteur and Médéric Loyez have been at the forefront of many important advances in biochemical sensors based on fiber-optic devices and are thus ideally positioned to write about these topics.

They have produced a comprehensive monograph that covers the fundamentals, the various implementations, and most recent advances in the field of fiber-optic sensors using plasmonic amplification. Since the intended audience includes researchers and students with widely different backgrounds, this challenging task includes brief tutorials on the most relevant aspects in photonics, chemistry or biochemistry, and physics, including references to textbooks for further information. Most importantly, this book provides a detailed description of selected specific examples (many from their own seminal contributions and those of some of their collaborators) to highlight the different approaches utilized in the field. Finally, most chapters end with a large reference list to guide readers in the vast literature on plasmonic fiber sensors and associated technologies. The authors also made an effort to present an overview of commercial developments in the field and to describe how the most recent findings in research may contribute to further practical applications. The book is also beautifully illustrated and organized.

For all these reasons, and because the information provided is currently dispersed and segmented in so many discipline-specific books and scientific journals, I think that this book will prove extremely useful as a single source of information for a wide variety of users, from students to senior researchers in academia and industry.

Professor Jacques Albert
Carleton University, Ottawa
April 2023

Acknowledgments

We wish to thank all of our many internal and external colleagues and collaborators who have contributed to the experimental research projects on plasmonic tilted fiber Bragg grating sensors over the years. Their main advances are summarized in this book.

We wholeheartedly thank Professor Jacques Albert from Carleton University of Ottawa in Canada and Professor Tuan Guo from the Jinan University in China for our long-lasting and fruitful international collaborations. Professor Albert has pioneered the development of plasmonic sensing with gold-coated tilted fiber Bragg gratings and has been an endless source of inspiration since our first contact in 2007.

We also sincerely thank Professor Patrice Mégret from the Electromagnetism and Telecommunication Department and Professor Ruddy Wattiez from the Proteomics and Microbiology Department at the University of Mons. They have supported us since the beginning of our research activities.

Particularly, we thank Hadrien Fasseaux from our research team for his essential and dedicated support on reviewing the physical considerations reported in this book.

We acknowledge the financial support of the F.R.S.-FNRS (Fonds de la recherche scientifique).

CHAPTER 1

Introduction

Plasmonic optical fiber sensors gather different miniaturized configurations usually based on centimeter-long sections of multimode or single-mode optical fibers that are locally modified and surrounded by a thin metal film. They constitute a transposition of commercially available prism configurations that offers easy light injection, remote operation, and *in situ* operation capabilities. While they were emerging almost three decades ago, these lab-on-fiber devices have been the subject of intense research and development efforts over the years. They have now matured to such a degree that they can be considered for versatile practical applications including environmental sensing and medical diagnosis where they can bring a unique added value. The literature on these thematics has flourished into hundreds of publications each year, together with several review papers, book chapters, and books on the subject. The goal of this introductory chapter is essentially to position the book content with respect to other relevant references in the field.

1.1 Rationale: Optical Fibers Versus Kretschmann Prism

An optical fiber is a low-loss cylindrical light waveguide made of at least two concentric layers, the core surrounded by the cladding. Usually, the former has a slightly higher refractive index than the latter so that light can be guided in the core by the mechanism of total internal reflections at the core-cladding interface, as depicted in Figure 1.1. The main role of optical fibers is to transmit digital information in the form of pulsed light signals at a very high speed and with a very small attenuation compared to copper cables. The optical attenuation is minimum at a wavelength of 1,550 nm and is close to 0.2 dB/km, enabling long-haul telecommunications.

Optical fiber sensors are almost as old as the first optical fibers developed for telecommunications. The idea to use optical fibers as sensors certainly occurred as a result of efforts to prevent them from being perturbed by unwanted temperature variations or micro-bending and macro-bending in telecommunication networks. Eminent scientists such as Rayleigh, Raman, Brillouin, and Bragg have given their names to the physical principles (and thereby to the different types of technologies) on which most optical fiber sensing modalities are based. Many optical fiber sensor configurations have matured into commercial products that are very competitive with other sensing technologies, in terms of both performance and cost. Civil engineering structures, industries, aircrafts, pipelines, or even volcanoes and ocean floors are now wired optically to monitor safety, maintenance, or operating parameters.

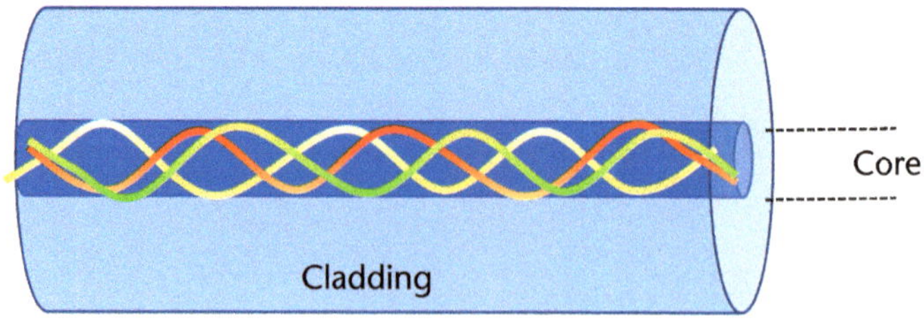

Figure 1.1 Light guidance mechanism in an optical fiber.

Medical applications have also strongly emerged in recent years, essentially for localized pressure and temperature measurements of body fluids.

These applications all benefit from the optical fiber's decisive advantages over electrical sensors. Hundreds of sensors can be multiplexed along a single optical wire (outer diameter of 250 μm with the usual polymer coating) and continuously interrogated in real time from a single end. There is no need for electrical plug-in at the sensor locations, which can therefore be placed in a harsh environment and will not be affected by electromagnetic interferences. These important practical assets and others have spawned a flourishing and rapidly expanding business, which continues to be supported by ever-growing research and development. There are now hundreds of companies worldwide that produce and sell physical (the measurand is either temperature or a mechanical parameter including strain, vibrations, tilt, pressure, and bending) optical fiber sensors.

The expansion of fiber sensor markets in biochemical or electrochemical sensing has been less straightforward, despite the very high volume of research publications per year. These sensors are based on absorption measurements, refractometry, or interferometry. Such applications generally require higher accuracies to be useful, and the measured parameters can be noisy on widely different timescales. When biosensing is involved, achieving the capability to detect molecules at concentrations of the order of picomolar levels or even less should be provided. People active in this field know that this happens to be more sophisticated in practice than measuring temperature variations of 0.1°C or so. The need for functional or sensitizing layers of material around the fibers brings inherent complications in controllable and repeatable fabrication, which has an impact on their reliability and cost. Also, cross-sensitivities to other parameters such as temperature and strain must be addressed, as they can affect the overall signal change and therefore yield a false response. These inherent difficulties require a collective effort to objectively improve this technology and turn these platforms into user-friendly and versatile interfaces. This is precisely what happens now at different levels of integration.

Such complexities are not present in physical optical fiber sensing and certainly explain why practical applications of functionalized optical fiber sensors have been currently much more limited. Strong progress is coming from the labs. In many implementations, optical fiber biochemical sensors have reached very high sensitivity and very low detection limits, which make them prone to compete with other technologies. Cross-sensitivity issues to temperature have also been solved, by separating its influence from the parameters to be measured. This is highly significant, because numerous forthcoming applications will require sensors located in nonstandardized environments, where temperature may impact the sensor's calibration or the chemical processes being monitored.

For biosensing, a rapid and accurate detection of analytes at low concentration (proteins, DNA, pathological markers, toxins) is crucial in numerous fields such as medical diagnosis, environmental monitoring, or quality control in the food industry. In the context of these applications, detection systems can be divided into two general categories: laboratory-based and field-based systems, where field is taken to mean detecting in samples where they happen to be located in contrast to having to bring samples back to a laboratory. A further distinction can be made between direct detection and labeled methods, whereby the latter requires some sort of tags added to the analyte in order to enable its detection. Direct detection methods are generally preferred over labeled approaches from cost and ease of use aspects (and for field use in particular). However, label-free detection is also generally less sensitive because labeling enables the use of additional selection and amplification methods that raise the signal level of very small concentrations over the background response of samples.

Plasmonic sensing remains a prominent example of direct detection method and optical fiber-based configurations hold a large part of the aforementioned sensing progress. Important research efforts on these structures has considerably raised their technology readiness level over the past decade, and results emerging now appear to confirm their status as possible game changers. Plasmonic optical fiber sensors encompass several different configurations that can be considered as transpositions of the Kretschmann prism implementation. The latter remains the gold standard for plasmonic sensing, also referred to as surface plasmon resonance (SPR) sensing, and is the basis of commercially available plasmonic devices that come along with microfluidic systems to bring the analytes (i.e., the chemical species to be detected) in contact with the sensor surface.

In the Kretschmann configuration, light is injected towards the metal-coated face of a glass prism at an angle superior to the critical angle of incidence. An evanescent wave extends in the metal film (whose mean thickness is of the order of 50 nm) as well as in the surrounding dielectric medium. Following the fulfillment of phase-matching conditions, the evanescent wave excites a plasmon wave, a collective excitation of free electrons within the metal layer, at the interface between the metal and the surrounding dielectric medium. This plasmon wave is extremely sensitive to changes of the refractive index of the surrounding medium at the sensor surface and can be used to monitor density fluctuations or thickness measurements. When bioreceptors (i.e., biomolecules that have a selective affinity with target molecules and can bind to them) are grafted on the metal surface, plasmonic substrates can be used for direct (i.e., label-free) biochemical sensing. Upon binding of the target molecules with the anchored bioreceptors, a shift of the surface plasmon resonance can be detected and correlated to analyte detection. The operating principle of this measurement platform is sketched in Figure 1.2.

Optical fibers allow devising miniaturized and integrated counterparts to the otherwise bulky prism configuration. To that aim, light is locally outcoupled from the optical fiber and brought into contact with the surrounding medium at a location where a thin metal film has been deposited over the fiber surface. There are many competing ways to do so, the most straightforward certainly being to bend

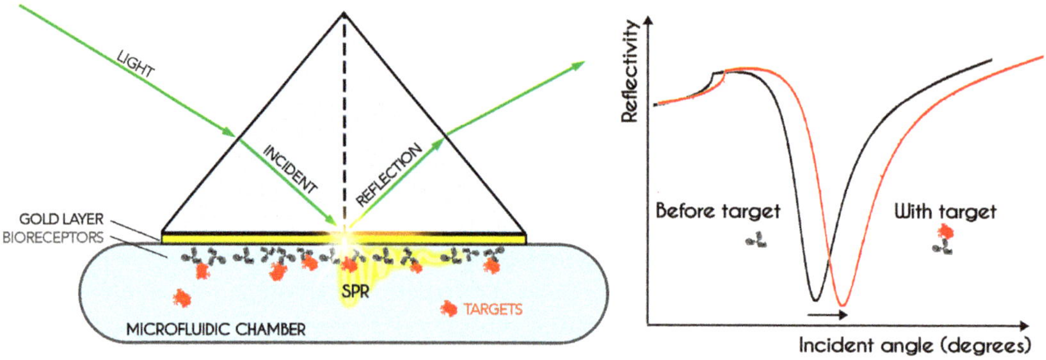

Figure 1.2 Sketch of the Kretschmann prism configuration and principle of detection based on the surface plasmon resonance shift.

the fiber in a U-shaped geometry, so as to bring an evanescent wave in contact with the outer medium. All configurations will be reviewed in this book. The use of optical-fiber devices such as plasmonic sensors presents many well-known desirable features (size, cost, light path control) for label-free detection. They can be inserted into small volumes of a sample either as being used as a handheld probe or as a set of remotely operated devices along a fiber-optic cable (in environmental monitoring applications, for instance). Fiber-based sensing solutions are not yet competitive with bulk optic laboratory instrumentation (like microtiter plate array systems) in applications for pharmaceutical research where a large number of tests need to be performed simultaneously. This may be no longer true in the future as the fiber Bragg grating technology, for instance, comes along with multichannel interrogation devices that have the possibility to process up to 64 channels in parallel. Aside from the pharmaceutical side, there are several industrial applications where fiber sensors' low cost and ease of use could lead to widespread deployment, such as screening of viral or bacterial infections, identifying specific toxins in food processing plants, and monitoring the water quality in urban water supply systems or in the environment surrounding toxic industries and resource extraction operations. In healthcare, plasmonic optical fiber sensors have also the unique potential to be used *in vivo*, certainly when inserted into a specific catheter or in the operating channel of an endoscope.

This important and quite rapid progress has been materialized by thousands of scientific publications accumulated over the years and essentially over the past decade, confirming the growing interest in this field. While fundamental aspects and basic applications have been mainly reported in seminal papers, recent publications have reported more advanced optical fiber configurations for use in relevant biochemical or environmental conditions. This comes along with an integration of the technology and its combined use with, for example, microfluidic systems, smart demodulation processes, and other physical principles. Besides this continually growing reservoir of scientific literature, companies have emerged and are committed to bring to the market their research and development efforts on plasmonic optical fiber sensors.

The purpose of this book is therefore to review these recent important advances and to draw up perspectives to bring even more original research interest and industrial applications in a common effort.

1.2 Positioning of the Book Content

With this book, we wish to provide an overview of the overall progress on plasmonic optical fiber sensing in their numerous fields of applications while bringing pedagogic information for people who are new to plasmonics and optical fiber sensors. This subject is intrinsically highly multidisciplinary and can be tackled by various scientific profiles. Hence, biochemists or biologists will certainly want to learn about how optical fiber configurations operate, how they couple light to the sensitive area, and how their signal can be interpreted in response to the analytes to detect. Considering their scientific background, engineers in optics or photonics will be more interested in information regarding coatings that can be deployed around an optical fiber and the way that receptors can be grafted on these coatings and interact with the species to detect. Other people will certainly require both kinds of information. The book content is therefore adapted to a large audience and will cover the different disciplines that pertain to the plasmonic optical fiber sensing field from an engineering point of view. The book is also organized and written to be complementary with other relevant references (especially those that provide a more physical coverage) on the subject, although background chapters will present strong similarities with already available literature. This content management comes along with the following objectives:

- Bringing an up-to-date and unified vision on all optical fiber configurations used for plasmonic sensing so far, focusing on the practical progress made over the last decade, and making the technology compatible for use in real industrial applications. In particular, the content will focus on the recent and emerging applications of plasmonic optical fiber biosensors such as the use of tilted fiber Bragg gratings, the integration of sensors *in situ*, and the use of smart interrogation techniques. To the best of our knowledge, these diverse applications have not been collectively addressed in a single reference dedicated to the technology.
- Bridging the gap between fundamental aspects and applications of plasmonic optical fiber biosensors and highlighting their main properties as a function of the design parameters (optical fiber configurations, coatings, functionalization). Understanding these key physical properties is of paramount importance for the efficient design of sensing platforms that can meet the target specifications.
- Demonstrating roadmaps for the design process and practical implementation of plasmonic optical fiber sensors. It will help the reader in understanding the role of the fiber configuration and the functional coating in the fundamental properties of the derived optrodes. This will make easier their practical implementation by interested researchers and developers.
- Gathering recent information and technological improvements, showing the maturity reached by the different subsequent technologies and their degree

of maturity to envisage practical implementations. It will also focus on metrological performances for practical applications.

1.3 Content Review

Based on the aforementioned considerations, the book content is more largely described hereafter. We wish to state that it was defined to bring a modern and unified review on recent trends in plasmonic optical fiber sensing from an engineering point of view while providing a historical perspective. The content is also such to minimize any redundancy with previously published books on the subject (that will be referred to the text whenever appropriate), except for the background information that we find relevant to recall for an overall understanding of the technology and in the aim to be pedagogic, as stated in Section 1.2.

Chapter 2 provides the useful physical background to get acquainted with surface plasmons and their use for sensing. This background information is essential to figure out how to achieve surface plasmon excitation with optical fibers.

Considering the bunch of optical fiber-based plasmonic sensing configurations and their huge difference in terms of implementation, spectral features, and relative performance, it was decided to split their presentation into three different chapters. Hence, Chapter 3 describes multimodal optical fiber configurations used for plasmonic sensing, while Chapter 4 summarizes single-mode optical fiber configurations. Chapter 5 targets the use of specialty optical fibers. Each chapter focuses on the typical spectral response obtained in each case and their subsequent metrological performance.

Following a similar organization, Chapters 6, 7, and 8 will be devoted to the presentation of the main strategies used to biofunctionalize plasmonic optical fiber sensors. Chapter 6 is dedicated to the use of antibodies and their derivates as bioreceptors to achieve optical fiber immunosensing. Immobilization strategies and sensing applications are described. Chapter 7 concerns the use of DNA-based or RNA-based molecules as bioreceptors. They can lead to the detection of a wide range of targets such as ions, DNA or RNA, proteins, and cells, among others. These applications and recent achievements are thoroughly covered in this chapter. Chapter 8 aims to describe other bioreceptors such as molecularly imprinted polymers, cells, and enzymes, among many others. Trends and emerging sensing strategies are also highlighted.

Plasmonic optical fiber sensors cover much more applications than biosensing. Chapters 9 and 10 will focus on these recent advances. Chapter 9 then covers all the recent forms of sensors that combine other physical principles along with plasmonics and plasmonics-related sensing features. Chapter 10 is about current and forthcoming developments on the technology towards integrated probes (e.g., point-of-care (POC) or *in situ* sensing).

Finally, Chapter 11 draws the general conclusions of the book, confirming the rise in technology-readiness level that accompanies the research and development on plasmonic optical fiber biosensing platforms.

To end this chapter, we find it relevant to provide very basic information for nontrained people who wish to understand typical practical considerations for the development of plasmonic optical fiber sensors.

1.4 Practical Considerations

This section refers to tips and tricks for the handling of both optical fibers and biochemical reagents and covers the needed basics to safely manipulate both fiber and biochemical approaches.

Getting started with optical fibers is not necessarily easy for anyone. So fragile at first sight but so robust finally, nontrained people will need to learn how to strip, cleave, splice, and practice to handle optical fibers and connectorize them. This step is essential to link the developed sensors with the read-out devices (optical source and detector) in order to optimize the signal and reduce the noise level. Figure 1.3 summarizes the main consumables or small devices that are required for optical fiber handling. Stripping a standard optical fiber is usually performed with a mechanical stripper. This is the first step of the splicing process. The fiber is then cleaved at the right angle using a dedicated ultra-precise device called a cleaver that uses a diamond wheel translated transversely to the optical fiber axis to cut it. The cleaved fiber is then inserted into a splicer, close to its central pair of electrodes, where an electric arc will happen and fuse both fibers to a single wire.

Note that it is important to maintain a clean and tidy work environment, especially to avoid dust and fiber residues. Optical fibers are so thin that, while they break, they can easily fall to the ground or stay on clothes. In addition, the quality of a splice or a connection requires rigor. Thoroughly cleaning the work surface as well as the optical fiber with clean wipes and acetone or isopropanol is mandatory. It is sometimes useful to check the end facet of the fiber using a dedicated microscope such as a fiberscope to ensure high coupling quality. We recommend using a connector cleaner and performing regular maintenance on cleavers and splicers to avoid any drop in quality. Unexpected drops of signal can sometimes be observed experimentally due to defective pigtails or patch cords. The use of a visible (green or red) laser can identify a possible break or loss along the fibers. Be careful about the storage of optical fiber samples and their identification. We recommend the use of dedicated holders enclosed in protective environments such as large petri dishes.

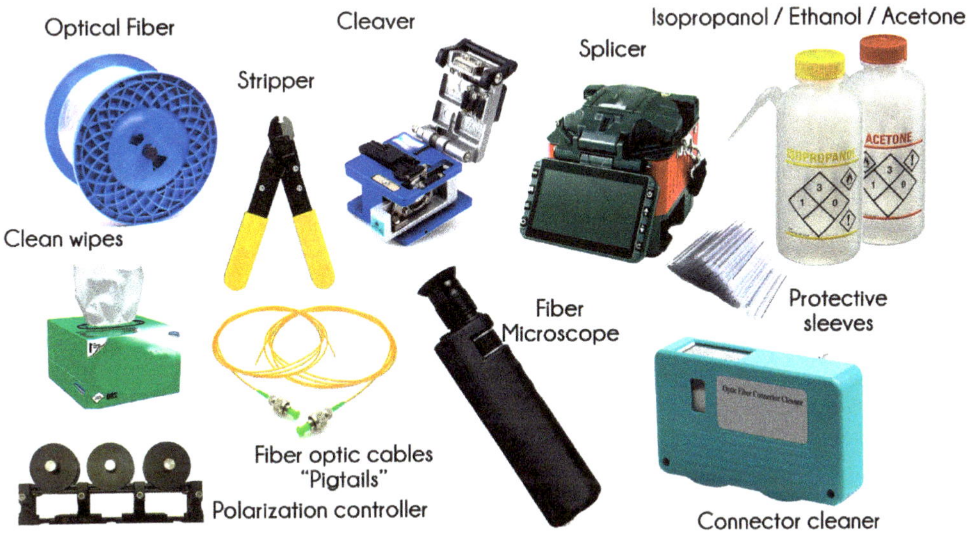

Figure 1.3 Getting started kit for optical fiber-based sensing.

Finally, as many types of optical fiber-based probes are sensitive to polarization and temperature changes, it is therefore essential to work in a controlled environment and to secure all the fibers with adhesive tape.

An important thing to mention, which may sound naïve for trained hands but is very important for novices, is the type of connection required. To make it simple, many optical fibers exist (e.g., single mode or multimode) and need to be correctly connected. There is a huge difference between fiber connectors, namely, angled physical contact (APC), physical contact (PC), or ultraphysical contact (UPC). These names refer to the polish style of the ferrules (often made of ceramic or stainless steel) inside the fiber connectors (Figure 1.4). The ferrule maintains the fiber in the center of the connector and is designed to connect with other ends of fibers, usually thanks to dedicated metallic connectors or specific adapters. PC connectors are referring to physical contact connectors and were the most commonly used connectors for multimode fibers. They were originally used to reduce the air gap between two fibers and therefore reduce the connection losses. Nowadays, this polishing is replaced by UPC connectors for ultraphysical contact. UPC can be considered as an improved version of PC connectors with higher efficiency (lower losses), but can be easily damaged by repeated cycles of connections and disconnections. UPC can also be distinguished from PC as it presents a slight curvature, while PC connectors are flatter.

Aside from these types of fiber ends, APC connectors refer to angled physical contact. These connectors present a polishing at a typical angle of 8° aiming at minimizing the back reflection. As a result, the reflected light leaks out into the cladding due to the induced angle, instead of traveling back into the core of the fiber. To avoid any damage, it is mandatory to connect the APC connectors with

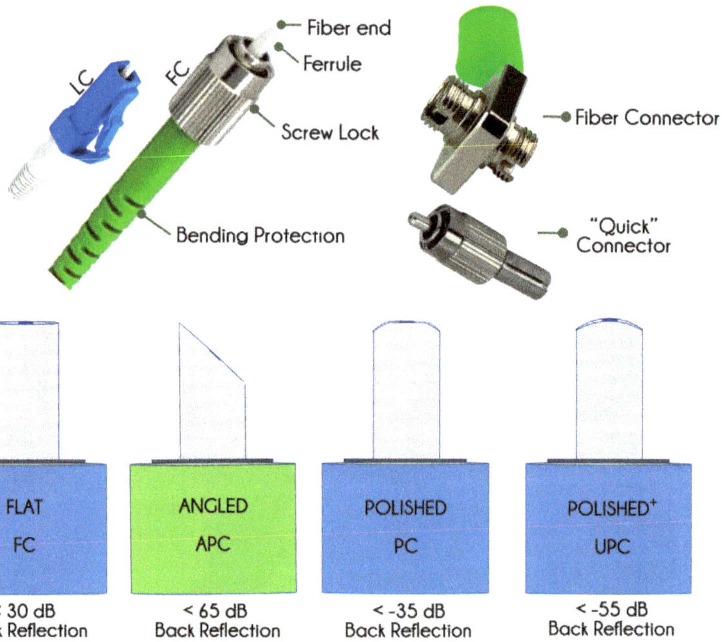

Figure 1.4 Types of optical fiber connectors.

other APC connectors, as their use with flat connectors could provoke high losses or damage the fiber end. The typical return loss of APC connectors is around 60 dB. Flat connectors (FCs) are usually designed for single-mode fibers or polarization-maintaining fibers and are exploited in telecommunications or measurement equipment. Be careful not to mistake it, because the nomenclature often uses FC for fiber connectors and for flat connectors. Therefore, an FC/APC connector means: fiber connector type APC. Some instruments require other types of connections such as LC or SC (LC stands for lucent connector, while SC is subscriber connector or standard connector). Generally, the standard color code for APC connectors is green, while flat ones are black or blue. These miniaturized connectors are usually simply connected by a click connection. Finally, temporary connectors, also called quick connectors, can be used instead of fiber patch cords or pigtails to directly connect a piece of fiber to an instrument or a secondary connector.

Once the optical fiber probes are ready, scientists often need to be inventive to find the appropriate containers and carry out the chemical or biological reactions. This is especially true when it comes to use fibers in transmission and to immerse a central portion of the fiber in a small volume. The required material must therefore be adapted and various containers such as pipette tips closed with parafilm or 3-D printed cassettes must sometimes be designed. For novices, a very important notion is the preservation of the samples. Often mentioned on a dedicated data sheet, it is important to respect the conditioning temperatures and avoid freeze and thaw cycles, especially with biomaterials. The dilutions, homogenization, and handling of biosamples are very important in biosensing. The use of a vortex, centrifuge, micro-pipettes, and basic knowledge in biochemical laboratory in terms of equipment and safety must be mastered (Figure 1.5). We recommend the use of a dedicated laboratory book and the tracking of samples using a specific notation. To improve data reproducibility, the use of timers and experimental plannings are obvious.

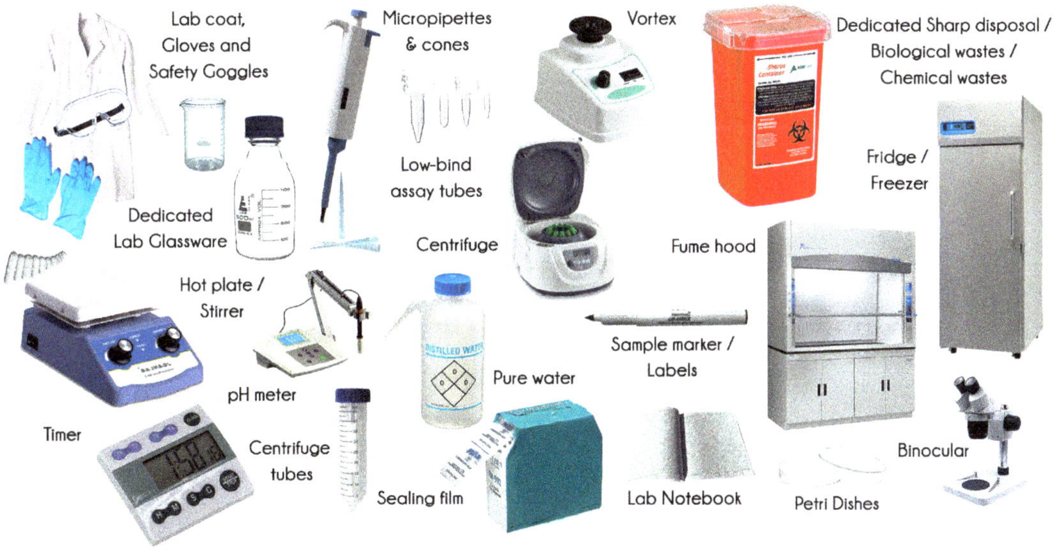

Figure 1.5 Getting started kit for biochemical sensing.

CHAPTER 2

Physical Concepts of Surface Plasmon Sensing

In metals, there are particular waves called plasma waves that correspond to an oscillation of the charge density. These waves have a longitudinal structure such that their wave vector is parallel to the electric field. They therefore cannot be generated optically given the transverse structure of the light wave. However, it is possible to remove this constraint at the interface between a metal and a dielectric, provided that an evanescent wave with a longitudinal component is generated at this interface. The mixed light/plasma oscillation mode thus generated constitutes the plasmon. In practice, the coupling between the plasma wave and the light wave is only possible by matching the phase velocities of both waves. This condition is obtained when the wave vectors are identical along the interface. A simple and efficient way to generate an evanescent wave suitable for coupling with the plasmon is to work in total internal reflection in a prism, one face of which is covered with a thin metal layer. When the incident light is coupled with the plasmon wave, there is no more reflected light since the light energy transferred to the plasmon is dissipated in the metal. This dissipation is related to the imaginary part of the dielectric constant of the metal and results in a certain resonance width. Because surface plasmon waves are very sensitive to changes in the refractive index of the external dielectric medium, they are naturally exploited to perform fine refractometry. Main applications include the measurement of dielectric constants of metals, the production of chemical and biochemical sensors, and spectroscopy. This physical principle has been extensively studied and documented. Many books have detailed the principle of operation of surface plasmon waves and their use. The objective of this chapter is to introduce the reader to the physical principle underlying the generation of surface plasmons so that it can be understood in the following chapters how to achieve that in practice with optical fibers.

2.1 Mathematical Formalism

The first documented observation of surface plasmon waves happened in 1902. It results from the work of Wood when he illuminated a metal diffraction grating with white (or polychromatic) light and found anomalies in the diffraction spectrum, taking the form of narrow dark bands [1]. A few decades later, the theoretical work of Fano led to the conclusion that these anomalies were associated with the excitation of surface electromagnetic waves on the edge of the diffraction grating [2]. In 1952,

Bohm and Pines introduced electron plasma in a metal solid to explain the energy losses due to fast electrons passing through a metal sheet. They called this excitation of a plasmon. Today, this type of excitation is rather called volume plasmon, to distinguish it from surface plasmons. Although surface electromagnetic waves were first discussed by Sommerfeld and Zenneck, Ritchie was the first to introduce the term "surface plasmons" in 1957 by extending the work of Bohm and Pines at metal interfaces [3]. In 1958, Turbadar observed a strong drop in reflectivity by illuminating a thin metal film deposited on a substrate [4]. However, he did not link this effect with surface plasmons. It was in 1968 that Otto explained the results of Turbadar by demonstrating that the loss of reflectivity was due to the excitation of surface plasmons [5]. The same year, Kretschmann and Raether demonstrated the excitation of plasmons on the surface of an illuminated prism at a critical angle to obtain total internal reflection [6]. This pioneering work has established a simple and effective method of excitation of surface plasmons, allowing them to be exploited for different purposes [7]. At the end of the 1970s, they were notably used for the characterization of thin films [8] and the study of chemical processes at the metal interface. They are now used for the production of sensors and biosensors [9], as well as for spectroscopy [10].

There are other terminologies for surface plasmons (SPs) in the literature. The term "polaritons" (surface plasmon polaritons (SPPs)) is commonly used and has the advantage of emphasizing the link between the electronic excitation in a solid that constitutes a plasmon and its associated electromagnetic field. The terms surface plasma waves (SPWs), surface plasma oscillations (SPOs), and surface electromagnetic waves (SEWs) are also encountered sometimes. Going forward, we will principally use SPs.

There are also several definitions for surface plasmons, most of which are incomplete. The suffix "on" indicates a connection to quantum mechanics and testifies that surface plasmons have properties of particles, thus including specific energies. A surface plasmon can therefore be defined as an excitation occurring at the interface between a first medium with negative permittivity and free charge carriers (a metal) and a second medium with positive permittivity. This excitation is a collective oscillation of surface charges and behaves like a particle exhibiting discrete energy. That said, the majority of the properties of surface plasmons can be derived from classical electromagnetic equations, as it will be summarized hereafter. It is indeed a fundamental electromagnetic mode at the interface between a material of negative permittivity and another of opposite permittivity, presenting a determined frequency and involving an electronic oscillation of surface charges. The question then arises of whether or not to use the classical description. If the materials supporting the surface plasmons are large enough to be described by a dielectric function (permittivity), the classical electromagnetic approach can be employed satisfactorily. In practice, it has been shown that a dielectric constant precisely describes media whose minimum dimensions are around 10 nm. For media whose sizes are between 1 and 2 nm, a size-dependent dielectric constant is used [11]. For a detailed discussion about size effects in small metallic aggregates, the reader is invited to consult [12, 13].

The mathematical formalism used to describe surface plasmons rely on electromagnetic wave propagation equations, also known as Maxwell equations. This formalism is detailed hereafter in the case of a metal-dielectric interface. For double interfaces, which are a little more sophisticated, the reader is invited to consult [14, 15].

2.1.1 Maxwell Equations

The electromagnetic field in vacuum is represented in terms of two vectors, the electric field $\boldsymbol{E}$ and the magnetic field $\boldsymbol{B}$. The presence of matter in the space occupied by these vectors requires the introduction of three other vectors, $\boldsymbol{D}$, $\boldsymbol{H}$, and $\boldsymbol{J}$, respectively, called electric displacement, magnetic field in matter, and electric current density. These vectors, whose components are expressed in an orthonormal Cartesian coordinate system (x, y, z), are linked by the following equations (two scalar equations and two vector equations)

$$\nabla \cdot D = \rho \tag{2.1}$$

$$\nabla \cdot B = 0 \tag{2.2}$$

$$\nabla \times E = -\frac{\partial B}{\partial t} \tag{2.3}$$

$$\nabla \times H = J + \frac{\partial D}{\partial t} \tag{2.4}$$

where ρ is the density of free electrical charges.

2.1.2 Constitutive Equations

The following relationships can be established

$$D = \varepsilon_0 \varepsilon_r E \tag{2.5}$$

$$B = \mu_0 \mu_r H \tag{2.6}$$

$$J = \sigma E \tag{2.7}$$

with ε_0 being the vacuum permittivity and ε_r the relative permittivity (also called the dielectric function). μ_0 is the vacuum permeability while μ_r is the relative permeability. σ is the conductivity.

2.1.3 Boundary Conditions

At the interface between two media where a sharp change in the relative permittivity occurs, a set of four boundary conditions have to be fulfilled. Let $\boldsymbol{N}_{12}$ be a unit vector pointing from a medium 1 to a medium 2 perpendicular to an infinitesimal section of the interface between these two media. The vector equations (2.8) and (2.9) express that the tangential component of the electric field is continuous across the section while the tangential component of the magnetic field is equal to the electric current density across the interface

$$\mathbf{N}_{12} \times \left(\boldsymbol{E}_2 - \boldsymbol{E}_1\right) = 0 \tag{2.8}$$

$$\mathbf{N}_{12} \times \left(\boldsymbol{H}_2 - \boldsymbol{H}_1\right) = \boldsymbol{J} \tag{2.9}$$

where subscripts 1 and 2 refer to each of the media.

Two other scalar relationships express that the normal component of the electric displacement is equal to the charge density at the surface while the normal component of the magnetic field remains continuous.

$$\mathbf{N}_{12} \cdot \left(\boldsymbol{D}_2 - \boldsymbol{D}_1\right) = \rho \tag{2.10}$$

$$\mathbf{N}_{12} \cdot \left(\boldsymbol{B}_2 - \boldsymbol{B}_1\right) = 0 \tag{2.11}$$

2.1.4 Wave Equations

To solve Maxwell equations in the case of surface plasmons, we consider two semi-infinite, homogeneous, linear, and isotropic media. They are nonmagnetic ($\mu_r = 1$) and in the absence of surface current ($\boldsymbol{J} = 0$) and surface charges ($\rho = 0$). One of them is a dielectric while the other is a metal. The waveguide thus considered is sketched in Figure 2.1. The metal has a complex permittivity ε_m, while the dielectric has a real permittivity ε_d. They have refractive indexes of n_m and n_d, respectively. The zOy plane of the configuration shown in Figure 2.1 corresponds to the interface. The surface plasmon propagates in the z-direction and the system is therefore invariant along the y-axis. Taking these conditions and the constitutive equations into account, Maxwell equations (2.1) to (2.4) become

$$\nabla \cdot \left(\varepsilon_0 \varepsilon_r \boldsymbol{E}\right) = 0 \tag{2.12}$$

$$\nabla \cdot \left(\mu_0 \boldsymbol{H}\right) = 0 \tag{2.13}$$

$$\nabla \times \boldsymbol{E} + \mu_0 \frac{\partial \boldsymbol{H}}{\partial t} = 0 \tag{2.14}$$

$$\nabla \times \boldsymbol{H} - \varepsilon_0 \varepsilon_r \frac{\partial \boldsymbol{E}}{\partial t} = 0 \tag{2.15}$$

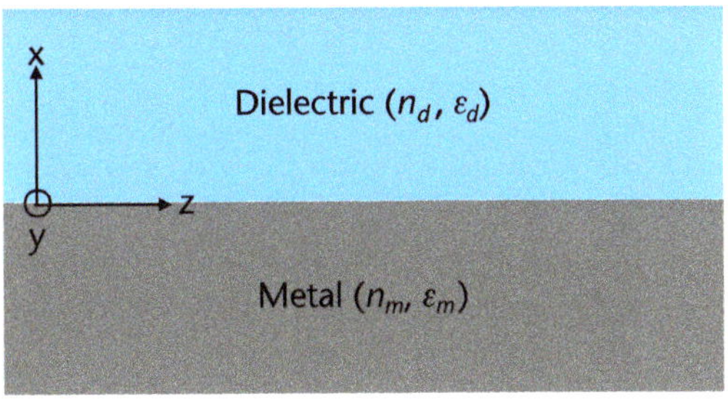

Figure 2.1 Schematic representation of the metal-dielectric interface considered for the study of surface plasmons (not to scale).

The fields propagating in the zOy plane are solutions of Maxwell equations that decay exponentially on both sides of the interface. The wave equations are obtained from relationships (2.5), (2.6), (2.14), and (2.15) and are written as

$$\nabla \times \nabla \times \boldsymbol{E} = -\mu_0 \frac{\partial^2 \boldsymbol{D}}{\partial t^2} \tag{2.16}$$

$$\nabla \times \nabla \times \boldsymbol{H} = -\varepsilon_0 \varepsilon_r \frac{\partial^2 \boldsymbol{B}}{\partial t^2} \tag{2.17}$$

Since $\nabla \times \nabla \times E \equiv \nabla(\nabla \cdot E) - \nabla^2 E$ and $\nabla \cdot (\varepsilon E) \equiv \boldsymbol{E} \cdot \nabla\varepsilon + \varepsilon\nabla \cdot \boldsymbol{E}$, considering homogeneity, (2.16) and (2.17) turn into:

$$\nabla^2 \boldsymbol{E} - \frac{\varepsilon_r}{c^2}\frac{\partial^2 \boldsymbol{E}}{\partial t^2} = 0 \tag{2.18}$$

$$\nabla^2 \boldsymbol{H} - \frac{\varepsilon_r}{c^2}\frac{\partial^2 \boldsymbol{H}}{\partial t^2} = 0 \tag{2.19}$$

with the speed of light in vacuum $c = \sqrt{1/\mu_0\varepsilon_0}$.

Let's now assume a harmonic temporal-dependent solution such as

$$\boldsymbol{F}(\boldsymbol{r},t) = \boldsymbol{F}_0 e^{i(\omega t - \boldsymbol{k}_0 \boldsymbol{r})} \tag{2.20}$$

where $\omega = 2\pi\nu$ and t are the angular frequency and the time, respectively, and $\boldsymbol{F}$ represents the electric or magnetic field. $\boldsymbol{F}_0$ is not time-dependent but depends on the position. $\boldsymbol{r}$ is a vector of position, while $\boldsymbol{k}_0$ is the complex wavevector perpendicular to the plane of constant phase of the propagating field. In this case, (2.18) and (2.19) lead to Helmholtz's equations

$$\nabla^2 \boldsymbol{E} + k_0^2 \varepsilon_{r,j} \boldsymbol{E} = 0 \tag{2.21}$$

$$\nabla^2 \boldsymbol{H} + k_0^2 \varepsilon_{r,j} \boldsymbol{H} = 0 \tag{2.22}$$

where $k_0 = \omega/c$ and j identifies the metal (m) or the dielectric (d). As the considered system does not vary along y, $\nabla^2 \boldsymbol{E} = \partial^2 \boldsymbol{E}/\partial x^2 + \partial^2 \boldsymbol{E}/\partial z^2$. In addition, as the wave propagates along the z-direction but remains invariable in the y-direction, the solution of (2.21) can be expressed by $\boldsymbol{E}(x, y, z) = \boldsymbol{E}(x)\exp(-i\beta z)$. Inserting this expression in (2.21) allows obtaining the following wave equation

$$\frac{\partial^2 \boldsymbol{E}(x)}{\partial x^2} + \left(k_0^2 \varepsilon_{r,j} - \beta^2\right) \boldsymbol{E}(x) = 0 \tag{2.23}$$

Similarly for the magnetic field, the expression $\boldsymbol{H}(x, y, z) = \boldsymbol{H}(x)\exp(-i\beta z)$ can be inserted in (2.22) to yield

$$\frac{\partial^2 \boldsymbol{H}(x)}{\partial x^2} + \left(k_0^2 \varepsilon_{r,j} - \beta^2\right) \boldsymbol{H}(x) = 0 \tag{2.24}$$

Equations (2.23) and (2.24) are the starting point for an analysis of guided modes in an electromagnetic waveguide. A thorough discussion of their properties and applications can be obtained in [18]. To use the wave equations to determine the spatial profile of the field, it is necessary to find explicit expressions of the different components of $\boldsymbol{E}$ and $\boldsymbol{H}$. This is possible from (2.14) and (2.15). Considering a harmonic time dependence ($\partial/\partial t = i\omega$), the following set of coupled equations can be obtained

$$\frac{\partial E_z}{\partial y} - \frac{\partial E_y}{\partial z} = -i\omega\mu_0 H_x \tag{2.25}$$

$$\frac{\partial E_x}{\partial z} - \frac{\partial E_z}{\partial x} = -i\omega\mu_0 H_y \tag{2.26}$$

$$\frac{\partial E_y}{\partial x} - \frac{\partial E_x}{\partial y} = -i\omega\mu_0 H_z \tag{2.27}$$

$$\frac{\partial H_z}{\partial y} - \frac{\partial H_y}{\partial z} = i\omega\varepsilon_0\varepsilon_r E_x \tag{2.28}$$

$$\frac{\partial H_x}{\partial z} - \frac{\partial H_z}{\partial x} = i\omega\varepsilon_0\varepsilon_r E_y \tag{2.29}$$

$$\frac{\partial H_y}{\partial x} - \frac{\partial H_x}{\partial y} = i\omega\varepsilon_0\varepsilon_r E_z \tag{2.30}$$

The propagation in the z-direction ($\partial/\partial z = -i\beta$) and the invariability in the y direction ($\partial/\partial y = 0$) yield the following simplifications

$$\beta E_y = -\omega\mu_0 H_x \tag{2.31}$$

$$i\beta E_x + \frac{\partial E_z}{\partial x} = i\omega\mu_0 H_y \tag{2.32}$$

$$\frac{\partial E_y}{\partial x} = -i\omega\mu_0 H_z \tag{2.33}$$

$$\beta H_y = \omega\varepsilon_0\varepsilon_{r,j} E_x \tag{2.34}$$

$$-i\beta H_x - \frac{\partial H_z}{\partial x} = i\omega\varepsilon_0\varepsilon_{r,j} E_y \tag{2.35}$$

$$\frac{\partial H_y}{\partial x} = i\omega\varepsilon_0\varepsilon_{r,j} E_z \tag{2.36}$$

This system can be separated into two sets of independent equations. The first set is composed by transverse magnetic (TM) modes, also referred to as P-polarized, for which $H_z = 0$ (and thus $E_y = 0$ and $H_x = 0$). In this case, the system of equations is reduced to the following

$$E_x = \frac{\beta}{\omega\varepsilon_0\varepsilon_{r,j}} H_y \tag{2.37}$$

$$E_z = -i\frac{1}{\omega\varepsilon_0\varepsilon_{r,j}}\frac{\partial H_y}{\partial x} \tag{2.38}$$

$$\frac{\partial^2 H_y}{\partial x^2} + \left(k_0^2\varepsilon_{r,j} - \beta^2\right) H_y = 0 \tag{2.39}$$

Thus, the knowledge of the H_y component is sufficient to find the electromagnetic field solution for TM modes. The second set of solutions is composed of transverse electric (TE) modes, also referred to as S-polarized, for which $E_z = 0$ (and thus $H_y = 0$ and $E_x = 0$). The following system is therefore obtained

$$H_x = -\frac{\beta}{\omega\mu_0} E_y \tag{2.40}$$

$$H_z = i\frac{1}{\omega\mu_0}\frac{\partial E_y}{\partial x} \tag{2.41}$$

$$\frac{\partial^2 E_y}{\partial x^2} + \left(k_0^2\varepsilon_{r,j} - \beta^2\right)E_y = 0 \tag{2.42}$$

So E_y is sufficient to find the electromagnetic field solution for TE modes. The wave equations (2.39) and (2.42) allow studying plasmons confined to an interface.

2.1.5 Plasmons at a Single Interface

Let's further consider the geometry sketched in Figure 2.1 and start with the solutions of the TM modes. Let's assume for $x > 0$ (dielectric medium) $H_y(x) = A_d\exp(-\gamma_d x) \equiv H_{d,y}(x)$ where A_d is the amplitude and γ_d is a positive attenuation constant. The minus sign expresses energy conservation avoiding infinite field amplitude at an infinite distance from the interface. Inserting this solution into (2.37) to (2.39), one can extract the whole components of $\boldsymbol{H}(x)$ and $\boldsymbol{E}(x)$ (see (2.43) to (2.45)). For $x < 0$ (in the metal), the solutions are assumed to be $H_y(x) = A_m\exp(+\gamma_m x) \equiv H_{m,y}(x)$ where A_m is the amplitude and γ_m is a positive attenuation constant. This yields:

For $x > 0$:

$$H_{d,y}(x) = A_d\exp(-\gamma_d x) \tag{2.43}$$

$$E_{d,z}(x) = iA_d\frac{1}{\omega\varepsilon_0\varepsilon_d}\gamma_d\exp(-\gamma_d x) \tag{2.44}$$

$$E_{d,x}(x) = A_d\frac{\beta}{\omega\varepsilon_0\varepsilon_d}\exp(-\gamma_d x) \tag{2.45}$$

For $x < 0$:

$$H_{m,y}(x) = A_m\exp(\gamma_m x) \tag{2.46}$$

$$E_{m,z}(x) = -iA_m\frac{1}{\omega\varepsilon_0\varepsilon_m}\gamma_m\exp(\gamma_m x) \tag{2.47}$$

$$E_{m,x}(x) = A_m\frac{\beta}{\omega\varepsilon_0\varepsilon_m}\exp(\gamma_m x) \tag{2.48}$$

where $\gamma_{d(m)}^2 = \beta^2 - k_0^2\varepsilon_{r,d(m)}$. As explained above, the boundary conditions ((2.8) and (2.9)) impose continuity of the parallel components to the interface. These equations can be rewritten for TM modes as $H_{d,y}(x = 0, y, z) = H_{m,y}(x = 0, y, z)$ and $E_{d,z}(x = 0, y, z) = E_{m,z}(x = 0, y, z)$ with $H_{d(m),y}(x, y, z) = H_{d(m),y}(x)\exp(-i\beta z)$ and $E_{d(m),z}(x, y, z) = E_{d(m),z}(x)\exp(-i\beta z)$. This leads to the conditions

$$A_m = A_d \tag{2.49}$$

$$\frac{\gamma_m}{\varepsilon_m} = -\frac{\gamma_d}{\varepsilon_d} \tag{2.50}$$

If we first consider a perfect metal without absorption, its complex permittivity reduces to its real part. In this case, since $\gamma_{d(m)} > 0$, the condition (2.50) imposes that the real parts of the permittivities are of opposite sign. This is the case of the system considered so far, comprising a conductor and an insulator. The dispersion relation of a surface plasmon is obtained using the expression of $\gamma^2_{d(m)}$ and (2.50). Indeed,

$$\gamma_d^2 = \beta^2 - k_0^2\varepsilon_d = \gamma_d\gamma_d = \gamma_d\left(-\frac{\gamma_m}{\epsilon_m}\epsilon_d\right) \tag{2.51}$$

and

$$\gamma_m^2 = \beta^2 - k_0^2\varepsilon_m = \gamma_m\gamma_m = \gamma_m\left(-\frac{\gamma_d}{\epsilon_d}\epsilon_m\right) \tag{2.52}$$

So

$$\gamma_m\gamma_d = \frac{\epsilon_m}{\epsilon_d}\left(k_0^2\epsilon_d - \beta^2\right) = \frac{\epsilon_d}{\epsilon_m}\left(k_0^2\epsilon_m - \beta^2\right) \tag{2.53}$$

The latter equality leads to

$$\beta = k_0\sqrt{\frac{\epsilon_m\epsilon_d}{\epsilon_m + \epsilon_d}} = \frac{\omega}{c}\sqrt{\frac{\epsilon_m\epsilon_d}{\epsilon_m + \epsilon_d}} \tag{2.54}$$

Surface plasmons are all guided modes that satisfy the eigenvalue equation (2.54). If we now consider a dissipative metal characterized by a nonzero imaginary part, the surface plasmon propagation constant is complex and given by

$$\beta = \beta' + i\beta'' = \frac{\omega}{c}\sqrt{\frac{\epsilon'_m\epsilon_d}{\epsilon'_m + \epsilon_d}} + i\frac{\epsilon''_m}{2\left(\epsilon'_m\right)^2}\frac{\omega}{c}\left(\frac{\epsilon_d\epsilon'_m}{\epsilon_d + \epsilon'_m}\right)^{\frac{3}{2}} \tag{2.55}$$

where notations ′ and ″ refer to the real and imaginary parts, respectively. Since $\beta' \in \mathbb{R}$ and $\epsilon'_m < 0$, it is necessary that $|\epsilon'_m| > \epsilon_d$.

As expressed by (2.43) to (2.48), the amplitude of the transversal field (perpendicular to the interface) exponentially decreased with the distance. We can define the penetration depth in the medium j corresponding to the distance where the field amplitude drops to $1/e$ such as

$$L_{x_j} = \frac{1}{\gamma_j} \tag{2.56}$$

Since $\gamma_j^2 = \beta^2 - k_0^2\varepsilon_{r,j}$, (2.54) and (2.56) lead to

$$L_{x_j} = \frac{\lambda}{2\pi}\left(-\frac{\epsilon_m' + \epsilon_j}{\epsilon_j^2}\right)^{1/2} \tag{2.57}$$

Using (2.45) to (2.48) and (2.54) and the definition of γ_j, we have

$$\left|\frac{E_{m,x}(x)}{E_{m,z}(x)}\right| = \left(\frac{\epsilon_d}{-\epsilon_m}\right)^{1/2} \tag{2.58}$$

and

$$\left|\frac{E_{d,x}(x)}{E_{d,z}(x)}\right| = \left(\frac{-\epsilon_m}{\epsilon_d}\right)^{1/2} \tag{2.59}$$

Since $|\epsilon_m'| > \epsilon_d$, the transverse (perpendicular to the interface) part of the field extended in the dielectric is higher than the transverse part of the field extended in the metal. The limit case where $-\epsilon_m \rightarrow \epsilon_d$ is the one of a pure surface wave, as we will see later.

Concerning the TE modes. Let's assume as previously that $E_y(x > 0) = B_d\exp(-\gamma_d x) \equiv E_{d,y}(x)$ and $E_y(x < 0) = B_m\exp(-\gamma_m x) \equiv E_{m,y}(x)$. Using (2.40) and (2.41), we have:

For $x > 0$:

$$E_{d,y}(x) = B_d\exp(-\gamma_d x) \tag{2.60}$$

$$H_{d,z}(x) = -iB_d\frac{1}{\omega\mu_0}\gamma_d\exp(-\gamma_d x) \tag{2.61}$$

$$H_{d,x}(x) = -B_d\frac{\beta}{\omega\mu_0}\exp(-\gamma_d x) \tag{2.62}$$

For $x < 0$:

$$E_{m,y}(x) = B_m \exp(\gamma_m x) \tag{2.63}$$

$$H_{m,z}(x) = iB_m \frac{1}{\omega\mu_0}\gamma_m \exp(\gamma_m x) \tag{2.64}$$

$$H_{m,x}(x) = -B_m \frac{\beta}{\omega\mu_0}\exp(\gamma_m x) \tag{2.65}$$

The boundary conditions (2.8) and (2.9) impose, respectively, the following relationships:

$$B_d = B_m \equiv B \tag{2.66}$$

$$B(\gamma_m + \gamma_d) = 0 \tag{2.67}$$

The solution of (2.67) implies that $B = 0$. Consequently, surface plasmons only exist for TM modes.

Based on these developments, the conditions to be respected to generate surface plasmons are therefore:

- Surface plasmons are described by the dispersion relationship (2.54).
- $\varepsilon'_m < 0$. This condition is verified by metals such as gold and silver.
- $\varepsilon_d < |\varepsilon'_m|$. This condition is verified with a metal-dielectric interface.
- Surface plasmons exist exclusively for the TM polarization state.

Properties of surface plasmons can now be examined based on their dispersion relationship.

2.2 Excitation of Surface Plasmons

As we will see in the next section, the most widespread approach for the excitation of surface plasmons consists in the exploitation of the attenuated total reflection when light is sent, at an incident angle larger than the critical one, through prism covered with a nanoscale layer of a noble metal (usually gold or silver). The evanescent field excites the surface plasmon when the component of its propagation constant parallel (or tangential) to the metal surface is equal to that of the plasmon. In practice, this condition implies to verify the following relationship

$$\beta_z = \beta_{SP} \tag{2.68}$$

where β_z is the tangential component of the propagation constant associated with the evanescent wave ($\beta = \sqrt{\beta_x^2 + \beta_z^2}$) and β_{SP} is the one of the plasmon wave. Figure 2.2 depicts a scheme of the considered system where the external dielectric medium is air. At the metal-prism interface, the tangential component is $\beta_z = \beta_p \sin\delta$ with δ as the incident angle of light in the prism, (2.68) turns into:

$$\beta_p \sin\delta = \beta_{SP} \tag{2.69}$$

From Maxwell equations, it is well known that the dispersion relations of a photon propagating through a medium like the prism or air are, respectively, given by

$$\beta_p = \frac{\omega}{c}\sqrt{\varepsilon_{\text{prism}}} = \frac{\omega}{c} n_{\text{prism}} \tag{2.70}$$

and

$$\beta_a = \frac{\omega}{c}\sqrt{\varepsilon_{\text{air}}} = \frac{\omega}{c} n_{\text{air}} \tag{2.71}$$

where $\boldsymbol{\varepsilon}_{\text{prism(air)}}$ and $\boldsymbol{n}_{\text{prism(air)}}$ are, respectively, the permittivity and refractive index of the prism (air). Because the prism is a dielectric medium, (2.54) allows us to write the dispersion relation at the prism-metal interface as well as at the air-metal one

$$\beta_{pm} = \frac{\omega}{c}\sqrt{\frac{\varepsilon_{\text{prism}}\varepsilon_{\text{metal}}}{\varepsilon_{\text{prism}} + \varepsilon_{\text{metal}}}} \tag{2.72}$$

$$\beta_{am} = \frac{\omega}{c}\sqrt{\frac{\varepsilon_{\text{air}}\varepsilon_{\text{metal}}}{\varepsilon_{\text{air}} + \varepsilon_{\text{metal}}}} \tag{2.73}$$

In practice, the metal permittivity appearing in these last two equations is a function of the frequency and wavevector, $\epsilon(\omega, k)$. However, the spatial dependence

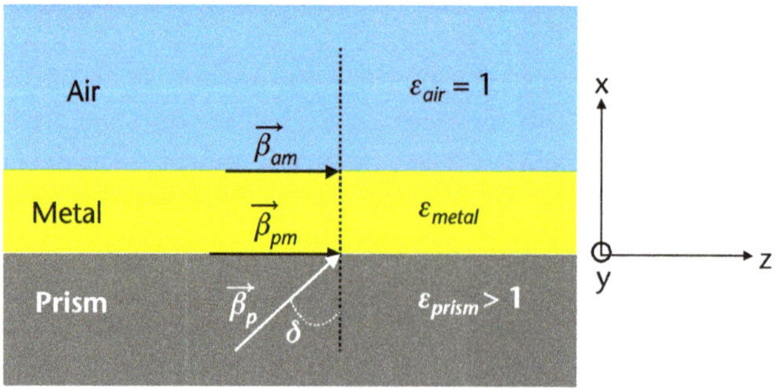

Figure 2.2 Configuration (not to scale) stating the conditions for surface plasmon wave excitation.

is usually unconsidered because the characteristic dimension of the system is not of the same order as the wavelength. The frequency dependency is usually explained using the Drude free electron model for an idealized metal.

$$\epsilon(\omega) = 1 - \frac{\omega_p^2}{\omega^2 + i\omega\gamma} \tag{2.74}$$

where γ is the damping constant and $\omega_p = \sqrt{Ne^2/m\epsilon_0}$ is the plasma frequency with m being the optical mass of electrons. The dispersion relations (2.72) and (2.73) are usually presented using the lossless Drude model (i.e., with no damping term: $\gamma = 0$) as depicted in Figure 2.3.

At $\omega << \omega_p$, the dispersion relation of SPs follows the one of the considered dielectric (prism for SPs at prism/metal interface and air for SPs at air/metal interface). The transversal component of the field dominates the longitudinal one (see (2.58) and (2.59) with $\epsilon_m << \epsilon_d$ from (2.74)) leading to a weakly bounded SP to the interface. The surface wave looks more like a grazing photon. As the energy of the incident light increases, the dispersion relation of SPs is getting further away from the light lines in the prism or in air. The longitudinal component increases while the transverse part decreases, so the SP becomes more bounded to the surface. At the asymptotic limit of $\beta \to +\infty$ (i.e., $\epsilon_m = -\epsilon_d$ in (2.54)), $\omega \to 1/\sqrt{1+\epsilon d}$ which is the angular frequency limit for a pure surface plasmon excitation. Above the gap, this is no more a surface plasmon and corresponds to the Brewster angle.

Let's consider an angular frequency ω_0 between the linear trend of the dispersion relation (at low frequency) and the surface plasma frequency (asymptotic limit).

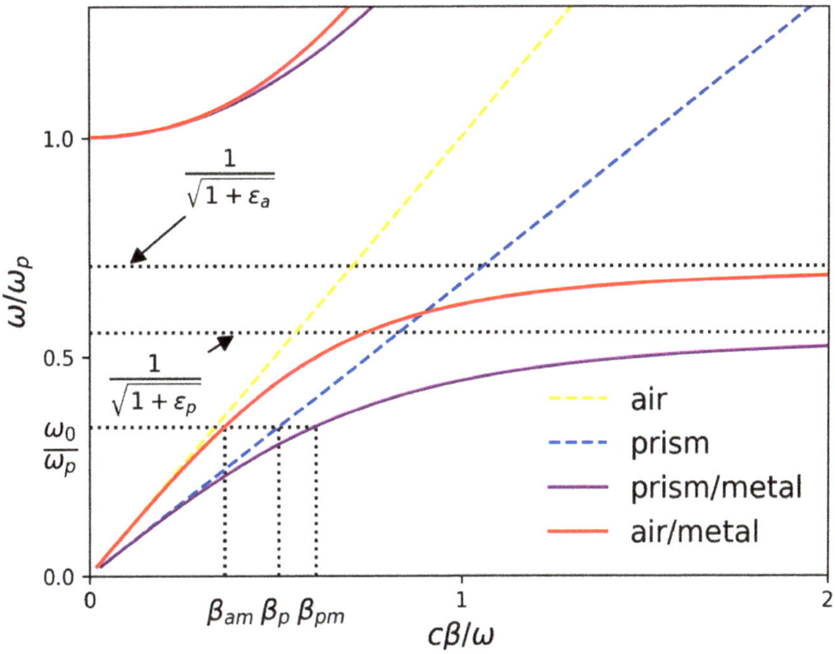

Figure 2.3 Dispersion relationships considering a lossless Drude free electron model.

One can observe that the SP propagation constant at the prism-metal interface (β_{pm}) is always higher than the propagation constant in the prism (β_p). Hence, whatever the incidence angle of light in the prism, the tangential component $\beta_p\sin(\delta)$ of the propagation constant β_p cannot be equal to β_{pm} at the prism-metal interface. Therefore, there is no surface plasmon excitation at this interface. Oppositely, the evolution of propagation constants at the air-metal interface (red curve) and in the prism (blue curve) possesses a crossing point. Using relationships (2.69) and (2.73), one can see that there is excitation and propagation of a surface plasmon wave when

$$\beta_{am} = \beta_p \sin\delta \tag{2.75}$$

which is equivalent to

$$\sqrt{\frac{\varepsilon_{\text{air}}\varepsilon_{\text{metal}}}{\varepsilon_{\text{air}} + \varepsilon_{\text{metal}}}} = \sqrt{\varepsilon_{\text{prism}}}\,\sin\delta \tag{2.76}$$

This condition is met at a given frequency (or wavelength) for a particular angle of incidence δ. This condition is necessary but not sufficient. As indicated before, light injected inside the prism should be TM polarized. Moreover, the metal thickness should range between 30 and 70 nm [16]. Indeed, it should be thin enough so that the evanescent wave can go through the metal layer and couple with the surface plasmon at the metal-dielectric interface. However, it should also be thick enough in order for the SP to "feel" almost infinity on both sides.

2.3 Long-Range Surface Plasmons

The single-interface surface plasmon is characterized by a high attenuation, caused by free-electron scattering in the metal and some surface roughness. This attenuation can be reduced by bounding the thin metal film on both sides using the same dielectric medium. In this case, the structure operates in the long-range surface plasmon polariton (LRSPP) mode and is usually referred to a symmetric metal slab (see Figure 2.4). Long-range means that the attenuation is at least a factor of 2 to 3 lower than that of the single-interface surface plasmon, which results in propagation over a longer distance. If the metal thickness is much greater than the penetration depth of the surface plasmon at each metal-dielectric interface, this type of waveguide supports two TM modes, corresponding to the two surface plasmons localized at the two metal-dielectric interfaces. When the metal layer decreases and becomes of the order of a few tens of nanometers, a coupling appears between the two surface plasmons. These modes can be obtained by solving the eigenvalue equation (2.54). In the case where the two dielectrics have the same refractive index (i.e., the two materials are the same), there are two coupled surface plasmons whatever the thickness of the metal in a range varying between 30 and 70 nm. These modes are qualified as symmetric and antisymmetric, taking into account the symmetry of the distribution of the electric field [17]. The symmetric surface plasmon exhibits an

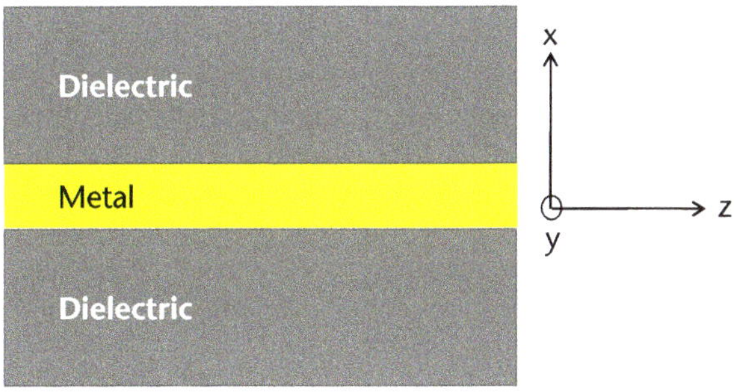

Figure 2.4 Waveguide configuration composed by a thin metal film sandwiched between two semi-infinite dielectric media.

effective refractive index and an attenuation that both increase with the thickness of the metal film. It is the opposite for the antisymmetric mode. In the case where the waveguide is asymmetrical (the two dielectrics have different refractive indices), the effective refractive index of the symmetrical surface plasmon decreases with the thickness of the metal layer and, for a thickness given, the surface plasmon is no longer guided. It is then called the cutoff mode. The symmetric surface plasmon exhibits lower attenuation than the antisymmetric mode. Its field also penetrates deeper into the dielectric than that of the antisymmetric mode.

Therefore, the symmetric electromagnetic field mode is called LRSPP, while the antisymmetric electromagnetic field mode is called short-range surface plasmon resonance (SRSPP). The long-range surface plasmon resonance is usually narrower than that of the short-range surface plasmon. Despite this practical advantage, long-range surface plasmon-based sensors remain by far less popular than their short-range counterparts, because of the need for an additional layer that complexifies the practical development. For more information on the subject, the interested reader is invited to consult [17, 18].

In the following section, the main configurations used for surface plasmon excitation with planar interfaces are described. The case of cylindrical interfaces is only briefly presented, as it will be the focus of Chapters 3 and 4.

2.4 Prism Configurations and Optical Fiber Counterpart

As mentioned in Section 2.1, the most spread configuration for the surface plasmon wave generation is sketched in Figure 2.5 and is called the Kretschmann prism. Light is injected from the bottom of the prism towards the upper flat face covered with a thin metal film. The angle of incidence θ is greater than the critical angle θ_c so that light is totally reflected at the interface. The reflected optical power is then detected at the prism output using a photodiode. The total internal reflection of light generates an evanescent wave outside of the prism. Its amplitude decreases exponentially with respect to the interface between the prism and the metal layer. Its penetration depth is such that this wave extends both in the metal medium and in the surrounding dielectric medium. This wave excites the surface plasmon at the outer metal-dielectric interface. When the light wave couples with the surface plasmon, it transfers part of its optical power to it, which is manifested by a decrease in reflectivity for a given angle of incidence (here θ_0), as illustrated on the right side of Figure 2.5. This is the surface plasmon resonance.

The propagation constant of the plasmon wave propagating along the thin metal layer is influenced by the dielectric at the opposite side of the metal and can be written as follows

$$\beta_{SP} = \beta_{SP,0} + \Delta\beta = \frac{\omega}{c}\sqrt{\frac{\varepsilon_d \varepsilon_m}{\varepsilon_d + \varepsilon_m}} + \Delta\beta \tag{2.77}$$

where $\beta_{SP,0}$ is the propagation constant of the plasmon wave propagating at the metal-dielectric interface in the absence of the prism and $\Delta\beta$ takes into account the finite metal thickness and the presence of the prism. To make possible the coupling between the evanescent wave and the surface plasmon, the propagation constant of the evanescent wave β_E should be equal to the one of the surface plasmon

$$\beta_E = k_z = \frac{2\pi}{\lambda} n_p \sin\theta = \mathrm{Re}\{\beta_{SP}\} = \mathrm{Re}\left\{\frac{2\pi}{\lambda}\sqrt{\frac{\varepsilon_d \varepsilon_m}{\varepsilon_d + \varepsilon_m}} + \Delta\beta\right\} \tag{2.78}$$

In terms of effective refractive index, the relationship (2.78) can be rewritten as

$$n_{\mathrm{eff},E} = n_p \sin\theta = n_{\mathrm{eff},SP} = \mathrm{Re}\left\{\sqrt{\frac{\varepsilon_d \varepsilon_m}{\varepsilon_d + \varepsilon_m}}\right\} + \Delta n_{\mathrm{eff},SP} \tag{2.79}$$

where $n_{\mathrm{eff},E}$ and $n_{\mathrm{eff},SP}$ are the effective refractive indices of the evanescent wave and the surface plasmon, respectively. Figure 2.6, adapted from [9], represents the coupling condition (2.79) in the case of a prism in borosilicate glass (Schott N-BK7, $n_p = 1.51$ at the wavelength of 600 nm) and a gold-water interface. This figure shows

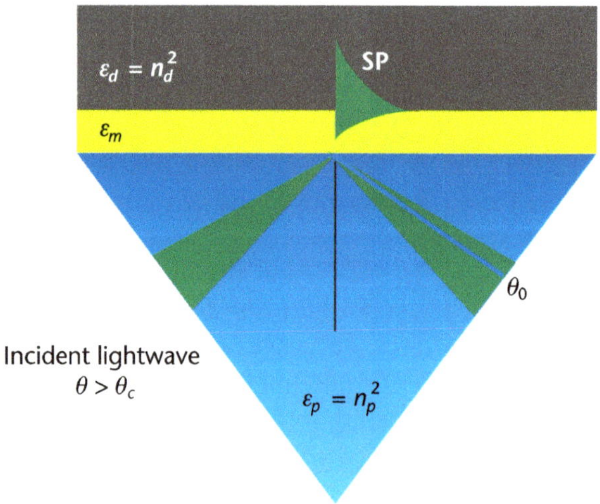

Figure 2.5 Sketch of the Kretschmann prism configuration.

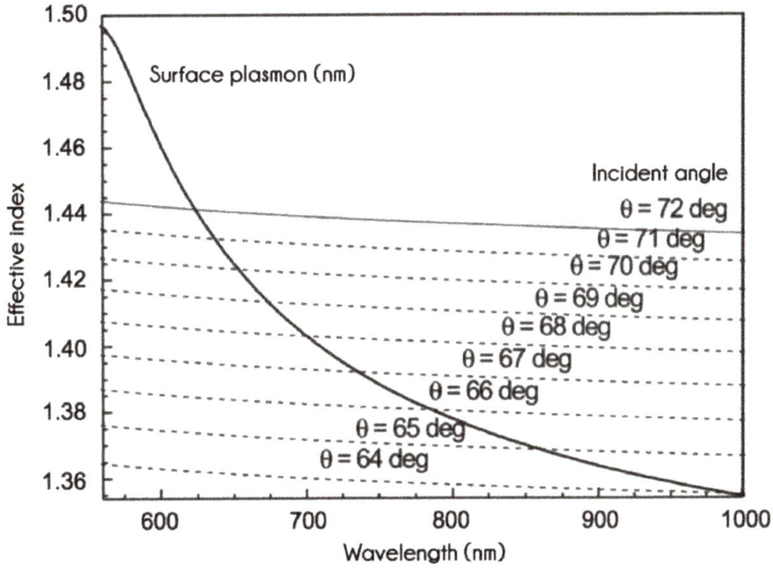

Figure 2.6 Spectral evolution of the effective refractive index of the surface plasmon wave at the gold-water interface and evolution of the effective refractive index of the evanescent lightwave produced in a N-BK7 glass prism for 9 different incidence angles (values displayed above the corresponding curve). (*Adapted from:* [9].)

that for each wavelength, the coupling condition is satisfied for a single incident angle that increases when the wavelength decreases.

This shows that there are two possible ways to determine the permittivity of the outer medium. Either using a spectral interrogation at fixed incident angle or using photodetector scanning the angles at fixed wavelength.

Figure 2.7 sketches another configuration called Otto prism. In this case, a dielectric medium (the one we want to know the refractive index) is first deposited on the flat surface of a prism. Then the thin metal layer surrounds this dielectric medium. Similar operating conditions as those of the Kretschmann prism can be

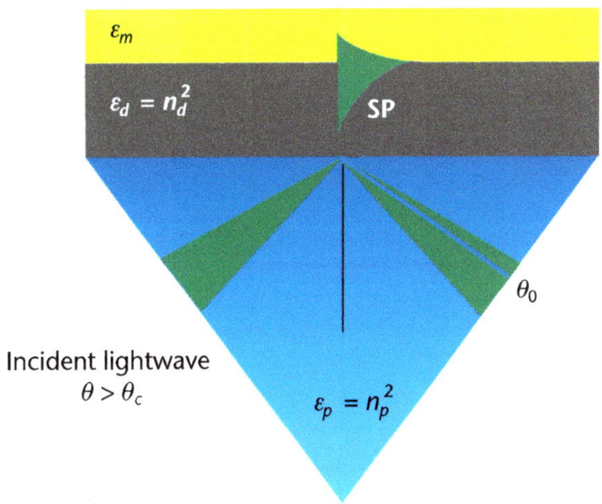

Figure 2.7 Sketch of the Otto prism configuration.

applied to this configuration for the generation of surface plasmon waves. This configuration will not be further described, as there is no counterpart with the optical fiber technology.

The interaction between an incident light wave and a surface plasmon in a total internal reflection method such as the use of prisms can also be studied using the Fresnel reflection approach. Assuming an infinite prism-metal-dielectric structure in the y-z plane and an incident plane wave as shown in Figure 2.8, the reflected light amplitude A_r can be expressed as

$$A_r = r_{pmd} A_i = \left| r_{pmd} \right| e^{i\phi} A_i \tag{2.80}$$

where A_i is the amplitude of the incident wave, r_{pmd} is the reflection coefficient, and ϕ is the phase. The reflection coefficient for the TM polarization state is:

$$r_{pmd} = \frac{r_{pm} + r_{md} \exp\left(-2ik_{mx}q\right)}{1 + r_{pm} r_{md} \exp\left(-2ik_{mx}q\right)} \tag{2.81}$$

where

$$r_{pm} = \frac{\epsilon_m k_{px} - \epsilon_d k_{mx}}{\epsilon_m k_{px} + \epsilon_p k_{mx}} \tag{2.82}$$

$$r_{md} = \frac{\epsilon_d k_{mx} - \epsilon_m k_{dx}}{\epsilon_d k_{mx} + \epsilon_m k_{dx}} \tag{2.83}$$

$$k_{p(m)x} = \sqrt{\left(\frac{2\pi}{\lambda}\right)^2 \epsilon_{p(m)} - k_z^2} \tag{2.84}$$

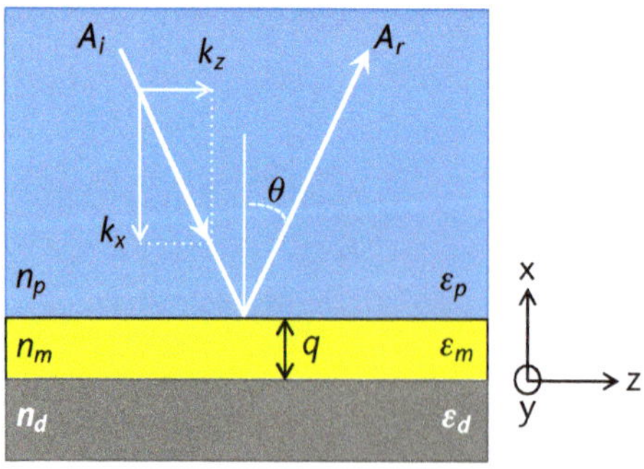

Figure 2.8 Light reflection in the Kretschmann prism configuration study with the Fresnel reflection approach.

$$k_{dx} = \sqrt{\left(\frac{2\pi}{\lambda}\right)^2 \epsilon_d - k_z^2} \tag{2.85}$$

with k_z given by the first part of (2.78), q is the thickness of the metal, and $\bar{w}$ stands for the complex conjugate of w. The reflectivity is finally given by

$$R = \left|r_{pmd}\right|^2 \tag{2.86}$$

Figure 2.9 represents the reflectivity R as a function of the angle of incidence for a TM wave calculated for 4 different thicknesses of gold and silver at 2 distinct wavelengths (600 nm and 800 nm). The prism considered is a N-BK7 glass (n_p = 1.51 at 589 nm). The external dielectric is water (n_d = 1.3332 at 589 nm). All values were corrected at the wavelengths used. Gold (silver) is characterized by $\epsilon_m^{\mathrm{Au},600} = -9.41 - 1.3i$ ($\epsilon_m^{\mathrm{Ag},600} = -13.0 - 0.8i$) and $\epsilon_m^{\mathrm{Au},800} = -24.10 - 1.55i$ ($\epsilon_m^{\mathrm{Ag},800} = -27.8 - 1.55i$) at 600 and 800 nm, respectively. The peak present in the angular evolution

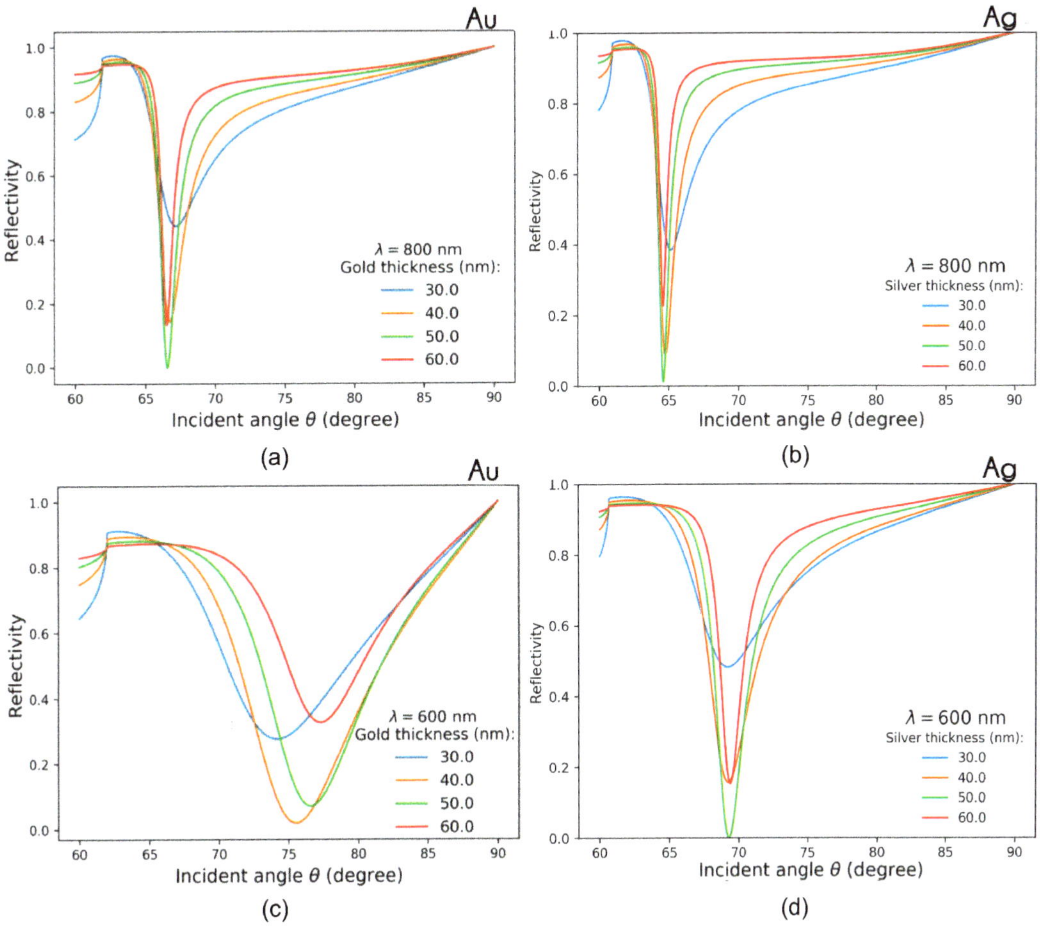

Figure 2.9 Metal thickness dependency of the reflectivity of TM light at 600 nm and 800 nm in stratified structure. The latter is composed of a metal layer (gold or silver) in between two semi-infinite dielectric media (glass and water).

materializes the transfer of energy between the incident light wave and the surface plasmon and its dissipation in the metal film. The resonance angle as well as the depth of the peak depends on the thickness of the metal film. The most pronounced excitation appears for a given thickness, around 50 nm for the geometry and the wavelength considered in these examples.

Figure 2.10 shows that the optimized gold thickness and the resulting FWHM depend on the wavelength used. While the FWHM decreases monotonically with wavelength, the gold thickness first increases to a maximum value of 50 nm around a wavelength of 700 nm before decreasing.

Another way to generate surface plasmons is through the use of a light diffraction grating. In this method, a light wave coming from a dielectric medium with dielectric constant ϵ_d and wave vector $\boldsymbol{k}$ is incident on the surface of a metal diffraction grating, with dielectric constant ε_m, and a period Λ, as shown in Figure 2.11.

Without loss of generality, let's consider a grating with a sine structure in the z-direction. Diffraction gives rise to a set of diffracted waves whose wave vectors in the z-direction are given by

$$\boldsymbol{k}_{m,z} = \boldsymbol{k}_z + m\mathbf{G} \tag{2.87}$$

where m is an integer representing the diffraction order and $\mathbf{G}$ is the diffraction grating vector, according to [19]. $\mathbf{G}$ is a vector in the plane (y-z plane) that is perpendicular to the lattice plane (i.e., in the z-direction) and with an amplitude inversely proportional to the period of the grating. $\mathbf{G}$ can therefore be expressed as follows

$$\mathbf{G} = \frac{2\pi}{\Lambda} z_0 \tag{2.88}$$

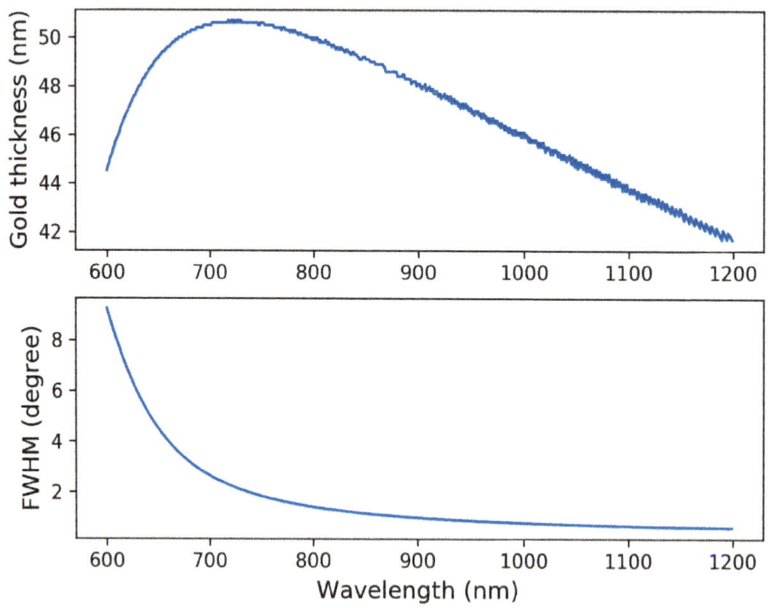

Figure 2.10 Gold thickness optimization for a minimum reflectivity and the corresponding width of the resonance.

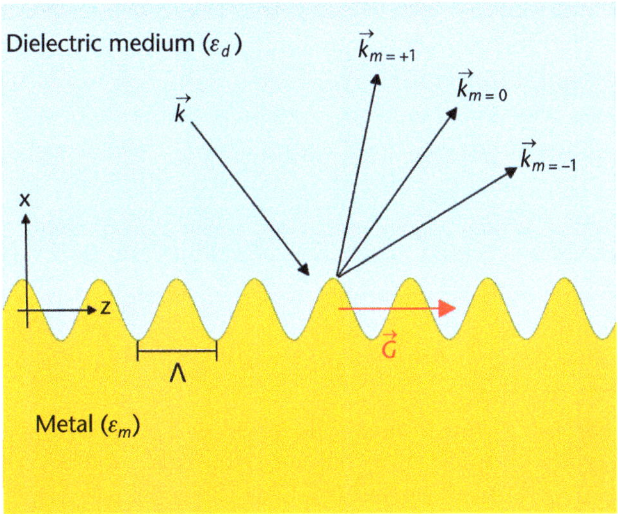

Figure 2.11 Grating coupling method.

where z_0 is the unit vector in the z-direction. The vectorial equation (2.87) leads to

$$k_{zm} = k_z + m\frac{2\pi}{\Lambda} \tag{2.89}$$

The diffracted wave couples with a surface plasmon when their propagation constant is equal, which is equivalent to verifying the following relationship using (2.79) for the wavevector in the dielectric media and (2.59) for the SP wavevector

$$k_{zm} = \frac{2\pi}{\lambda}n_d \sin\theta + m\frac{2\pi}{\Lambda} = \pm\mathrm{Re}\{\beta_{SP}\} \tag{2.90}$$

where β_{SP} is defined by relationship (2.77) with $\beta_{SP,0}$ the surface plasmon propagation constant at the smooth interface between a semi-infinite dielectric and a metal, $\Delta\beta$ taking into account the presence of the lattice. The coupling condition (2.90) can be expressed in terms of effective refractive index using (2.79)

$$n_d \sin\theta + m\frac{\lambda}{\Lambda} = \pm\mathrm{Re}\left\{\sqrt{\frac{\varepsilon_d\varepsilon_m}{\varepsilon_d + \varepsilon_m}}\right\} + \Delta n_{\mathrm{eff},SP} \tag{2.91}$$

This condition can be verified for different combinations of the incident angle, grating period and diffraction order. For positive diffraction orders, the coupled wavelength increases with a decrease in the angle of incidence while the trend is opposite for negative orders.

The reflection spectrum is again characterized by a stopband resonance whose reflectivity depends on the depth of the grating. For a diffraction grating immersed in water, characterized by a pitch of the order of 500 nm and illuminated by light with a wavelength close to 650 nm, the reflectivity reaches 0 for a grating depth of about 30 nm. Both the width and the asymmetry of the stopband resonance increase with the depth of the grating.

Surface plasmons can also be excited by the modes of a dielectric waveguide. When a propagating mode of the dielectric waveguide enters the region of the thin metal layer, it couples with a plasmon at the outer surface of the metal when the propagation constants are equal. As the plasmons are much more dissipative than the dielectric waveguide modes, the coupling condition is verified only for a small range of wavelengths. As a result, surface plasmon excitation is spectrally manifested as a narrow extinction band in the narrow light spectrum. The strength of the coupling depends on the thickness of the metal and the physical length of the interaction region.

Similarly to planar or channel waveguides, optical fibers also enable the excitation of surface plasmon waves, along cylindrical interfaces, this time. This technology allows for a small sensing element to probe a very tiny sample volume of the order of 1 microliter or so. It comes along with a simplified optical design and brings the potential use for disposable fiber-optic sensors and the capability for remote sensing. The following chapters will focus on the different configurations that can be produced based on optical fibers. The latter can be derived from multimode or single-mode optical fibers. All the configurations share the common point to locally (typically over a length of a few centimeters) enable guided light to be totally internally reflected at the external surface of the optical fiber, by removing the cladding or by extracting the light from the core and let it be guided by the cladding, over small distances where the interaction is to take place. Unlike traditional SPR measurements, which employ a discrete excitation wavelength while modulating the angle of incidence, the optical fiber-based SPR sensors use a broadband optical source, resulting in a large range of excitation wavelengths. The range of incident angles is limited to those that propagate within the optical fiber. An SPR resonance is obtained in the transmitted or reflected amplitude spectrum of the optical fiber. Its analysis allows using such configurations for biochemical sensing.

Before going deeper into the SPR excitation in the context of optical fiber configurations, it is worthwhile to briefly mention that plasmon mode can also be supported by metallic nanostructure.

2.5 Localized Surface Plasmons

In the case of a metal nanosphere, the electrons at the surface of a metal particle can oscillate in phase when they are excited by an incident electromagnetic wave, as depicted in Figure 2.12. This quasi-point electron density variation has the symmetry of an oscillating dipole. The incident wave can then enter into resonance with the plasmon, whatever the angle of incidence. In the same way, when it is excited, the localized surface plasmon can emit as long as the symmetry of its oscillating electronic cloud is mainly dipolar and thus acts as an oscillator.

The size of the particles obviously has a significant influence on the extinction spectrum (absorption + elastic light scattering) of nanoparticles of a given metal. When the metallic nanosphere embedded in dielectric medium has dimensions much

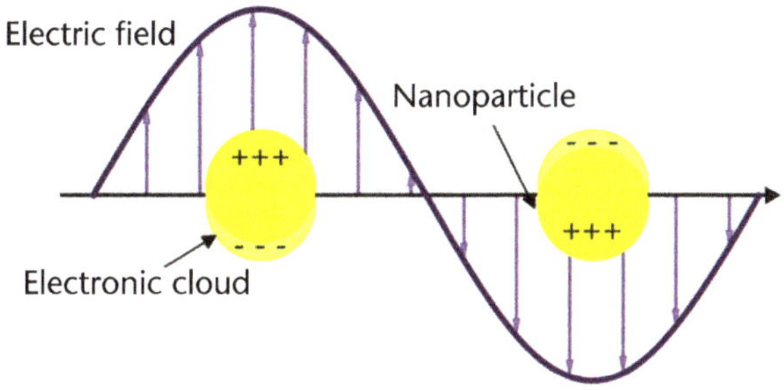

Figure 2.12 Sketch of the localized surface plasmon excitation in metal nanoparticles due to the application of an oscillating electrical field.

smaller than the wavelength of the incident plane wave, its extinction cross-section is only dependent on the dipole oscillation and it is given by [6]:

$$C_{ext} = \frac{18\pi\epsilon_d^{3/2}V}{\lambda}\frac{\epsilon''(\lambda)}{\left[\epsilon'(\lambda)+2\epsilon_d\right]^2+\epsilon''(\lambda)} \tag{2.92}$$

where $\epsilon'(\lambda)$ and $\epsilon''(\lambda)$ are, respectively, the real and imaginary parts of the metal, V is the volume of the sphere, and ϵ_d the dielectric constant of the surrounding medium. This equation can be used to obtain the extinction spectrum once the values of the dielectric constant of the metal is known for the considered spectral range. One can see that a strong extinction appears on the condition $\epsilon'(\lambda) = -2\epsilon_d$ (i.e., when the denominator of (2.92) is minimal), which is the condition of localized surface plasmon resonance (LSPR). This explains the sensitivity of the external permittivity as well as the specific red wine color of the gold nanosphere in water. Indeed, for gold nanoparticles immersed in water ($\epsilon_d \simeq 1.777$), the resonant condition leads to $\epsilon' \simeq -3.555$, which is fulfilled for a wavelength around 525 nm.

In practice, for particles of a size less than 10 nm and therefore of limited volume, it is the diffusion due to collisions between the electrons and the walls of the particle that takes over. It gains in importance when the diameter of the particles is much smaller than the mean free path of the electrons inside the metal (which is about 38 nm in the case of gold [20]). This phenomenon has the effect of broadening the resonance peak. Note the brownish tint taken by the suspensions of gold particles, whose size is less than 4 nm.

For a size greater than 10 nm, the position and the width of the plasmon peak remain relatively constant as a function of the size over a range that can extend up to a few tens of nanometers. The optical behavior of particles whose size is in this range can be described by the Rayleigh scattering model [21]. It considers that all the free electrons of the particle oscillate in phase, thus forming only one resulting dipole. This quasi-static approximation remains valid as long as the diameter of the particles is less than one-tenth of the incident wavelength ($\lambda/10$). According to this model, the absorbed light is elastically reemitted by the oscillation of the excited dipole.

When the size of the particle exceeds $\lambda/10$, the conduction electrons coming from different zones of the particle are not all subjected to the same local electric field at time t. The resulting phase shift leads to the appearance of higher multipole modes whose resonant frequency differs from that of the dipole. These higher modes absorb but do not reemit. They dampen the resonance and help to broaden and redshift the plasmon peak.

The most important factor in the optical extinction spectrum of a particle is its structure. A modification of shape as small as it is compared to the sphere has a considerable impact on the plasmon bound to the particle. In that case of nanosphere, LSPR does not have any polarization dependency. When dealing with other types of nanoparticles (i.e., if the spherical symmetry is broken), the extinction cross-section needs to be numerically determined. However, in the cases of prolate or oblate spheroids, the extinction cross-section is determined by using the Gans equation [12]:

$$C_{ext} = \frac{2\pi\epsilon_d^{3/2}V}{3\lambda}\sum_j \frac{\epsilon''/P_j^2}{\left\{\epsilon' + \left[\frac{1-P_j}{P_j}\right]\epsilon_d\right\}^2 + \epsilon''^2} \tag{2.93}$$

where P_j is the depolarization factor for the 3 axes A, B, and C of the spheroid (with $A > B = C$) and is defined by

$$P_A = \frac{1-e^2}{e^2}\left[\frac{1}{2e}\ln\left(\frac{1+e}{1-e}\right) - 1\right] \tag{2.94}$$

$$P_B = P_C = \frac{1-P_A}{2} \tag{2.95}$$

where

$$e = \sqrt{1-\left(\frac{B}{A}\right)^2} = \sqrt{1-\left(\frac{1}{R}\right)^2} \tag{2.96}$$

and R is the aspect ratio. From (2.93), one can see that the resonant condition is now

$$\epsilon' = -\left[\frac{1-P_j}{P_j}\right]\epsilon_d \tag{2.97}$$

which is dependent of the aspect ratio through the weighted factor of ϵ_d. Using spheroids instead of a sphere-shaped nanoparticle leads to two extinction maxima instead of one. One is due to the transverse mode and is located around the maximum for nanospheres, while the other corresponds to the longitudinal mode. This factor, $(1 - P_j)/P_j$, increases with the aspect ratio, which can be higher than 2, as is the case for the nanosphere. Thus, increasing the aspect ratio redshifts the longitudinal mode.

The effective polarizability of the particle's surface electrons (i.e., those that contribute most to the plasmon resonance) is affected by the polarizability of the surrounding medium. Because of this external contribution to polarization, a transparent matrix with a higher refractive index than a vacuum will attenuate the net charge of the particle surface. This decrease in the net surface charge has the consequence of reducing the restoring force of the oscillator, thus causing a negative displacement of the resonant frequency.

Lastly, let's mention that when two metal nanoparticles are sufficiently close to each other, their resonant plasmonic modes can interact, thus giving rise to new coupled modes whose frequency and intensity differ from the individual mode. This coupling of plasmons strongly affects the extinction spectrum of aggregated particles.

For more details on nanoparticles and the generation of surface plasmons for various structures, the reader is invited to consult [15, 17].

2.6 Plasmonic Materials

Most often, gold and silver are used as a nanoscale metal layer for the generation of surface plasmons. It is often considered that gold exhibits a greater resonance shift than silver in response to a change in refractive index and that silver provides a narrower resonance, resulting in better detection accuracy (better signal-to-noise ratio (SNR)). This essentially depends on the experimental conditions, in particular the film thickness and the excitation wavelength, but also on the dielectric constant of the metal. To estimate the dielectric constant of the metal ϵ_{DC}, one can use the Drude-Lorentz model (a correction of the idealized metal considered by the Drude model; see Section 2.2) through the following equation [7, 8]:

$$\varepsilon_{DC}(\omega) = \varepsilon_{\infty} - \underbrace{\frac{\omega_D^2}{\omega(\omega + i\gamma_D)}}_{\text{Drude model}} - \underbrace{\frac{\Delta\varepsilon\omega_L^2}{\omega^2 - \omega_L^2 + i\gamma_L\omega}}_{\text{Lorentz model}} \tag{2.98}$$

where ε_{∞} represents the relative permittivity for high frequencies where electrons cannot follow excitation and is usually equal to 1 for the Drude model only (see (2.74)); i is the pure imaginary number such as $i^2 = -1$; γ_D and γ_L are the frictional forces; ω_L is the plasma frequency associated with interband transmissions (absent in the Drude model), while ω_D is the plasma frequency; and $\Delta\varepsilon$ is the oscillator strength. Using this model, it is possible to estimate both the real and imaginary values of the dielectric constant for both gold and silver (Figure 2.13), as a function of the wavelength range. It was obtained using the following values from the experimental result reported in Table 2.1.

Gold remains the most used metal for plasmonics, as it does not oxidize and allows the grafting of bioreceptors through thiol bonds. Silver (Ag) remains a choice, as it is associated with sharp (see Figure 2.13) and deep (see Figure 2.14(c–e))

Table 2.1 Dielectric Constant Fitted Parameters for Au and Ag

	Au	*Ag*
ε_∞	5.967	4.6
ω_D	1.3262×10^{16} rad/s	1.3673×10^{16} rad/s
γ_D	9.8752×10^{13} rad/s	1.0635×10^{14} rad/s
$\Delta\varepsilon$	1.09	1.1
ω_L	4.077×10^{15} rad/s	7.4444×10^{15} rad/s
γ_L	6.5784×10^{14} rad/s	62.8231×10^{14} rad/s

Source: [22].

plasmonic resonances offering high resolution at a lower cost than gold films. This behavior is due to its dielectric constant, with a larger absolute value for the real part (ε'_m) and a smaller absolute value for its imaginary part (ε''_m) around the commonly used visible spectral domain and with respect to other metal films such as copper or gold [23–25]. Silver is therefore often considered as the best plasmonic material, with the lowest ohmic losses at optical frequencies. Some have described this layer as a way to achieve "dream films" for the highest Q-factor plasmonic devices [26]. The counterpart of this advantage is that silver oxidizes quickly and is therefore more difficult to handle for biodetection. In addition to the metal thickness dependency (see Figure 2.9), this behavior is also highly dependent on the wavelength range used to excite the plasmon, as shown in Figure 2.14.

Due to high oxidation rate, thin silver films are less investigated for biofunctionalization. It also appears that surface roughness or the presence of metal clusters can influence the functionalization and should therefore be precisely controlled [27]. Thus, silver nanoparticles (SNPs) are commonly used as strong plasmonic amplifiers and show interesting properties when they are implemented as electrochemical/redox tags, essentially under localized plasmonic enhancements [28]. This surface

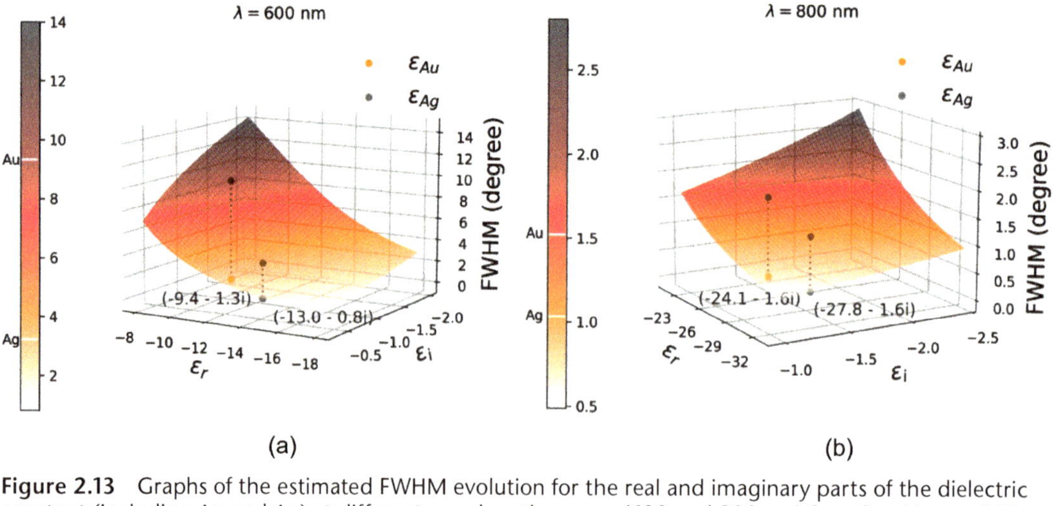

Figure 2.13 Graphs of the estimated FWHM evolution for the real and imaginary parts of the dielectric constant (including Au and Ag) at different wavelength ranges (600 and 800 nm) for a fixed layer of 50 nm.

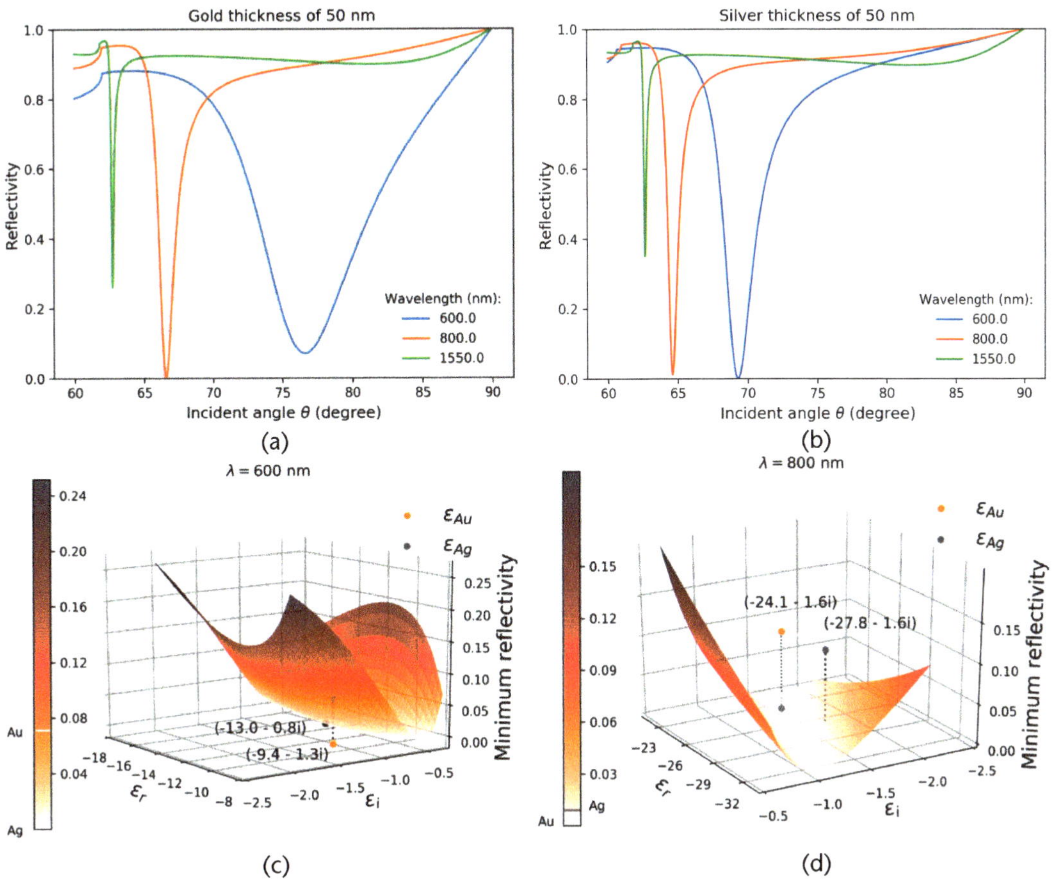

Figure 2.14 (a, b) Plasmonic resonances for gold and silver at different wavelengths (600 and 800 nm) for a thickness of 50 nm. (c, d) 3-D plots of the estimated evolution of the minimum reflectivity with permittivity (including gold and silver) at different wavelengths.

shape consideration is of paramount importance for all surface coatings, because both synthesis techniques and detailed surface analyses are needed to qualify and quantify the properties carried out (homogeneity, roughness, thickness, nanoparticle shapes). Dealing with these parameters needs time to adapt and optimize until reaching the suited conditions for any specific device.

Palladium (Pd) also achieves excellent sensing properties, but is often limited to research use because it is expensive to implement in commercial devices. Its applications are often related to hydrogen detection because of intense sorption of H_2 in Pd layers. This element presents electrical and optical changes under exposure to specific gases [29]. Moreover, some of its physical properties enable the generation of specific inverted surface plasmon resonance (ISPR) observed by internal reflectivity of a bilayer of two thin films and displaying a maximum in place of a minimum resonance, which can be of interest to further develop some biosensing strategies. Pd films are also described with combinations of graphene layers, and are suited for thiol chemistry such as for gold and silver [30]. Proofs of concept and prototypes using palladium as probe layers already exist but are still limited to very specific applications [30].

Platinum (Pt) layers are also widely present in optoelectronic devices and can also be exploited as plasmonic materials. Pt is often coupled with other composites such as nanoparticles or into multiple-layer configurations, such as those made with indium tin oxide (ITO) or tungsten trioxide (WO_3). ITO is considered as a transparent conductor that has high transmissivity for the visible light, with very low electrical resistivity. It is therefore often used in combination with other materials such as noble metals and Pt to improve and fine-tune the sensing performances [31]. Pt/WO_3 layers are mostly used for hydrogen sensing [32]. Bio-immobilization techniques on Pt are roughly the same as those used with gold or silver, through self-assembled alkane thiol monolayers and thiolated receptors. Pt is also used on optical fiber for high temperature and gas sensing because of its physical properties and important stability while used in nanometric layers.

Other metals can also be used for sensing purposes, such as copper (Cu). Copper nanoparticles and thin films have therefore been studied for high-throughput biosensing [33]. It has also been added on optical fiber sensors for SPR and could help to improve the SNR, but it seems to show slightly lower sensitivities than gold or silver when used at the same thicknesses. The main advantage of Cu is mainly its cost and wide availability, because Ag and Au are approximately 100 to 10,000 times more expensive. However, as for silver, copper is prone to oxidation. The inability of copper to withstand the environmental factors in which it operates is probably the main reason why it is considered as a low-tier nanomaterial for plasmonic applications. A promising solution proposed for both nanoparticles and continuous layers made of oxidative materials is the use of protective coatings, such as those made of polymeric gels, graphene, aluminum, or hafnium (Hf) oxides, originally found in zirconium ores [34, 35]. Thanks to these protective layers, Cu sensors are better adapted to sustain in acids, bases, water, and air and resist high temperatures [36, 37]. The encapsulation of Cu materials inside capping layers also presents the advantages to tune and enhance surface sensitivity and to improve the adhesion and the amount of bioreceptors trapped all along the sensing surface. These layers also influence the properties of localized SPR (LSPR) and light scattering effects, leading to interesting resources. The selection of protective layers must therefore be done with two objectives: to protect the surface, and to play an active role in the plasmonic amplification or the bioreceptor's immobilization. In addition to this, metals can also be investigated as a way to achieve electro-plasmonics or to implement microcircuits directly on the biosensor surface.

Another metal of interest is tungsten (W) because it can be deposited on optical fibers for plasmonic sensing as well. It is often used as a metal dichalcogenide material (MDC) as, for example, under the form WS_2. It is described forming highly sensitive and protective layers, which can further tune the plasmonic resonance wavelength. Tungsten disulfide is often employed as nanosheets and its immobilization on optical fibers often leads to granular morphologies and rough surfaces. It can be combined with other metal films such as gold or silver precoatings. Bio-immobilization on WS_2 can be performed using chitosan or other intermediate linkers to create strong anchoring bonds. Nanocomposites also often call on AuNPs to create a physical link between amino groups (from the surface) and thiolated groups (from receptors) [38].

Finally, metal oxides, nitrides, and metal oxide nanoparticles and their composites can also be implemented as biosensors' matrices. These components possess a large surface area with a high surface free energy so they can adsorb biomolecules. Many types of metal oxides exist and can be exploited for biosensing purposes, such as Cu_2O and Fe_3O_4 nanoparticles [39, 40] titanium nitride (TiN), zirconium dioxide (ZrO_2) films, and many hybrid structures coupling metal oxides or nitrides with different kinds of polymers. These promising materials present unusual properties, such as the ones highlighted for titanium nitride and zirconium oxide, which could outperform the ones of gold layers or nanoparticles in many plasmonic assays [41]. These components possess a dielectric permittivity close to the one of Au making them plasmonic in the visible and near infrared (NIR) range. They also have high melting points and are thus adapted for high temperature sensing. However, their main advantage is probably their large (bio)compatibility and their field enhancement in the telecommunication window, which is an important benefit when using optical fiber-based technologies [42].

As far as metal deposition methods are concerned, the different means encountered in practice have been validated on optical fibers. These techniques are cathodic sputtering, evaporation, plasma deposition, or electro or electroless plating.

2.7 Signal Analysis and Performance Indicators

2.7.1 Signal Analysis

Figure 2.15 presents the typical response of a prism-based SPR sensor for angular interrogation when the value of the surrounding refractive index is slightly shifted. The angular interrogation when the Kretschmann prism is used for fine refractometry consists in using monochromatic light and precisely scanning incidence angles of light beyond the critical angle. The sensitivity of a plasmonic sensor relying on the angular interrogation is defined as follows [16]

$$S_n = \frac{\delta\theta_0}{\delta n_s} \tag{2.99}$$

where $\delta\theta_0$ is the shift in resonance angle due to a change in refractive index of the outer dielectric by δn_s. The detection accuracy of a plasmonic sensor therefore depends on its ability to detect changes in resonance angle. The narrower the resonance, the greater the detection accuracy. According to Figure 2.15, if $\delta\theta_{0.5}$ corresponds to the width of the plasmon resonance at mid-height, the detection accuracy of the sensor is inversely proportional to this parameter. The SNR of the plasmonic sensor interrogated angularly, which qualifies the accuracy of the sensor, is therefore defined by [16]

$$\mathrm{SNR} = \frac{\delta\theta_0}{\delta\theta_{0.5}} \tag{2.100}$$

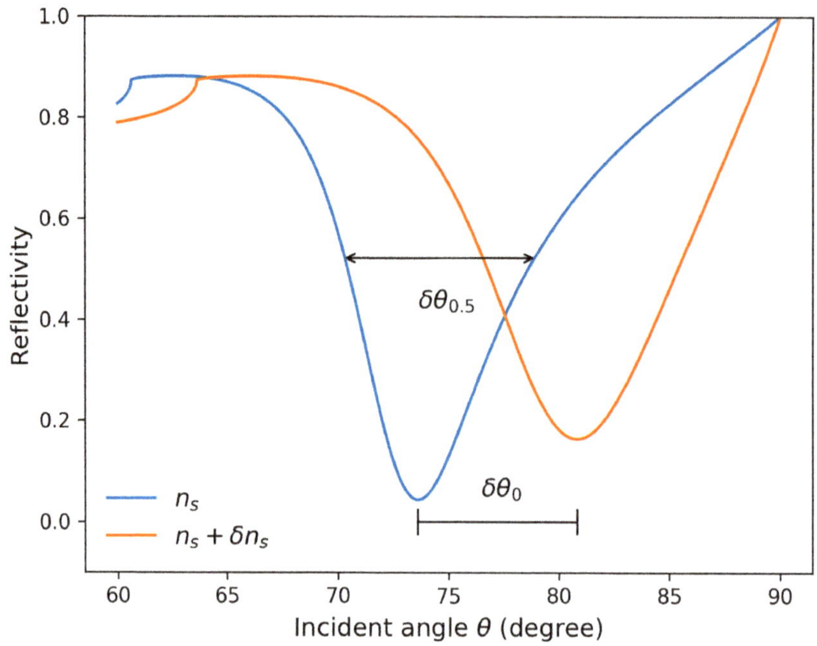

Figure 2.15 Evolution of the SPR in the case of the angular interrogation for two different values of the surrounding refractive index.

Note that there is another possibility of interrogation, called a spectral technique, which consists in using a fixed angle of incidence and modifying the wavelength of the incident light. This technique is intrinsically used by plasmonic optical fibers sensors, as will be shown in the following chapters.

The main applications of this technique are:

- Measurement of density fluctuations of the dielectric in contact with the noble metal layer;
- Measurement of changes in thickness in contact with the metal surface;
- Molecular adsorption measurement. This possibility, which remains the most widely used, requires grafting onto the gold layer receptacles (often called bioreceptors or ligands) on which the molecules to be detected (analytes) are bonded. This measurement scheme relies on affinity mechanisms and leads to a localized refractive index change, detectable by the surface plasmon.

Next we recall the main performance indicators, especially when such sensors are used for biodetection.

2.7.2 Performance Indicators

Plasmonic biosensors are often judged and appreciated by their limit of detection. However, this is often a shortcut because it also heavily depends on the conditions of the experiments: the medium used for its characterization (water, blood, food sample) and the biotarget itself (protein, DNA, cells) [43, 44].

The main performances indicators are:

- Sensitivity;
- Limit of detection (LOD)/limit of quantification (LOQ);
- Specificity/accuracy;
- Repeatability versus reproducibility;
- Figure of merit (FOM) and Q-factor.

These parameters are mainly depending on:

- Bioavailability of the target in the sensing medium;
- Experimental conditions (temperature, volumes, resolution of the measurement device, microfluidics);
- Efficiency of the biofunctionalization process;
- Penetration depth of the plasmonic field.

As we will consider spectral interrogation instead of an angle one in optical fiber next, we will define those indicators in terms of wavelength shift of the plasmonic resonance. The sensitivity of a sensor corresponds to the ratio of the signal response versus the change of the measurand (presence/absence, density, concentration).

$$\text{Sensitivity} = \frac{\Delta\,\lambda}{\Delta\ \text{analyte}} \tag{2.101}$$

The limit of detection of a sensor is expressed in units of concentration and, following the International Union of Pure and Applied Chemistry (IUPAC) definition, indicates "the smallest analyte concentration that a given analytical system is able to distinguish with reasonable reliability from a sample without analyte." The SNR is therefore an important parameter to optimize. Despite this, the LOD often leads to unrealistically low estimations because it is often presented indirectly through a linear regression performed on a set of measurements, assuming that the sensitivity is quasi-constant and the sensor in under optimal conditions. However, it may also take into account the dynamic range of the interactions and be experimentally verified [45, 46]. There are three main methods for the determination of the LOD:

1. Use the standard deviations at as low a concentration as the LOD. This can also be done by a curve-fitting to determine a theoretical LOD from points obtained experimentally. It can be limited to a theoretical LOD or can become experimental LOD if the reported values are efficiently measured with the device after its theoretical determination (Figure 2.16).
2. This approach is based on the calibration curve of the biosensor and is the method recommended by the IUPAC. It is based on the addition of 3 times the maximum standard deviation obtained along all the measurements at low concentrations (Figure 2.17).
3. This strategy is to calculate a quotient that relies on a value obtained with a certain experimental concentration, reported to the biosensor surface and its surface density against the target analyte. For example, let's consider

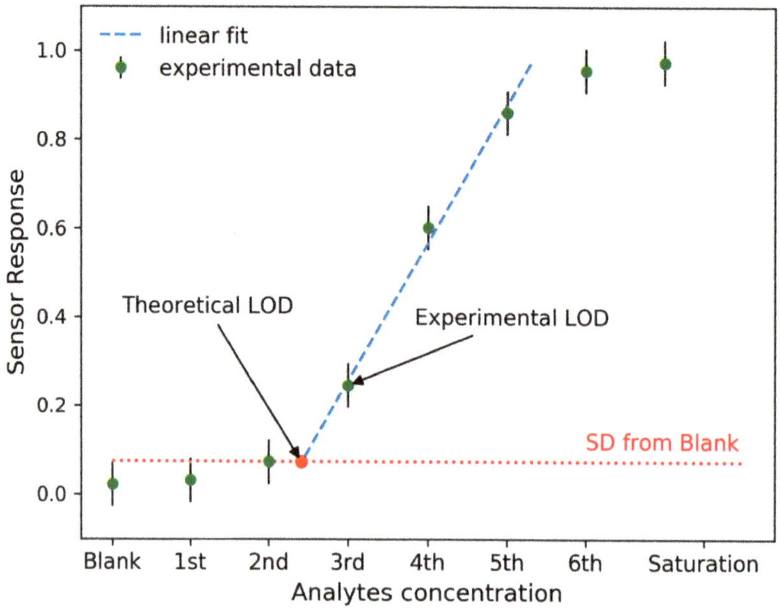

Figure 2.16 The first method applied on fictive data for the LOD determination relying on standard deviation of blank buffers and curve-fitting.

that x receptors can be grafted on the sensors surface. We can assume that if the saturation is reached, all receptors are bound (i.e., x analytes). Then the LOD in that case will be the shift of the sensor response divided by x. Given that it relies on a theoretical calculation in ideal conditions, the two firsts are the most widespread methods among the literature.

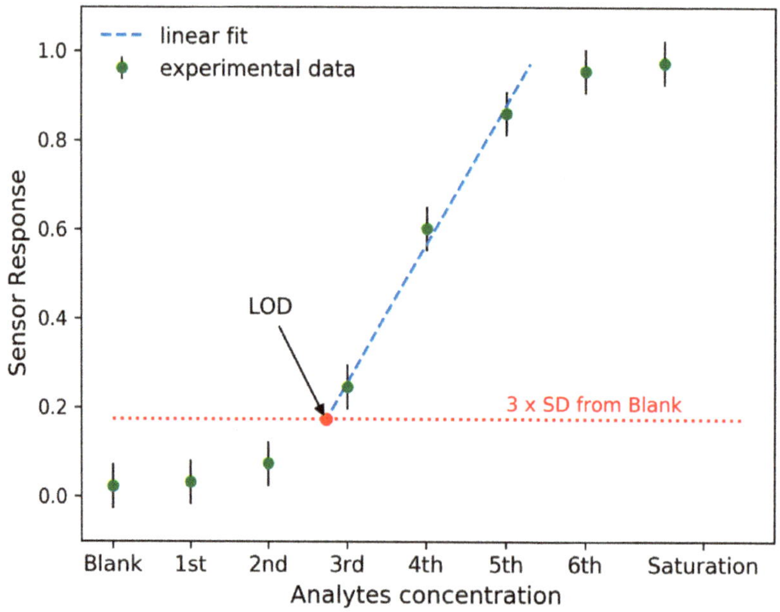

Figure 2.17 The second method applied on fictive data for the LOD determination relying on 3× standard deviation from blank buffer and curve-fitting.

It has to be said that, for most biomedical sensors, the LOD is also given as the first concentration detected with at least a probability of 95% confidence, which makes the method much more reliable. Although the two previous LOD methods are far more optimistic compared to their experimental reality, they remain extensively used by the academic community because they allow highlighting better performances on paper, which are, however, often difficult to reach in practice. It is important to mention that a LOD needs to be calculated carefully to effectively present the performances of the device tested [44, 47]. For instance, the calibration curve needs sufficient data points to fit a reliable linear or sigmoidal curve in order to estimate its value.

Another parameter that usually follows the LOD is the LOQ, for limit of quantification. It is defined as the lowest level that an analyte can be quantitated with any degree of certainty.

The determination of the LOQ is similar to the one of the LOD so they are often reported together. One common method consists of reporting the SNR of 10:1 or to find an approximation by multiplying the LOD by 3. By extension, the LLOQ (lowest calibration standard on the calibration curve) can be mentioned. The latter is reported as the detection response at least 5 times over the blank measurement.

The accuracy is the degree (the percentage) to which the sensor readout value corresponds to the actual value of the measurand (analyte). To be clear, it represents the deviation between the mean value achievable from a series of measurements (the data set) and the true value of the measurand assumed as a reference value (Figure 2.18). There is a difference between the accuracy and the precision: it represents the level of dispersion of the data set (e.g., the variance) with respect to its mean value. It shows how every single measurement agrees with the other. A measurement can therefore be precise but not accurate, and the opposite is also true.

ACCURACY	PRECISION
MEASURING NEAR THE TRUE VALUE	GETTING CONSISTENT RESULTS

Example:

	Biosensor Response		
Concentration 1 ng/mL	TEST 1: 0.98 TEST 2: 1.01 TEST 3: 0.99	Correspond to the concentration	accurate and precise The biosensor is well calibrated
Concentration 1 ng/mL	TEST 4: 0.99 TEST 5: 0.62 TEST 6: 0.02	No conclusion	not accurate and not precise The biosensor needs improvements
Concentration 1 ng/mL	TEST 7: 0.50 TEST 8: 0.51 TEST 9: 0.50	Do not correspond to the concentration	precise, but not accurate The biosensor needs calibration
Concentration 1 ng/mL	TEST 10: 0.95 TEST 11: 1.05 TEST 12: 0.93	Correspond to the concentration	accurate but not precise The biosensor needs improvements

Figure 2.18 Accuracy versus precision example for a biosensing response.

The specificity is the ability of the sensor to only recognize the targets and avoid nonspecific detection. A number of false-positive or false-negative responses versus true-positive and true-negative responses need to be highlighted to obtain both sensitivity and specificity ratios. To be determined, it often relies on an arbitral threshold. If the sensing response is over this threshold in a medium with nontarget molecules, the sensor is not specific enough. A good balance between the noise level, the sensitivity, and the specificity needs to be adjusted, especially in complex media.

The repeatability of the sensor is its ability to reproduce the same response under the same conditions. It is only possible to define a range of repeatability after many tests ($n > 10$) and is often calculated in percentage. While repeatability involves the same conditions (location, measurement procedure, observer, instrument), reproducibility refers to results obtained at different locations, with different instruments and conducted by different individuals. It therefore measures the ability to replicate research from one group to others. Reproducibility is one of the most important key parameters in terms of experimental quality. In order to improve its success rate, the experimental conditions must be clearly enumerated and the material completely listed. The descriptions must be precise enough to be able to transpose the technique easily to other research groups or industries.

The figure of merit for a wavelength shift is proportional to the ratio between the sensitivity (nanometers) and the linewidth of the resonance, considering that it is easier to measure the exact location of a narrow resonance (some optical fiber configurations are presented in the following chapters) than a broad resonance (Kretschmann prism or multimode unclad fiber).

$$FOM = \frac{\text{Sensitivity}}{FWHM} \tag{2.102}$$

A supplementary parameter called Q-factor (quality factor) can also be found. It is the quotient between the initial wavelength of the resonance and the full width at half-maximum or minimum (FWHM). A high Q-factor indicates that the FWHM is small in comparison with its central wavelength, so there is a good capability to distinguish close SRI values (small analytes' concentration changes). A low Q-factor means that it is more difficult to distinguish close SRI changes.

$$Q\ \text{Factor} = \frac{\lambda_0}{FWHM} \tag{2.103}$$

All these parameters qualifying the sensor itself have to be distinguished from the parameters of the sensing method, which relies on the number of true or false positives or negatives (especially for medical diagnosis).

$$\text{Sensitivity of the bioassay} = \frac{\text{True Positives}}{\text{True Positives} + \text{False Negatives}}$$

$$\text{Specificity of the bioassay} = \frac{\text{True Positives}}{\text{True Negatives} + \text{False Positives}}$$

$$\text{Precision of the bioassay} = \frac{\text{True Positives}}{\text{True Positives} + \text{False Positives}}$$

$$\text{Accuracy of the bioassay} = \frac{\text{True Positives} + \text{True Negatives}}{\text{True and False Positives} + \text{True and False Negatives}}$$

The bioavailability of the target molecule is also highly important. For instance, targets need to be homogeneously distributed in the sensing medium. If the target is complexed by other molecules or cells, precipitated, or denatured, the ability for the sensor to detect the target is reduced. If it is the case, the biosensing response does not represent the initial concentration of the total targets in the medium but only an available fraction for interactions. Moreover, the volume in contact with the probe is important. It is the reason why microfluidics systems can help to avoid the presence of a dead volume that never interacts with the sensor surface and aims to increase the contact (and the number) of targets onto the biosensor surface. The efficiency of the functionalization has to be well controlled to ensure sufficient and repeatable interactions with the targets. The strength of plasmonic sensors is their ability to monitor all the biofunctionalization steps in real time, which is not the case for other point-of-care techniques such as paper-based tests or microtiter plate-based platforms, which require additional validation assays for each batch prepared.

Finally, plasmonic sensors are driven by the depth and quality of the plasmonic field. This can induce high variability because small molecules can lead to high resonance shifts while bigger structures can lead to less efficient responses. This parameter depends on the metal layer thickness (and nature) but also the power and wavelength range of the incident light source. The coupling efficiency between incident light and the metal surface also has to be optimized [47, 48]. The manifold ways to determine these parameters make the readability and comparison of the experimental results from one study to another a tough task. For instance, low LOD values are often preferred for their attractiveness in scientific publications to the detriment of their real scope of application. This trend must change in order to increase both relevance and credibility to the field and give these new technologies a real chance to rise on the biosensing market. The weaknesses of certain sensors should definitely not be masked by selective calculation methods or seen as an obstacle to research opportunities but as a boon to improve them in the future, and to improve knowledge for a collective benefit. The fear of obtaining poor performance in order to publish high-impact articles must also take part at the level of publishers and actors in peer-reviewing process and no longer be an argument for pressure in terms of scientific quality.

In conclusion, this chapter reviewed the theoretical considerations required to understand the physical mechanisms of plasmonic chemical and biochemical sensors. The latter usually make use of a thin metal layer deposited on a dielectric medium. Based on the electromagnetism theory and the Kretschmann prism configuration, the mathematical formalism allowing supporting the excitation of plasmon waves using P-polarized light waves has been derived. The difference between standard plasmon waves and localized ones has been highlighted.

References

[1] Wood, R. W., "On a Remarkable Case of Uneven Distribution of Light in a Diffraction Grating Spectrum," *Proc. Phys. Soc. London*, Vol. 18, No. 1, 1902, pp. 269–275.

[2] Fano, U., "The Theory of Anomalous Diffraction Gratings and of Quasi-Stationary Waves on Metallic Surfaces (Sommerfeld's Waves)," *J. Opt. Soc. Am.*, Vol. 31, No. 3, 1941, pp. 213–222.

[3] Ritchie, R. H., "Plasma Losses by Fast Electrons in Thin Films," *Phys. Rev.*, Vol. 106, No. 5, 1957, pp. 874–881.

[4] Turbadar, T., "Complete Absorption of Light by Thin Metal Films," *Proc. Phys. Soc.*, Vol. 73, No. 1, 1959, pp. 40–44.

[5] Otto, A., "Excitation of Nonradiative Surface Plasma Waves in Silver by the Method of Frustrated Total Reflection," *Zeitschrift für Phys. A Hadron. Nucl.*, Vol. 216, No. 4, 1968, pp. 398–410.

[6] Kretschmann, E., and H. Raether, "Radiative Decay of Non Radiative Plasmons Excited by Light," *Zeitschrift Naturforsch A*, Vol. 23, 1968, pp. 2135–2136.

[7] Maier, S. A., *Plasmonics: Fundamentals and Applications*, New York: Springer, 2007.

[8] Pockrand, I., and J. D. Swalen, "Anomalous Dispersion of Surface Plasma Oscillations," *Journal of the Optical Society of America*, Vol. 68, 1978, pp. 1147–1151.

[9] Homola, J., *Surface Plasmon Resonance Based Sensors*, New York: Springer-Verlag, 2006.

[10] Fleischmann, M., P. J. Hendra, and A. J. McQuillan, "Raman Spectra of Pyridine Adsorbed at a Silver Electrode," *Chemical Physics Letters*, Vol. 26, 1974, pp. 163–166.

[11] Coronado, E., and G. Shatz, "Surface Plasmon Broadening for Arbitrary Shape Nanoparticles: A Geometric Probability Approach," *Journal of Chemical Physics*, Vol. 119, 2003, pp. 3926–3934.

[12] Kreibig, U., and M. Vollmer, *Optical Properties of Metal Clusters*, New York: Springer-Verlag, 1995.

[13] Hovel, H., et al., "Width of Cluster Plasmon Resonances: Bulk Dielectric Functions and Chemical Interface Damping," *Physical Review B*, Vol. 48, 1993, pp. 18178–18188.

[14] Sarid, D., and W. Challener, *Modern Introduction to Surface Plasmons: Theory, Mathematical Modeling and Applications*, New York: Cambridge University Press, 2010.

[15] Brongersma, M. L., and P. G. Kik, *Surface Plasmon Nanophotonics*, New York: Springer, 2007.

[16] Sharma, A. K., J. Rajan, and B. D. Gupta, "Fiber-Optic Sensors Based on Surface Plasmon Resonance: A Comprehensive Review," *IEEE Sensors Journal*, Vol. 7, 2007, pp. 1118–1129.

[17] Berini, P., "Long-Range Surface Plasmon Polaritons," *Advances in Optics and Photonics*, Vol. 1, 2009, pp. 484–588.

[18] Yanai, A., and U. Levy, "The Role of Short and Long Range Surface Plasmons for Plasmonic Focusing Applications," *Optics Express*, Vol. 17, 2009, pp. 14270–14280.

[19] Khurgin, J. B., and A. Boltasseva, "Reflecting upon the Losses in Plasmonics and Metamaterials," *MRS Bull.*, Vol. 37, No. 8, 2012, pp. 768–779.

[20] Mayer, K. M., and J. H. Hafner, "Localized Surface Plasmon Resonance Sensors," *Chem. Rev.*, Vol. 111, No. 6, 2011, pp. 3828–3857.

[21] Rayleigh, L., "On the Scattering of Light by Small Particles," *Philosophical Magazine A*, Vol. 41, 1871, pp. 447–454.

[22] Barchiesi, D., and T. Grosges, "Fitting the Optical Constants of Gold, Silver, Chromium, Titanium, and Aluminum in the Visible Bandwidth," *Journal of Nanophotonics*, Vol. 8, 2014, p. 083097.

[23] Yang, R., and Z. Lu, "Subwavelength Plasmonic Waveguides and Plasmonic Materials," *Int. J. Opt.*, 2012.

[24] Rodrigo, S. G., F. J. García-Vidal, and L. Martín-Moreno, "Influence of Material Properties on Extraordinary Optical Transmission Through Hole Arrays," *Phys. Rev. B—Condens. Matter Mater. Phys.*, Vol. 77, No. 7, 2008, pp. 1–8.

[25] Zainuddin, N. A. M., et al., "Investigation of Cladding Thicknesses on Silver SPR Based Side-Polished Optical Fiber Refractive-Index Sensor," *Results Phys.*, Vol. 13, June 2019.

[26] Baburin, A. S., et al., "Silver-Based Plasmonics: Golden Material Platform and Application Challenges," *Opt. Mater. Express*, Vol. 9, No. 2, 2019, pp. 611–642.

[27] Kumar, K. S., and R. Naaman, "Quantitative Analysis and Characterization of Self-Assembled DNA on a Silver Surface," *Langmuir*, Vol. 28, No. 41, 2012, pp. 14514–14517.

[28] Kashefi-Kheyrabadi, L., and M. A. Mehrgardi, "Aptamer-Conjugated Silver Nanoparticles for Electrochemical Detection of Adenosine Triphosphate," *Biosens. Bioelectron.*, Vol. 37, No. 1, 2012, pp. 94–98.

[29] Zuccon, S., et al., "Functional Palladium Metal Films for Plasmonic Devices: An Experimental Proof," *J. Opt. (United Kingdom)*, Vol. 16, No. 5, 2014.

[30] Zuppella, P., et al., "Palladium on Plastic Substrates for Plasmonic Devices," *Sensors (Switzerland)*, Vol. 15, No. 1, 2015, pp. 1138–1147.

[31] Shah, K., N. K. Sharma, and V. Sajal, "SPR Based Fiber Optic Sensor with Bi Layers of Indium Tin Oxide and Platinum: A Theoretical Evaluation," *Optik (Stuttg.)*, Vol. 135, 2017, pp. 50–56.

[32] Jin, W., et al., "Gas Detection with Micro- and Nano-Engineered Optical Fibers," *Opt. Fiber Technol.*, Vol. 19, No. 6, Part B, 2013, pp. 741–759.

[33] Stebunov, Y. V., et al., "Superior Sensitivity of Copper-Based Plasmonic Biosensors," Langmuir, Vol. 34, No. 15, 2018, pp. 4681–4687.

[34] Preston, A. S., et al., "Plasmonics Under Attack: Protecting Copper Nanostructures from Harsh Environments," *Chem. Mater.*, Vol. 32, No. 15, 2020, pp. 6788–6799.

[35] Kravets, V. G., et al., "Graphene-Protected Copper and Silver Plasmonics," *Sci. Rep.*, Vol. 4, No. 1, 2014, p. 5517.

[36] Albrecht, G., et al., "Comprehensive Study of Plasmonic Materials in the Visible and Near-Infrared: Linear, Refractory, and Nonlinear Optical Properties," *ACS Photonics*, Vol. 5, No. 3, 2018, pp. 1058–1067.

[37] Wang, X., C. Santschi, and O. J. F. Martin, "Strong Improvement of Long-Term Chemical and Thermal Stability of Plasmonic Silver Nanoantennas and Films," *Small*, Vol. 13, No. 28, 2017, p. 1700044.

[38] Huang, K. J., et al., "Signal Amplification for Electrochemical DNA Biosensor Based on Two-Dimensional Graphene Analogue Tungsten Sulfide-Graphene Composites and Gold Nanoparticles," *Sensors Actuators B Chem.*, Vol. 191, 2014, pp. 828–836.

[39] Zhu, H., J. Wang, and G. Xu, "Fast Synthesis of Cu2O Hollow Microspheres and Their Application in DNA Biosensor of Hepatitis B Virus," *Cryst. Growth Des.*, Vol. 9, No. 1, 2009, pp. 633–638.

[40] Sanaeifar, N., et al., "A Novel Electrochemical Biosensor Based on Fe3O4 Nanoparticles-Polyvinyl Alcohol Composite for Sensitive Detection of Glucose," *Anal. Biochem.*, Vol. 519, 2017, pp. 19–26.

[41] Naik, G. V., et al., "Epitaxial Superlattices with Titanium Nitride as a Plasmonic Component for Optical Hyperbolic Metamaterials," *Proc. Natl. Acad. Sci.*, Vol. 111, No. 21, 2014, pp. 7546–7551.

[42] Guler, U., V. M. Shalaev, and A. Boltasseva, "Nanoparticle Plasmonics: Going Practical with Transition Metal Nitrides," *Mater. Today*, Vol. 18, No. 4, 2015, pp. 227–237.

[43] Caucheteur, C., T. Guo, and J. Albert, "Review of Plasmonic Fiber Optic Biochemical Sensors: Improving the Limit of Detection," *Anal. Bioanal. Chem.*, Vol. 407, No. 14, 2015, pp. 3883–3897.

[44] Lavín, Á., et al., "On the Determination of Uncertainty and Limit of Detection in Label-Free Biosensors," *Sensors*, Vol. 18, No. 7, 2018.

[45] Socorro-Leránoz, A. B., et al., "Trends in the Design of Wavelength-Based Optical Fibre Biosensors (2008–2018)," *Biosens. Bioelectron.*, Vol. X, No. 1, 2019, p. 100015.

[46] Loock, H. P., and P. D. Wentzell, "Detection Limits of Chemical Sensors: Applications and Misapplications," *Sensors Actuators B Chem.*, Vol. 173, 2012, pp. 157–163.

[47] Chiavaioli, F., et al., "Towards a Uniform Metrological Assessment of Grating-Based Optical Fiber Sensors: From Refractometers to Biosensors," *Biosensors*, Vol. 7, No. 2, 2017, p. 23.

[48] Bhalla, N., et al., "Introduction to Biosensors," *Essays Biochem.*, Vol. 60, 2016, pp. 1–8.

CHAPTER 3

Multimode Optical Fiber Platforms

Optical fibers were developed for telecommunication purposes, with the objective to transmit numeric information in the form of light pulses at very high bit rates and with a very low attenuation. There are two main classes of optical fibers: multimode and single-mode optical fibers. The latter propagate only one light mode, while multimode fibers enable light propagation into different modes. Historically, the first plasmonic optical fiber configurations were derived from multimode optical fibers. This chapter will concentrate on such configurations. Section 3.1 will provide the reader with basic knowledge to explain light propagation in optical fibers, which we believe will be useful for people who do not have a scientific background in photonics. Using both geometric and electromagnetism approaches, the light propagation will be reviewed, especially in the context of step-index optical fibers, as they are largely used for plasmonic optical fiber sensing. The main plasmonic configurations derived from multimode optical fibers will then be reviewed, with a historical perspective and a focus on their typical performance in terms of surrounding refractive index sensitivity.

3.1 Light Propagation in Optical Fibers

As introduced in Chapter 1, optical fibers are cylindrical waveguides of light made of two concentric layers: the core in the center where the vast majority of the light power propagates, surrounded by the cladding. The refractive index of the core is usually slightly higher than the one of the cladding to enable light propagation in the core. Materials used to produce optical fibers have to possess several features:

- Enabling the possibility to develop long, thin, and flexible guides;
- Transparency in a broad range of wavelengths to enable light propagation;
- Compatibility between materials for the core and the cladding.

Materials that fulfill these requirements are glass and polymers. Optical fibers made of glass are the most spread and are usually composed of silica (SiO_2). The refractive index of silica at 850 nm is 1.458. To produce two materials with slightly different refractive index values for the core and the cladding, dopants such as oxides and fluorine are added to silica. The addition of GeO_2 or P_2O_5 increases the refractive index, while the addition of B_2O_3 or fluorine decreases the value. The following compositions are possible [1]:

1. Core in $SiO_2 + GeO_2$ with pure SiO_2 cladding;
2. Core in $SiO_2 + P_2O_5$ with pure SiO_2 cladding;
3. Core in SiO_2 with cladding in $SiO_2 + B_2O_3$.

The main advantages of silica-based optical fibers are:

- Very low attenuation;
- Transparency in the visible and near-infrared wavelength ranges;
- Immunity against electromagnetic interferences;
- No noise;
- High resistance to chemical corrosion, high temperatures, and thermal shocks;
- Lightweight, thin, and easy to handle;
- Sand is the raw material and is largely available.

The growing demand in telecommunications for optical fiber installation directly at the end-user location has encouraged manufacturers to develop cost-effective solutions in polymer optical fibers, usually made of poly(methylmetacrylate) (PMMA) to cover distances up to a few hundred meters. Such fibers are thicker (diameter approximately 1 mm), which makes their manipulation and connection easier. In the following, unless otherwise specified in the text, we will consider glass optical fibers.

The propagation in optical fibers can be studied as a particular case of dielectric waveguides, thanks to Maxwell equations and propagation modes. However, multimode optical fibers are largely over-dimensioned with respect to the operating wavelength and the geometric approach (ray optics) can then be used equivalently. Rays have the advantage of simplicity but some phenomena such as losses in the cladding require the electromagnetism approach, as in single-mode optical fibers. Let us start this review on light propagation with the geometric approach and remember the reflection and refraction laws at the interface of two media of the refractive index n_1 and n_2, respectively.

3.1.1 Geometrical Approach

Let us consider an incident light ray that hits the interface of two perfectly insulating dielectric media with an angle θ_1, as sketched in Figure 3.1. In general, it creates a reflected ray and a refracted ray whose propagation directions are defined according to Snell laws:

$$\theta_1 = \theta_1, \text{ and } n_1 \sin\theta_1 = n_2 \sin\theta_2 \tag{3.1}$$

When n_1 is higher than n_2, θ_2 is higher than θ_1, according to (3.1). There is an incidence angle called the critical angle, which noted θ_c, for which θ_2 is equal to $\pi/2$. The incident wave is therefore subject to a total internal reflection. This phenomenon explains the propagation of light rays within an optical fiber.

To understand how light propagates within an optical fiber, one can consider the sketch of Figure 3.2 that represents a symmetric planar waveguide composed of a central layer of thickness d (with $d = 2a$ where a is the core radius) comprised between two confining layers of refractive index n_2 such that $n_2 < n_1$ that form the

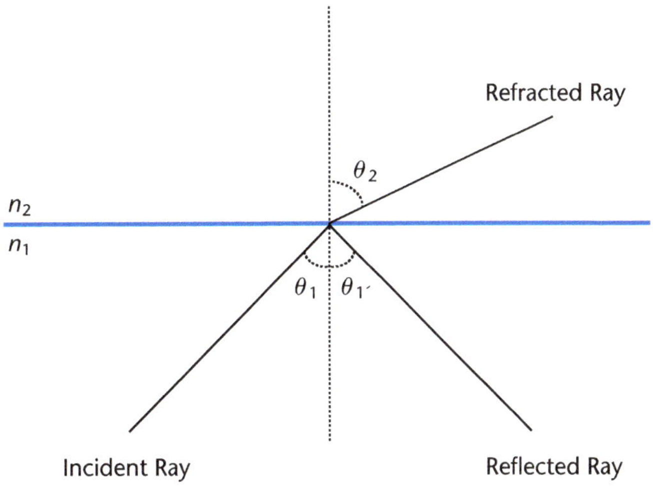

Figure 3.1 Light refraction and reflection at the interface between two materials of different refractive indices.

cladding. When a light ray propagates inside the core, it will be totally reflected at one of the core-cladding interfaces when the incident angle θ is superior to the critical angle θ_c. The following reflections will happen in the same way. Light energy is trapped within the core in the vertical plane and will then propagate along the horizontal axis. From the condition $\theta > \theta_c$, there will be ray propagation when the half-aperture angle θ_i is inferior to θ_0 with:

$$\theta_0 = \frac{\pi}{2} - \theta_c \tag{3.2}$$

The condition $\theta_i < \theta_0$ can also be written as $\cos\theta_0 > \cos\theta_0 = \sin\theta_c$ such that:

$$\cos\theta_i > \frac{n_2}{n_1} \tag{3.3}$$

This imposes the existence of an acceptance cone at the fiber entrance such that all rays comprised in this cone propagate within the core. Other rays situated outside (corresponding to half-aperture angles superior to θ_0) will be partially reflected and refracted at the interfaces and will be lost in the cladding. These rays will not be guided and will induce loss during propagation.

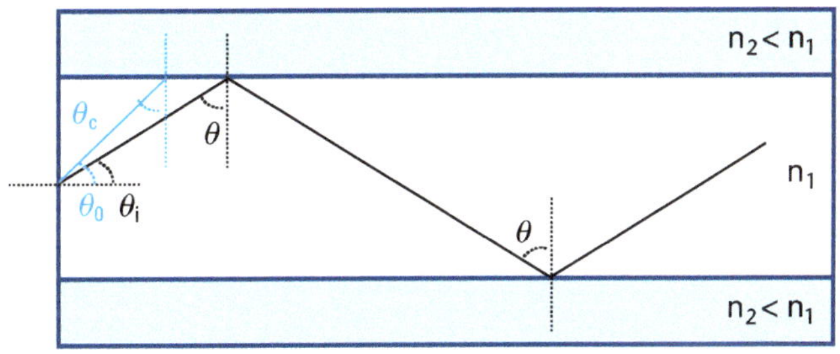

Figure 3.2 Light guidance by the mechanism of total internal reflections.

If θ_0 is now defined as the angle of acceptance of the waveguide for rays coming from air, one can define the numerical aperture NA as the sine of the acceptance angle:

$$\mathrm{NA} = \sin\theta_0 = n_1 \cos\theta_c \tag{3.4}$$

which can be rewritten based on further derivations [1]:

$$\mathrm{NA} = \sqrt{n_1^2 - n_2^2} = n_1\sqrt{2\Delta} \tag{3.5}$$

where

$$\Delta = \frac{1}{2}\frac{n_1^2 - n_2^2}{n_1^2} \cong \frac{n_1 - n_2}{n_1} \text{ if } \Delta \ll 1 \tag{3.6}$$

A plane wave can be associated to each ray, as shown in the left part of Figure 3.3. Accounting for interference effects due to the phase of this plane wave, one can show that the condition for guided propagation $\theta > \theta_c$ is not sufficient. Another condition is necessary such that all rays can be guided: all points situated on the same wavefront must be in phase. This means that only rays characterized by well-defined values of the incidence angle can propagate within the fiber.

Let us now consider the right part of Figure 3.3 showing two rays associated to the same plane wave (or same wavefront). The condition stating that all points situated on the same wavefront must be in-phase results in the fact that the difference between the phase change of the first ray ($\Delta\phi_1$) when it propagates between points A and B and the one of the second ray ($\Delta\phi_2$) between C and D should be an entire multiple of 2π. This condition can then be written as:

$$\Delta\phi_1 = \Delta\phi_2 + 2m\pi \tag{3.7}$$

where m is an integer. According to the sketch of Figure 3.3, this relationship can be rewritten as follows:

$$|\mathrm{AB}|kn_1 + 2\phi_r = |\mathrm{CD}|kn_1 + 2m\pi \tag{3.8}$$

where $|.|$ means the distance, k is equal to $2\pi/\lambda$ with λ as the wavelength, and ϕ_r is the phase retardance due to the reflection at the interfaces. In addition:

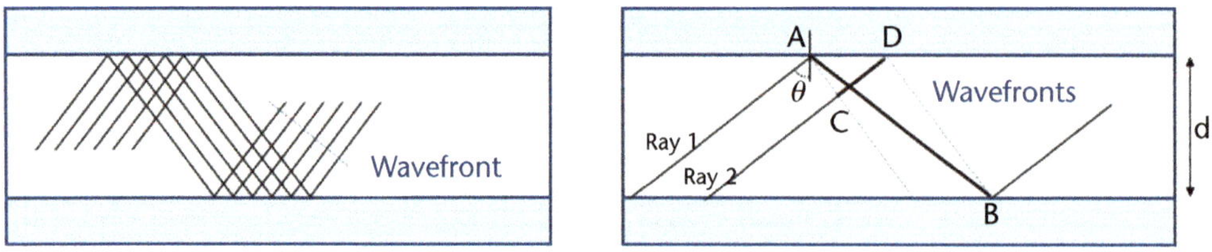

Figure 3.3 Sketch of (left) a wavefront and (right) a phase condition.

$$|AB| = \frac{d}{\cos\theta} \tag{3.9}$$

$$|CD| = \frac{d}{\cos\theta}\left(\sin^2\theta - \cos^2\theta\right) \tag{3.10}$$

Thereby, (3.8) becomes:

$$\frac{d}{\cos\theta}\frac{2\pi}{\lambda}n_1 + 2\phi_r = \frac{d}{\cos\theta}\left(\sin^2\theta - \cos^2\theta\right)\frac{2\pi}{\lambda}n_1 + 2m\pi \tag{3.11}$$

which, after simplification, can be reformulated as:

$$\frac{2\pi}{\lambda}n_1 d\cos\theta + \phi_r = m\pi \tag{3.12}$$

Assuming that the electric field is perpendicular to the incidence plane (TE wave), one can show that:

$$\phi_r = -2\arctan\frac{\sqrt{\sin^2\theta - n_2^2/n_1^2}}{\cos\theta} \tag{3.13}$$

and (3.8) finally becomes:

$$\tan\left(\frac{\pi}{\lambda}n_1 d\cos\theta - \frac{m\pi}{2}\right) = \frac{\sqrt{n_1^2\sin^2\theta - n_2^2}}{n_1\cos\theta} \tag{3.14}$$

Equation (3.14) is called the modal equation. Its numerical resolution in θ (for $m = 0, \ldots, N - 1$ with m as the number of propagated TE modes) leads to a discretization of the possible propagation angles associated to the TE_m modes or to propagation constants:

$$\beta_m = n_1 k \sin\theta_m = k n_{\mathrm{eff}_m} \tag{3.15}$$

with n_{eff} the effective refractive index of the mode. β_b is the component along the direction of propagation of the propagation constant $n_1 k$.

Rays that propagate within multimode optical fibers can be classified into two categories:

- Meridional rays that pass by the axis of the cylinder;
- Helical rays that do not pass by the axis but describe a spiral when they propagate.

Let us now compute the number of TE modes (N) that an optical fiber can support. To compute N, we have to determine the number of θ angles that satisfy

(3.14) for θ values in the range $[0 - \theta_c]$. Using (3.13) and since $\sin\theta_c = n_2/n_1$ (see (3.1), when $\theta_1 = \theta_c$, $\theta_2 = \pi/2$), one can write:

$$\frac{2\pi}{\lambda} n_1 d \cos\theta_c = (N - 1)\pi \tag{3.16}$$

Since $k = 2\pi/\lambda$ and considering that a is the core radius ($a = d/2$), one can write:

$$N - 1 = \frac{2n_1 ka \cos\theta_c}{\pi} \tag{3.17}$$

which, using (3.4) to (3.6), turns into:

$$N - 1 = \frac{2n_1 ka\sqrt{2\Delta}}{\pi} = \frac{2V}{\pi} \tag{3.18}$$

with

$$V = n_1 ka\sqrt{2\Delta} = ka\mathrm{NA} \tag{3.19}$$

V is a geometric parameter without dimension called the normalized frequency. It allows defining the number of modes that can be guided as follows:

$$N = 1 + \mathrm{EP}\left(\frac{2V}{\pi}\right) \tag{3.20}$$

where EP means the entire part. This relationship means that the number of guided modes decreases with the core thickness. It also decreases when the difference between n_1 and n_2 decreases. A waveguide will support a single propagation mode when $V < \pi/2$.

3.1.2 Electromagnetism Approach

The electromagnetism approach to study light propagation in optical fibers consists in solving Maxwell equations for cylindric waveguides. In the following, the case of step-index optical fibers will be considered. A more exhaustive study can be found in [2].

In metal waveguides, only transverse electric (TE) and transverse magnetic (TM) modes can propagate. In optical fibers, the boundary conditions at the core-cladding interface lead to the existence of hybrid modes called HE or EH whether the transverse electric field (E) or transverse magnetic field (H) is the most important for this mode.

In general, the difference between the core refractive index n_1 and the one of the cladding n_2 is very small and the theoretical study can be simplified. It leads to the definition of linearly polarized (LP) modes, as shown in Section 3.1.3.

Let us first qualitatively analyze the electric field distribution in the fiber cross-section for TE modes of orders 0, 1, and 2 (TE_0, TE_1, and TE_2). As observed in Figure 3.4, the order of a mode is equal to the number of times that the electric field crosses 0. The order of a mode is also linked to the angle that the corresponding ray makes with the fiber axis. Electric fields of the guided modes are not fully confined in the center of the guide but partially extend in the cladding. They harmonically evolve in the fiber core and exponentially decrease in the cladding. For low-order modes, the electric field is mainly concentrated in the center of the waveguide. For higher-order modes, the field is more distributed close to the core-cladding interface.

Solving Maxwell equations confirms that guided modes do not represent the whole set of solutions. Radiative modes are also solutions of the equations system. They are not guided by the core but result from optical power situated outside the acceptance cone, which is therefore refracted in the cladding. As the cladding has a limited thickness, part of this optical power remains trapped. A coupling between core modes and cladding modes can then be observed during their propagation.

There exists a third category of modes, called leaky modes. They are partially confined in the core and are continuously attenuated by radiating their optical power outside of the core when they propagate.

A mode stays guided provided that the following relationship is respected:

$$n_2 k < \beta < n_1 k \tag{3.21}$$

where β is the propagation constant of the mode. This relationship can also be written as:

$$n_2 < n_{\text{eff}} < n_1 \tag{3.22}$$

Maxwell equations have already been defined and used in Chapter 2. For the sake of consistency, they will be redefined in the following for a linear dielectric medium, without charge and current:

$$\nabla \cdot \boldsymbol{D} = 0 \tag{3.23}$$

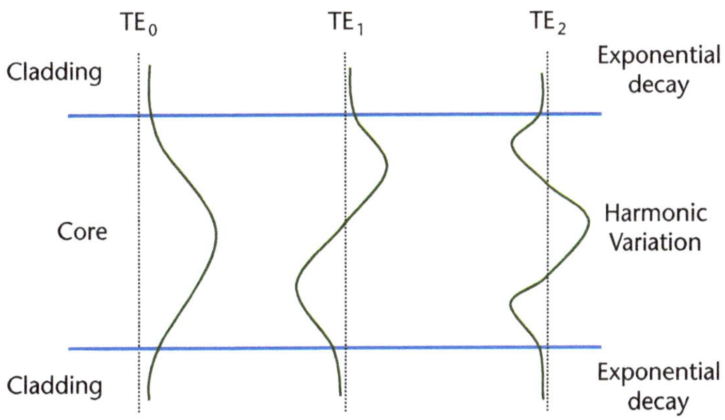

Figure 3.4 Distribution in the fiber cross-section of the electric fields for TE modes of orders 0, 1, and 2. (*Adapted from:* [1].)

$$\nabla \cdot B = 0 \tag{3.24}$$

$$\nabla \times E = -\frac{\partial B}{\partial t} \tag{3.25}$$

$$\nabla \times H = \frac{\partial D}{\partial t} \tag{3.26}$$

with

$$D = \varepsilon E \text{ and } B = \mu H \tag{3.27}$$

where ε and μ are the permittivity and permeability of the dielectric, respectively. An important relationship describing the wave phenomena of electromagnetic fields can be derived from Maxwell equations. Using the rotational of (3.25) and the relationships (3.26) and (3.27), one can get:

$$\nabla \times (\nabla \times E) = -\varepsilon\mu \frac{\partial^2 E}{\partial t^2} \tag{3.28}$$

Using the following identity:

$$\nabla \times (\nabla \times E) = \nabla(\nabla \cdot E) - \nabla^2 E \tag{3.29}$$

and (3.23), the relationship (3.29) becomes:

$$\nabla^2 E = \varepsilon\mu \frac{\partial^2 E}{\partial t^2} \tag{3.30}$$

Similarly, computing the rotational of (3.26), one can get:

$$\nabla^2 H = \varepsilon\mu \frac{\partial^2 H}{\partial t^2} \tag{3.31}$$

These two equations are called wave equations.

Let us now consider electromagnetic waves propagating along a cylindrical optical fiber. A cylindrical coordinates system (r, , z) can be defined with the z-axis corresponding to the fiber propagation axis. Waves propagate according to the z-direction and harmonically vary as a function of time t and distance z:

$$E = E_0(r,\phi)e^{j(\omega t - \beta z)} \tag{3.32}$$

$$\boldsymbol{H} = \boldsymbol{H}_0(r,\phi)e^{j(\omega t - \beta z)} \tag{3.33}$$

β is the propagation constant along the z-direction and can be determined from the boundary conditions at the core-cladding interface. Substituting (3.32) into (3.25) allows deriving the following relationships:

$$\frac{1}{r}\left(\frac{\partial E_z}{\partial \phi} + jr\beta E_\phi\right) = -j\omega\mu H_r \tag{3.34}$$

$$j\beta E_r + \frac{\partial E_z}{\partial r} = j\omega\mu H_\phi \tag{3.35}$$

$$\frac{1}{r}\left(\frac{\partial}{\partial r}\left(rE_\phi\right) - \frac{\partial E_r}{\partial \phi}\right) = -j\omega\mu H_z \tag{3.36}$$

Similarly, substituting (3.33) into (3.26), one can get:

$$\frac{1}{r}\left(\frac{\partial H_z}{\partial \phi} + jr\beta H_\phi\right) = j\omega\varepsilon E_r \tag{3.37}$$

$$j\beta H_r + \frac{\partial H_z}{\partial r} = -j\omega\varepsilon E_\phi \tag{3.38}$$

$$\frac{1}{r}\left(\frac{\partial}{\partial r}\left(rH_\phi\right) - \frac{\partial H_r}{\partial \phi}\right) = j\omega\varepsilon E_z \tag{3.39}$$

These equations can be rewritten in such a way that E_r, E_ϕ, H_r, and H_ϕ can be determined when E_z and H_z are known. For instance, E_ϕ can be eliminated from (3.34) and (3.38) such that H_r can be determined from E_z and H_z. This allows getting:

$$E_r = -\frac{j}{q^2}\left(\beta\frac{\partial E_z}{\partial r} + \frac{\omega\mu}{r}\frac{\partial H_z}{\partial \phi}\right) \tag{3.40}$$

$$E_\phi = -\frac{j}{q^2}\left(\frac{\beta}{r}\frac{\partial E_z}{\partial \phi} - \omega\mu\frac{\partial H_z}{\partial r}\right) \tag{3.41}$$

$$H_r = -\frac{j}{q^2}\left(\beta\frac{\partial H_z}{\partial r} - \frac{\omega\varepsilon}{r}\frac{\partial E_z}{\partial \phi}\right) \tag{3.42}$$

$$H_\phi = -\frac{j}{q^2}\left(\frac{\beta}{r}\frac{\partial H_z}{\partial \phi} + \omega\varepsilon\frac{\partial E_z}{\partial r}\right) \tag{3.43}$$

with

$$q^2 = \omega^2 \varepsilon\mu - \beta^2 = k_n^2 - \beta^2 \tag{3.44}$$

The following wave equations in cylindrical coordinates can be obtained by mixing (3.42) and (3.43) with (3.39) as well as (3.40) and (3.41) with (3.16):

$$\frac{\partial^2 E_z}{\partial r^2} + \frac{1}{r}\frac{\partial E_z}{\partial r} + \frac{1}{r^2}\frac{\partial^2 E_z}{\partial \phi^2} + q^2 E_z = 0 \tag{3.45}$$

$$\frac{\partial^2 H_z}{\partial r^2} + \frac{1}{r}\frac{\partial H_z}{\partial r} + \frac{1}{r^2}\frac{\partial^2 H_z}{\partial \phi^2} + q^2 H_z = 0 \tag{3.46}$$

These two equations only contain E_z or H_z. This means that the longitudinal components of the electric and magnetic fields are independent and can be arbitrarily chosen provided that they fulfill (3.45) and (3.46). However in practice, E_z and H_z are linked by boundary conditions, as will be shown later. There exist solutions or modes to (3.40) through (3.43) for which $E_z = 0$ or $H_z = 0$. When $E_z = 0$, modes are called TE, while they are TM when $H_z = 0$. There also exist hybrid modes for which E_z and H_z are not zero. These modes are labeled HE or EH when H or E provides the most important contribution to the transverse field.

We will now use the previous derivations to determine the guided modes in step-index optical fibers. A current method to solve equations such as (3.45) consists in considering a solution that takes the form of:

$$E_z = AF_1(r)F_2(\phi)F_3(z)F_4(t) \tag{3.47}$$

where A is a constant. The dependency in t and z is given by:

$$F_3(z)F_4(t) = e^{j(\omega t - \beta z)} \tag{3.48}$$

as it is a sine wave that propagates in the z-direction. Because of the circular symmetry of the optical fiber, each component of the field does not vary when 2π is added to the ϕ coordinate. This is equivalent to:

$$F_2(\phi) = e^{j\nu\phi} \tag{3.49}$$

ν is an integer number that can be positive or negative. Considering (3.47) through (3.49), the wave equation (3.45) turns into:

$$\frac{\partial^2 F_1}{\partial r^2} + \frac{1}{r}\frac{\partial F_1}{\partial r} + \left(q^2 - \frac{v^2}{r^2}\right)F_1 = 0 \tag{3.50}$$

An identical relationship can be derived for H_z. The previous equation has a well-known form for which the solutions are Bessel functions.

In the case of step-index optical fibers, we consider a homogeneous core of refractive index n_1 and radius a surrounded by an infinite cladding of refractive index n_2, assuming that an infinite cladding results from the exponential decay of the guided modes outside of the core, with negligible values at the cladding-air interface.

Equation (3.50) must be solved both in the core and cladding regions. In the core, solutions are Bessel functions of the first kind $J_v\,(ur)$ (as solutions must stay finite, also when $r \to 0$) with:

$$u^2 = k_1^2 - \beta^2 \tag{3.51}$$

k_1 is the propagation constant in the core and is equal to $2\pi n_1/\lambda$. In the core, E_z and H_z thus become:

$$E_z(r < a) = AJ_v(ur)e^{jv\phi}e^{j(\omega t - \beta z)} \tag{3.52}$$

$$H_z(r < a) = BJ_v(ur)e^{jv\phi}e^{j(\omega t - \beta z)} \tag{3.53}$$

where B is also a constant. In the cladding, solutions are modified Bessel functions of the second kind $K_v\,(wr)$ (as solutions must tend to 0 when $r \to \infty$) with:

$$w^2 = \beta^2 - k_2^2 \tag{3.54}$$

k_2 is the propagation constant in the cladding and is equal to $2\pi n_2/\lambda$. In the cladding, E_z and H_z then become:

$$E_z(r > a) = CK_v(wr)e^{jv\phi}e^{j(\omega t - \beta z)} \tag{3.55}$$

$$H_z(r > a) = DK_v(wr)e^{jv\phi}e^{j(\omega t - \beta z)} \tag{3.56}$$

where C and D are constants. In practice, $K_v\,(wr) \to e^{-wr}$ when $wr \to \infty$. Since K_v (wr) must tend to 0 when $r \to \infty$, one can conclude that $w > 0$, which means that $\beta \geq k_2$. Within the core, u must be real for F_1 to be as well, which yields $k_1 \geq \beta$. Hence:

$$k_2 \leq \beta \leq k_1 \tag{3.57}$$

We now have to consider the boundary conditions at the core-cladding interface to determine β and the different constant A, B, C, and D. They state that the tangential components of $\boldsymbol{E}$ and $\boldsymbol{H}$ must be equal at each side of the interface, so for $r = a$. For the electric field, this is mathematically represented by:

$$E_{z1} - E_{z2} = 0 \tag{3.58}$$

$$E_{\phi 1} - E_{\phi 2} = 0 \tag{3.59}$$

with $E_{z1}(E_{\phi 1})$ and $E_{z2}(E_{\phi 2})$ the components of the electric field along $z(\phi)$ at the core-cladding interface, at the core side and the cladding side, respectively. Using (3.52) and (3.55), (3.58) becomes:

$$E_{z1} - E_{z2} = AJ_{\nu}(ua) - CK_{\nu}(wa) = 0 \tag{3.60}$$

Substituting (3.52) and (3.53) into (3.41) to determine $E_{\phi 1}$ and, similarly, substituting (3.55) and (3.56) to determine $E_{\phi 2}$, the condition (3.59) becomes:

$$\begin{aligned} E_{\phi 1} - E_{\phi 2} &= -\frac{j}{u^2}\left(A\frac{j\nu\beta}{a}J_{\nu}(ua) - B\omega\mu u J'_{\nu}(ua)\right) \\ &-\frac{j}{w^2}\left(C\frac{j\nu\beta}{a}K_{\nu}(wa) - D\omega\mu w K'_{\nu}(wa)\right) = 0 \end{aligned} \tag{3.61}$$

The symbol ′ represents the derivative according to the argument. As both media (core and cladding) are dielectric, we assume that $\mu_1 = \mu_2 = \mu$. Similar developments can be made for the components of the magnetic field to obtain the following relationships:

$$H_{z1} - H_{z2} = BJ_{\nu}(ua) - DK_{\nu}(wa) = 0 \tag{3.62}$$

$$\begin{aligned} H_{\phi 1} - H_{\phi 2} &= -\frac{j}{u^2}\left(B\frac{j\nu\beta}{a}J_{\nu}(ua) + A\omega\varepsilon_1 u J'_{\nu}(ua)\right) \\ &-\frac{j}{w^2}\left(D\frac{j\nu\beta}{a}K_{\nu}(wa) + C\omega\varepsilon_2 w K'_{\nu}(wa)\right) = 0 \end{aligned} \tag{3.63}$$

Equations (3.60) through (3.63) form a set of 4 equations with 4 unknowns (A, B, C, and D). A solution exists when the following determinant is null:

$$\begin{vmatrix} J_{\nu}(ua) & 0 & -K_{\nu}(wa) & 0 \\ \frac{\nu\beta}{au^2}J_{\nu}(ua) & \frac{j\omega\mu}{u}J'_{\nu}(ua) & \frac{\nu\beta}{aw^2}K_{\nu}(wa) & \frac{j\omega\mu}{w}K'_{\nu}(wa) \\ 0 & J_{\nu}(ua) & 0 & -K_{\nu}(wa) \\ -\frac{j\omega\varepsilon_1}{u}J'_{\nu}(ua) & \frac{\nu\beta}{au^2}J_{\nu}(ua) & -\frac{j\omega\varepsilon_2}{w}K'_{\nu}(wa) & \frac{\nu\beta}{aw^2}K_{\nu}(wa) \end{vmatrix} = 0 \tag{3.64}$$

The computation of the determinant yields:

$$\left(\mathcal{J}_\nu + \mathcal{K}_\nu\right)\left(k_1^2\mathcal{J}_\nu + k_2^2\mathcal{K}_\nu\right) = \left(\frac{\nu\beta}{a}\right)^2\left(\frac{1}{u^2} + \frac{1}{w^2}\right)^2 \tag{3.65}$$

with

$$\mathcal{J}_\nu = \frac{J_\nu'(ua)}{uJ_\nu(ua)} \text{ and } \mathcal{K}_\nu = \frac{K_\nu'(wa)}{wK_\nu(wa)} \tag{3.66}$$

Equation (3.65) is called a characteristic equation. Its resolution in β yields all characteristics of guided modes. One can show that only some discrete β values are allowed in the interval defined by (3.57). Upon the determination of β, the components of the electric and magnetic fields in the core and in the cladding can be computed from (3.52) through (3.56) and (3.40) through (3.43).

Considering the oscillatory behavior of J_ν functions, m roots can be found for (3.65) for a given value of ν. These roots are identified by $\beta_{\nu m}$ and the corresponding modes are identified by $TE_{\nu m}$, $TM_{\nu m}$, $EH_{\nu m}$, or $HE_{\nu m}$. In optical fibers, all modes are hybrid except when $\nu = 0$. In this case, the characteristic equation becomes:

$$\left(\mathcal{J}_0 + \mathcal{K}_0\right)\left(k_1^2\mathcal{J}_0 + k_2^2\mathcal{K}_0\right) = 0 \tag{3.67}$$

$$\text{If } \mathcal{J}_0 + \mathcal{K}_0 = 0 \text{ then } \frac{J_1(ua)}{uJ_0(ua)} + \frac{K_1(wa)}{wK_0(wa)} = 0 \tag{3.68}$$

This case corresponds to TE_{0m} modes ($E_z = 0$).

$$\text{Else if } k_1^2\mathcal{J}_0 + k_2^2\mathcal{K}_0 = 0 \text{ then } k_1^2\frac{J_1(ua)}{uJ_0(ua)} + k_2^2\frac{K_1(wa)}{wK_0(wa)} = 0 \tag{3.69}$$

This case corresponds to TM_{0m} modes ($H_z = 0$).

When $\nu \neq 0$, numerical methods are required to solve the equations. The distribution of the electric field for some of the first guided modes is shown in Figure 3.5.

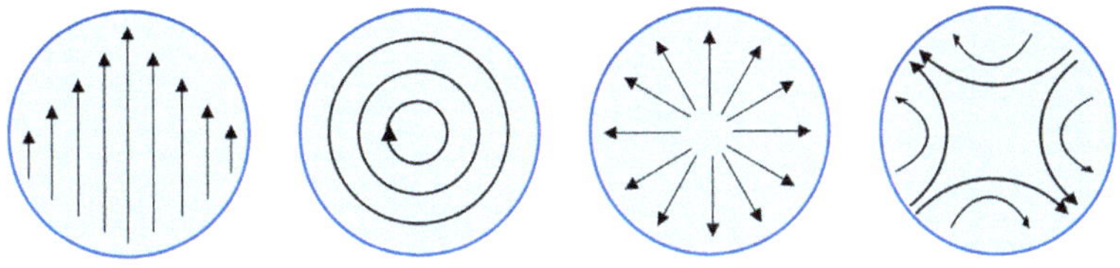

Figure 3.5 Sketch of the distribution of the electric field in the fiber cross for the first four guided modes in a step-index optical fiber. (*Adapted from:* [1].)

Let us now focus on the cutoff condition for guided modes. According to (3.57), the cutoff of a mode happens when it is no longer guided by the core, thus when its associated electric field do not longer decrease in the cladding. This happens when $\beta \leq n_2 k$. The cutoff conditions of the different modes can be studied by solving (3.65) when $\omega^2 \to 0$. This is rather complex and will not be recalled here. An important parameter linked to the cutoff condition is the normalized frequency redefined below:

$$V^2 = (ka\text{NA})^2 = \left(u^2 + w^2\right)a^2 \tag{3.70}$$

This parameter defines the number of modes that a fiber can support. The number of guided modes as a function of V can be obtained from the normalized propagation constant defined by:

$$b = \frac{a^2 w^2}{V^2} = \frac{(\beta/k)^2 - n_2^2}{n_1^2 - n_2^2} \approx \frac{n_{\text{eff}} - n_2}{n_2 \Delta} \tag{3.71}$$

Considering the evolution of b as a function of V (as depicted in Figure 3.6), each mode exists only for V values beyond a certain value. The cutoff of the modes is observed when $\beta/k = n_2$. The HE_{11} mode stops existing when the core diameter tends to zero. Single-mode operation is based on that observation. By choosing the a, n_1, and n_2 parameters such that $V < 2.405$, only the HE_{11} mode is guided and the fiber is single-mode. This happens for a core diameter of 8.2 μm, at a wavelength higher than the cutoff wavelength. Above this wavelength, the optical propagation is monomode.

Now that the propagation in step-index optical fibers has been reviewed, let's focus on plasmonic configurations derived from multimode optical fibers.

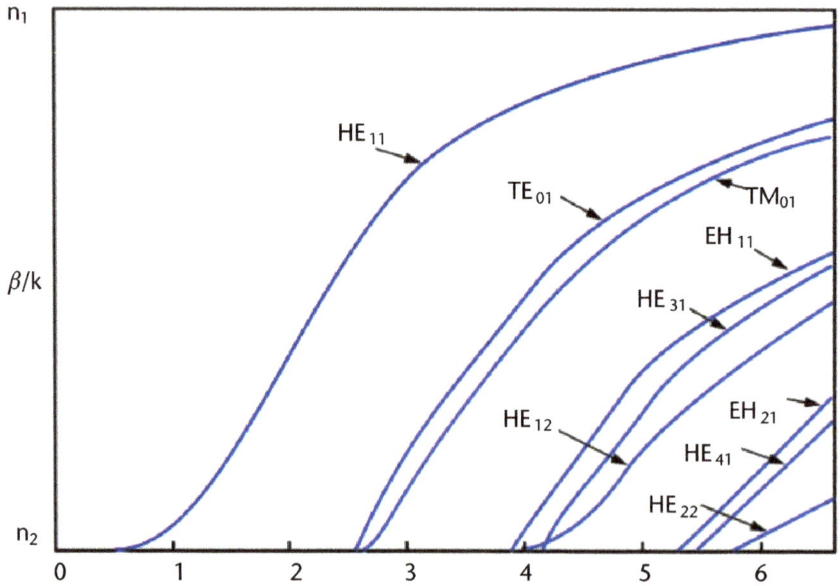

Figure 3.6 Evolution of b as a function of V for the first guided modes. (*Adapted from:* [1].)

3.2 Overview of Multimode Optical Fibers

People active in optical fiber telecommunications know that standard multimode optical fibers have a core diameter ranging between 50 and 100 microns surrounded by a cladding of 125 microns. Both materials are made from silica and the core region is slightly doped to allow light propagation, as reviewed before. These fibers are privileged over single-mode optical fibers for short-distance applications (especially in the fiber-to-the-home (FTTH) concept), because of more easy connection and light injection with light-emitting diodes (LEDs) or vertical-cavity surface-emitting lasers (VCSELs) at 850 nm and 1,300 nm.

There are two types of multimode optical fibers, depending on the transition of refractive index between the core and the cladding. When it is sharp, the fiber is called step index, while when it is gradual, the fiber is the graded index. In this case, the refractive index value decreases with increasing radial distance from the optical fiber axis, as sketched in Figure 3.7. The advantage of graded-index optical fibers compared to step-index optical fibers is the important decrease in modal dispersion, a parameter that plagues the communication as the signal is spread in time because the propagation velocity of the optical signal is not the same for all modes. Modal dispersion limits the bandwidth of multimode fibers. For instance, a 50-μm core step-index fiber would be limited to approximately 20 MHz for a 1-km length, which yields a bandwidth of 20 MHz·km. Modal dispersion can be strongly reduced, but never fully eliminated, with graded-index optical fibers. Hence, multimode graded-index fibers with bandwidths exceeding 3.5 GHz·km at 850 nm are common for use in 10-Gbps data links.

Modal dispersion should not be confused with chromatic dispersion, a distortion that results from the differences in propagation velocity of different wavelengths of light. Modal dispersion occurs even when a monochromatic light source is used.

It is also important to state that while the most straightforward configuration to outcouple light from the fiber core to the cladding is certainly by bending an optical fiber [3, 4], the vast majority of plasmonic multimode optical fiber sensors are derived from unclad configurations, where the fiber cladding is partially or totally removed. Such unclad sensors are most often used at operating wavelengths between 500 and 800 nm and have been made from very large core fibers, in the range from 200 μm to 600 μm instead of smaller core fibers, such as those used

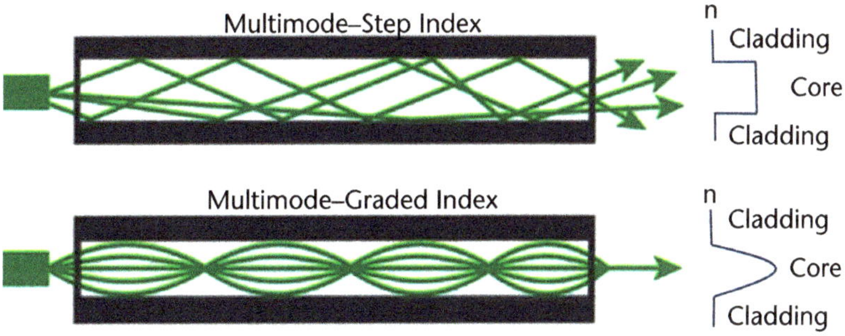

Figure 3.7 Sketch of the fiber cross-section of the step-index and graded-index multimode optical fibers.

in communication systems, as cladding removal alters the mechanical resistance of pristine fiber surfaces [5, 6]. Such fibers are very specific and, considering the usually very short distances required for sensing, they appear to be a much more pragmatic choice for sensor implementation. Indeed, their core is made in pure silica (step-index), while their cladding is often made with a material called Technology Enhanced Clad Silica (TECS) hard cladding. It is a hard fluoropolymer developed by 3M™ and now manufactured by Thorlabs™. Both materials are surrounded by a coating in Tefzel that can be readily removed with a mechanical stripper. The hard cladding is usually removed by immersion of the fiber in a solvent such as acetone, which appears to be a much easier process than the removal of a silica cladding. The latter requires either chemical etching (with very dangerous hydrofluoric acid) or fine mechanical polishing and is therefore much more sophisticated.

Looking at the standard specifications provided by the manufacturers of such fibers, Table 3.1 summarizes the main features for different geometries of step-index multimode optical fibers characterized by a numerical aperture equal to 0.39.

The performance of plasmonic sensors derived from such multimode optical fibers depends on intrinsic features such as the core diameter, numerical aperture, sensing layer, dopants in the fiber core, and the used metal layer. These effects are studied in [5, 7–16] and extensively reviewed in [17] and the reader is invited to refer to these references for additional considerations to the summary that is provided hereafter.

3.3 Unclad or Etched Configurations

The development of plasmonic optical fiber sensing began in the early 1990s. Among the first reports, we can quote the work of Villuendas and Pelayo [18]. They used optical fibers to inject and collect light to and from a mirrored cylindrical surface on which a metal film was deposited. The first intrinsic all-fiber plasmonic sensing configuration can be attributed to Jorgenson and Yee and dates back to 1993 [19]. Because it is the seminal contribution that has opened the path to thousands of contributions in the following years, we find relevant to remind here the abstract of their paper [19]:

> "A fiber-optic chemical sensor is presented which utilizes surface plasmon resonance excitation. The sensing element of the fiber has been fabricated by removing a section of the fiber cladding and symmetrically depositing a thin layer of highly reflecting metal onto the fiber core. A white-light source

Table 3.1 Typical Characteristics of Multimode Optical Fibers Used in Plasmonic Sensing

Core Diameter	*Cladding Diameter*	*Coating Diameter*	*Max. Attenuation at ~800 nm*
200 ± 5 μm	225 ± 5 μm	500 ± 30 μm	14 dB/km
400 ± 8 μm	425 ± 8 μm	730 ± 30 μm	14 dB/km
600 ± 10 μm	630 ± 10 μm	1,040 ± 30 μm	12 dB/km

is used to introduce a range of wavelengths into the fiber optic. Changes in the sensed parameters (e.g., bulk refractive index, film thickness and film refractive index) are determined by measuring the transmitted spectral-intensity distribution. Experimental results of the sensitivity and the dynamic range in the measurement of the refractive indices of aqueous solutions are in agreement with the theoretical model of the sensor."

Looking at the paper content, a "novel fiber-optic SPR sensing configuration without the required light-coupling prism" is presented. This configuration allows for a small sensing element and sample volume and a simplified optical design and brings the potential use for disposable and remote sensing. In this seminal paper [19], the sensing element was a multimode optical fiber section for which the cladding has been removed and a 550-nm-thick silver film was deposited all around the fiber core, using an electron-beam evaporation process. The authors specified that the length of the removed cladding was chosen to optimize the average number of reflections of propagating rays of light and modes in the fiber. We will discuss the meaning of that sentence next. The authors also stated that their sensing configuration relied on a white-light source, thereby resulting in a large range of excitation wavelengths. This is an important difference with respect to traditional SPR measurements based on prisms that use a discrete excitation wavelength and modulate the incidence angle. The authors finally underlined that the sensed parameters could be determined from the measured resonance spectrum in the transmitted spectral intensity distribution.

The authors used a 400-μm core diameter, a 600-μm cladding diameter, and a 760-μm coating diameter multimode optical fiber with a numerical aperture of 0.36 that can support internal propagating angles of light from 90.0° to 78.5°, with respect to the perpendicular to the core-cladding interface. The cladding and coating layers were removed using a hobby torch, following a method reported in [20]. They have produced sensing regions of 6, 10, and 18 mm in lengths that were coated with silver. The fibers were rotated during the metal deposition to ensure a good uniformity of the metal thickness. This yields the configuration sketched in Figure 3.8.

A refractive index sensitivity ranging between 4.5 10^{-4} and 7.5 10^{-5} RIU refractive index unit (RIU) is reported, which is comparable to the sensitivity of SPR bulk-optic systems at that time.

In terms of optical equipment (see the sketch of Figure 3.9), light from a tungsten-halogen lamp was focused in the optical fiber through a mode scrambler to populate all modes of the fiber core. A spectrograph encompassing a 1,024-pixel charge-coupled device (CCD) linear array detector was used to record the spectrum in the wavelengths range between 500 and 900 nm.

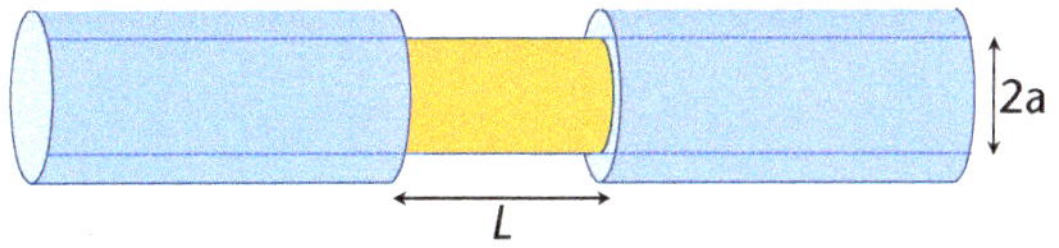

Figure 3.8 Schematic diagram of a metal-coated, unclad, multimode optical fiber section.

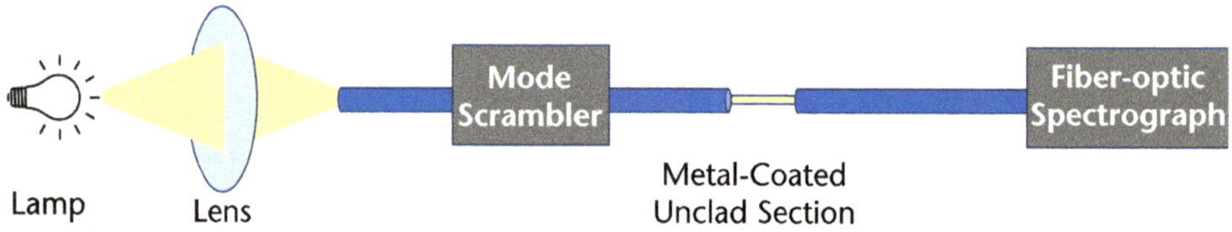

Figure 3.9 Sketch of plasmonic, unclad, multimode optical fiber working in transmission. (*Adapted from:* [19].)

The year after their seminal contribution, Jorgenson and Yee demonstrated that the dynamic range can be tuned between 1.00 and 1.40 RIU with the addition of thin high refractive index overlays [21]. This means that both aqueous and gaseous media can be probed with such a configuration. The upper limit of the dynamic range to the surrounding refractive index was even extended to 1.70, thanks to a fiber core in sapphire. To prepare the sensing region, a chemical etching with hot sulfuric acid was performed to remove the fiber cladding over a length of 10 mm.

In the early and rather complete studies of Jorgenson and Yee, a very good adequacy is also demonstrated between theoretical considerations and actual experiments and has paved the way to numerous research and development efforts on unclad multimode optical fiber configurations, supported by the advent of even more effective fiber implementations and optical devices. Progressively, the multimode optical fibers mentioned in Section 3.2 have emerged and have outclassed all-silica fibers, due to the relative easiness in removing their polymer-based cladding. Also, configurations in the reflection mode have become much more popular than the transmission operation sketched in Figure 3.9. In this case, the tip of a multimode optical fiber is unclad and cut at the right angle using a dedicated cleaver. The unclad section has a typical length between 0.5 cm and a couple of centimeters. For metal deposition, these fiber optrodes are usually positioned vertically in front of the metallization source so that both the unclad fiber section and the fiber end face are coated with the metal in a single run, as depicted in Figure 3.10. The metal deposited on the cleaved fiber end face acts as a mirror, allowing this configuration to operate in reflection mode.

A bifurcated optical fiber assembly, two fibers in the common end that break out into two legs at the other end, is then used to send light from the source to the optrode and collect the reflected signal towards a spectrometer. The resulting experimental setup is represented in Figure 3.11.

In the following, we derive the mathematical formalism that allows to explain the excitation of SPR with unclad multimode optical fibers and that is based on a geometric approach. Let us then consider a step-index MMF with a core radius a illuminated by a diffused (or Lambertian) source (such as the one of light-emitting diode) where the radiant intensity can be modeled by

$$I(\theta) = I_0 \cos(\theta) \tag{3.72}$$

where $0 \leq \theta \leq \pi/2$ and I_0 is a constant. Each differential element dS of that kind of source emits light in all directions.

Figure 3.10 (a) Multimode optical fiber tips deposited in a sputtering chamber for gold deposition from the top, (b) a cleaver is used to cut the multimode optical fibers, and (c) a gold-coated unclad multimode optical fiber tip inserted in a dedicated quick connector.

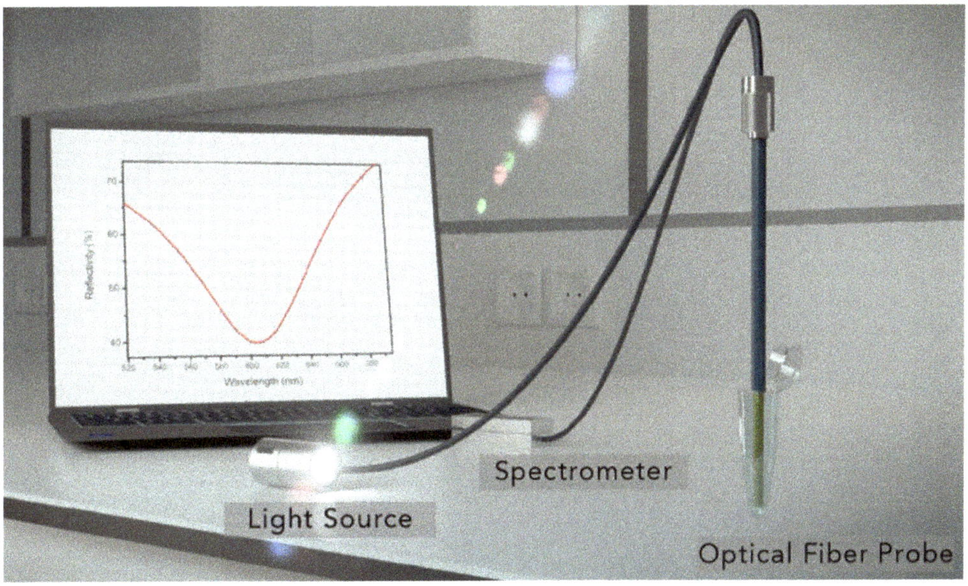

Figure 3.11 Artistic view of a plasmonic, unclad, multimode optical fiber operating in reflection mode.

Let us now assume an extended source of light placed at the fiber input. Considering the angles schemed in Figures 3.12 and 3.13, the power element dP radiated by an elementary source area dS through a solid angle $d\Omega$ is given by [2]:

$$dP = I(\theta) dS \, d\Omega \tag{3.73}$$

where $dS = r\, dr\, d\phi$, $d\Omega = \sin\theta\, d\theta d\varphi$, $r \in [0, a]$, $\phi \in [0, \pi]$, $\theta \in [0, \pi/2]$, and $\varphi \in [0, 2\pi]$. Then the total power radiated into the core of the optical fiber is

$$P_{\text{tot}} = \int_0^a dr \int_0^{2\pi} d\phi \int_0^{\pi/2} d\theta \int_0^{2\pi} d\varphi I_0 \cos\theta \sin\theta r = \pi^2 a^2 I_0 \tag{3.74}$$

However, only part of the source rays can propagate as guided rays. Indeed, θ has a threshold θ_m such that $0 \le \theta \le \theta_m < \pi/2$ and satisfying the Snell-Descartes law recalled hereafter:

$$n_0 \sin\theta_m = n_{co} \sin\theta_c \tag{3.75}$$

where n_0 is the refractive index of the medium where the source radiates, n_{co} is the one of the core, and n_{cl} is used for the cladding. θ_c is the complementary critical angle such that:

$$\sin\theta_c = \sqrt{1 - \left(\frac{n_{cl}}{n_{co}}\right)^2} \tag{3.76}$$

Equation (3.74) with the maximum angle θ_m yields to the total guided ray power

$$P_{gr} = \int_0^a dr \int_0^{2\pi} d\phi \int_0^{\theta_m} d\theta \int_0^{2\pi} d\varphi \; I_0 \cos\theta \sin\theta r = \frac{\pi^2 I_0 \text{NA}^2 a^2}{n_{0^a}} \tag{3.77}$$

where NA is the numerical aperture defined by (3.5). So the bound ray power increases with the square of the numerical aperture. At this stage, we can determine

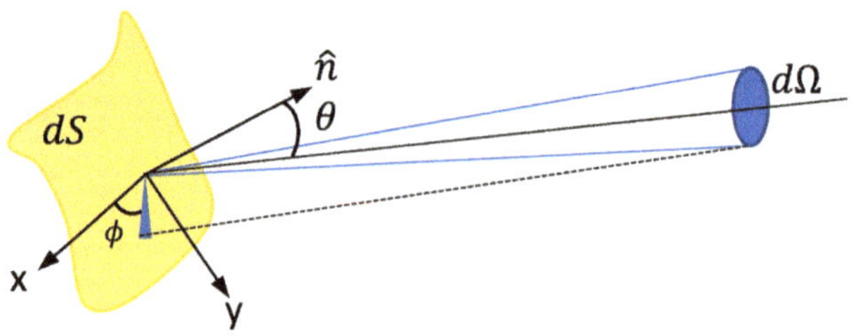

Figure 3.12 Optical source covering a surface dS emitting light through a solid angle $d\Omega$. $\hat{n}$ is a unit vector normal to the source surface.

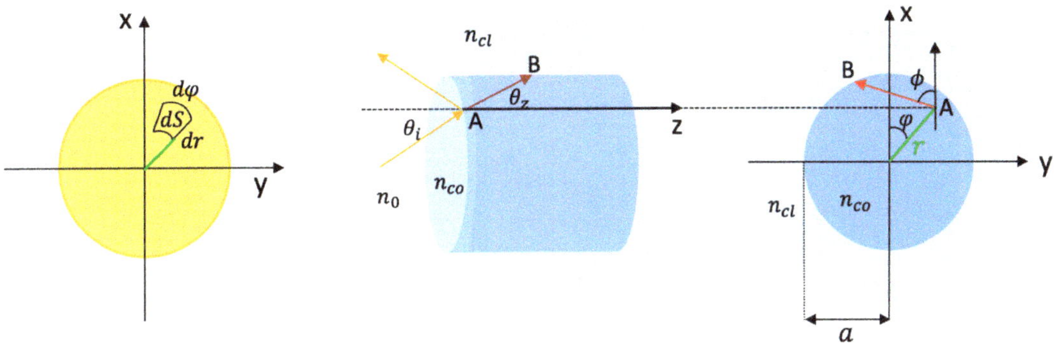

Figure 3.13 Extended optical source whose radius matches the one of the fiber core (left), incident ray of light impinging the core of an optical fiber (middle; the red arrow represents the transmitted ray through the core), and projection to the fiber cross-section (right).

the efficiency of the source, which is defined as the portion of the total power that is converted to bound rays. Making the ratio between (3.77) and (3.74), one can get

$$\varepsilon = \frac{P_{gr}}{P_{\text{tot}}} = \frac{\text{NA}^2}{n_{0^a}} \tag{3.78}$$

Because the refractive index n_0 is often the one of air, the efficiency of a diffused source to convert light to bound ray depends only on the numerical aperture. Now let us focus our attention to the bound ray propagation in the core of a step-index optical fiber.

Let us consider a guided ray satisfying (3.21) and propagating through the fiber core from α to β, as displayed in Figure 3.14. The distance L_p corresponds to the length between two reflections and a straightforward geometrical analysis yields:

$$L_p = \frac{2a\sin\phi}{\sin\theta_z} \tag{3.79}$$

Thus, we have the number of reflections per unit of length given by

$$N = \frac{1}{L_z} = \frac{\tan\theta_z}{2a\sin\phi} \tag{3.80}$$

where $L_z = L_p\cos\theta_z$ is the distance covered by the guided ray along the fiber axis (i.e., between α and γ). If we consider a diffused source, emitting a ray from a medium of refractive index n_0, the refracted ray through the fiber core satisfies the Snell-Descartes law such as

$$n_0\sin\theta = n_{co}\sin\theta_z \tag{3.81}$$

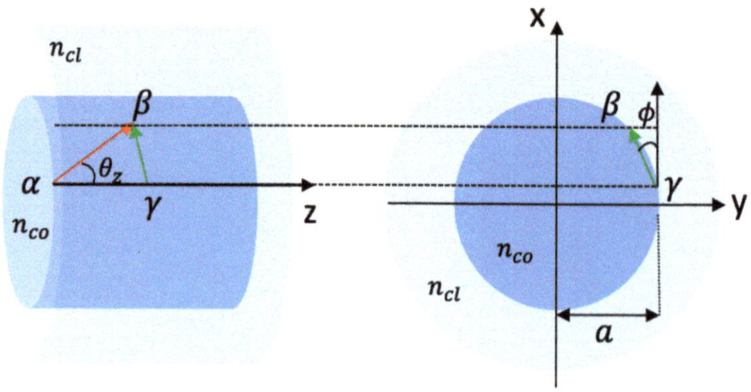

Figure 3.14 Sketch of a guided ray propagating from α to β (left) and projection onto the XY-plane (right).

Using (3.81) and (3.77), we can write:

$$P_{gr} = \int_0^a dr \int_0^{2\pi} d\phi \int_0^{\theta_c} d\theta_z \int_0^{2\pi} d\varphi \; I_0 \left(\frac{n_{co}}{n_0}\right)^2 \cos\theta_z \sin\theta_z r \tag{3.82}$$

For convenience, the light source is assumed to emit from air ($n_0 = 1$). After each reflection, the transmitted power is altered by the reflectance R. Consequently, the normalized power transmitted through the fiber after a propagation length along the fiber axis L_z is given by the ratio

$$P = \frac{\int_0^a dr \int_0^{2\pi} d\phi \int_0^{\theta_c} d\theta_z \int_0^{2\pi} d\varphi \, R^{NL_z} I_0 n_{co}^2 \cos\theta_z \sin\theta_z r}{\int_0^\rho dr \int_0^{2\pi} d\phi \int_0^{\theta_c} d\theta_z \int_0^{2\pi} d\varphi \, I_0 n_{co}^2 \cos\theta_z \sin\theta_z r} \tag{3.83}$$

where R is the reflectance at the core-cladding interface given by

$$R = \frac{R_p + R_s}{2} \tag{3.84}$$

where the subscripts p and s account for the polarization of light. In unclad plasmonic MMF configurations, the cladding along a length L has been removed and replaced by a metal layer and immersed into a solution of refractive index n_{ext}. Following the consideration of Chapter 2, R_p is given by (2.86) and R_s is obtained using the same development for TE modes. Figure 3.15 shows the resulting transmitted spectrum obtained using (3.83) in the case of a silver-coated fiber, as reported in the seminal paper of Jorgenson and Yee [19]. The dielectric constant parameters used are the same as in Chapter 2 (see Table 2.1) while other fiber parameters are grouped in

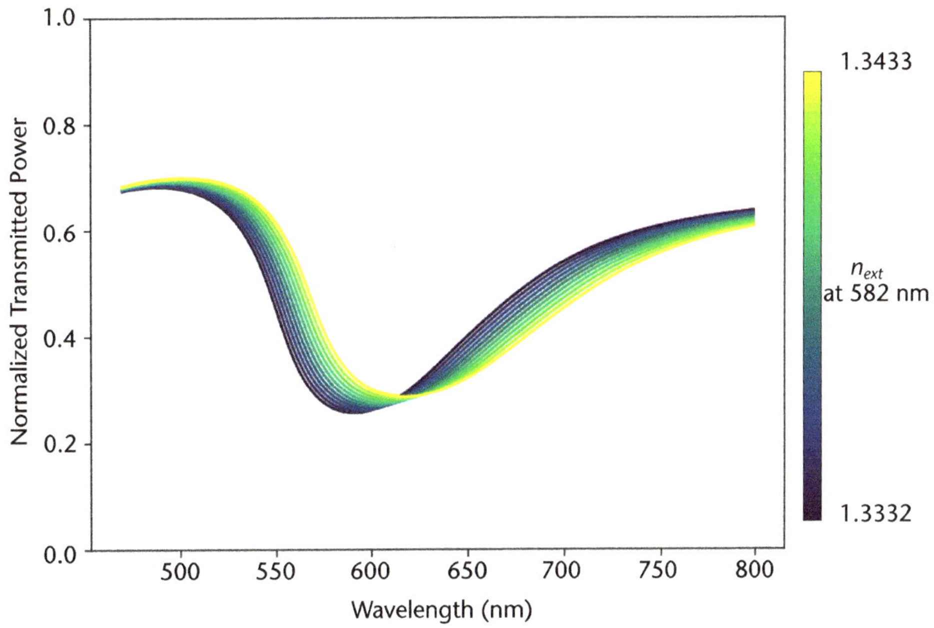

Figure 3.15 Simulation of a silver-coated, unclad, multimode fiber transmitted power. The refractive index of the fluid range from the one of water (1.3332 measured at 582 nm) to 1.3433.

Table 3.2. Using this simple, theoretical approach, one can see the characteristic attenuation related to SPR. While the refractive index of the surrounding medium increases, the spectrum is shifted while the FWHM increases.

According to (3.83), the number of reflections of a ray in the fiber core decreases as the core diameter increases. The effect of the sensing length is the opposite. For an increase of length from 2 mm to 10 mm, a decrease of approximately 5% of the refractometric sensitivity was reported, accompanied by an increase of the resonance width of approximately 50%. Looking at the numerical aperture parameter, its increase from 0.17 to 0.25 yields both an increase of the refractometric sensitivity and the SPR width. The metal and, to a lesser extent, the dopants also influence the refractometric performance, while the presence of an additional coating overlay can significantly improve the sensitivity [17]. One can consider that the ultimate bulk refractometry of these plasmonic unclad multimode optical fiber configurations can reach approximately 4,000 nm/RIU, with a figure of merit (FoM) of approximately

Table 3.2 Fiber Parameters Used for the Simulation Displayed in Figure 3.15

Parameters	*Values*
Thickness of the silver layer	50 nm
Core radius	200 μm
Numerical aperture	0.39
Fiber axis length of the metal layer	1 cm

40 [5, 6, 22, 23]. In [24], silver was used for the overlay and detections of refractive index variation as low as 5 10^{-5} RIU were reported.

Figure 3.16 shows the typical SPR spectrum recorded for a gold-coated, unclad, 400-μm-core multimode optical fiber optrode. One can see that the resonance is centered around 600 nm and its FWHM is approximately 100 nm.

Figure 3.17 displays the wavelength shift of the SPR spectrum of a gold-coated, unclad, 400-μm-core, multimode optical fiber when it is immersed in an aqueous solution whose refractive index is modified by adding small quantities of salt (here LiCl) in the solution. Tracking the minimum of the resonance allows measuring the surrounding refractive index shift, after calibration. The obtained sensitivity reaches 1,300 nm/RIU in this example. As discussed before, there is a room for optimization depending on the actual physical parameters of the gold-coated unclad fiber section.

Plasmonic, unclad, multimode optical fibers are one of the most studied and developed configurations, certainly due to their easy manufacturing. Since their first report in 1993, these configurations have known an important rise in maturity such that they became commercially available. As a very prominent example, the FOx Biosystems company in Belgium has developed a robust, low-maintenance, and easy-to-use benchtop device based on the plasmonic unclad multimode optical fiber technology that can be used for quantification, affinity, and binding interaction studies between biomolecules.

Let us finally mention that there are variants to this approach. For instance, we can quote [25], which consists in making a narrow trench in the optical fiber cladding with a femtosecond laser source, thereby exposing a narrow strip of the core, where metal nanoparticles can be deposited to form a localized SPR sensor.

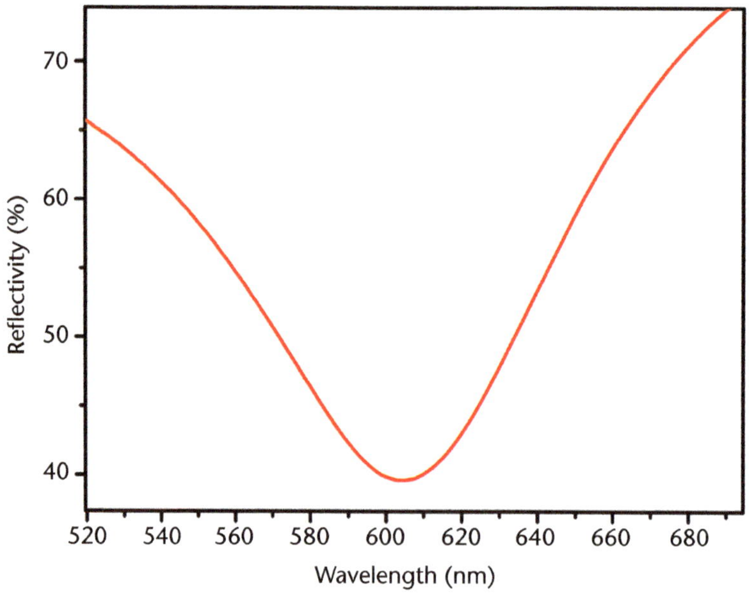

Figure 3.16 SPR spectrum of a gold-coated, unclad, 400 μm-core, multimode optical fiber (experimental data).

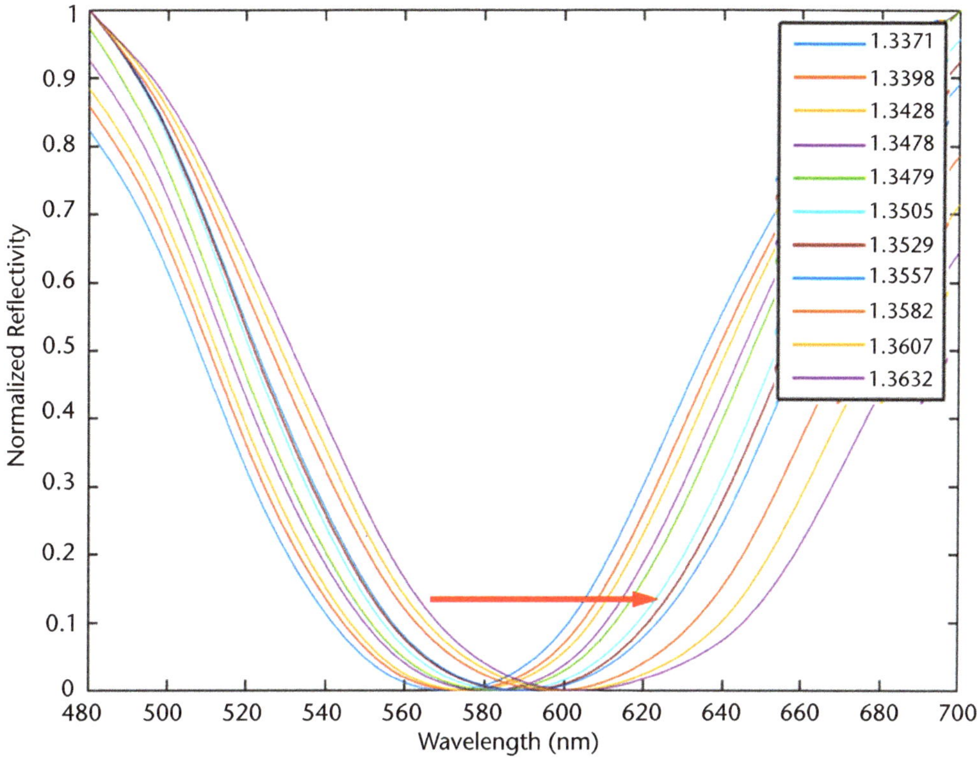

Figure 3.17 Wavelength shift of the normalized SPR spectrum of a gold-coated, unclad, multimode optical fiber optrode immersed in different calibrated refractive index solutions (experimental data).

3.4 Tapered Configurations

Tapered configurations are more sophisticated than unclad ones. A tapered optical fiber has a diameter that decreases gradually or abruptly to a certain value before increasing again to reach the initial fiber diameter. The narrowest segment of a tapered fiber is called the waist.

An optical fiber can be tapered by heating and pulling. A tapering station must entail a heat source (flame, mini-oven, CO_2 laser electric arc) to soften the glass and translation stages to pull the optical fiber in a controlled manner. Some fusion splicers have special programs to taper optical fibers and capillaries. Commercially available machines can taper optical fibers with literally any diameter and assemblies of optical fibers inside a capillary tube, in a controlled and reproducible manner. The taper profile can be linear, exponential, or parabolic. Therefore, the technology to fabricate tapered fibers can be considered as mature. Figure 3.18 depicts the sketch of a tapered optical fiber section.

The benefits of plasmonic sensors based on tapered optical fibers have been first theoretically studied using the ray approach [8, 11, 26]. In [27], an experimental validation was provided based on a standard graded-index multimode optical fiber with 62.5-μm core diameter and 125-μm cladding diameter. Using a tapering machine, a taper region (waist) with a small and uniform diameter was created between two narrowed transition regions with a gradually decreasing diameter (down-taper) and an increasing diameter (up-taper) adjacent to two untapered regions, as sketched in Figure 3.18. The whole tapered region was coated with a thin layer of gold following a sputtering process. The uniform waist length was 6 mm, and the up-taper and down-taper regions were 2 mm each.

As for unclad optical fiber configurations, it can be shown that the spectral width of the SPR curve increases with the number of reflections undergone by a ray. In this case, one can write:

$$N_{\text{ref}}(\theta,z) = \frac{L_w}{2a(z)\tan(\theta + \Omega)} \tag{3.85}$$

where L_w is the taper-waist length, $a(z)$ is the fiber taper radius at distance z from the input of the optical fiber section, θ remains the angle between the ray and the perpendicular to the surrounding medium interface, and Ω is the taper angle that can be expressed by:

$$\Omega = \tan^{-1}\left[\frac{a - a_0}{L_t}\right] \tag{3.86}$$

where a_0 is the radius of the waist-tapered fiber and L_t is the transition region length.

For measurements, such sensors are also connected to a white light source and a spectrometer. Figure 3.19 shows their typical transmitted amplitude spectrum. As for unclad fibers, an increase of the surrounding refractive index value leads to a red shift of the SPR curve, together with a change in amplitude.

The effect of the waist diameter on the refractometric sensitivity was experimentally studied in [27]. It was shown that a decrease of the taper-waist diameter from 85 μm to 25 μm resulted in an increase of the bulk refractometric sensitivity from 1,395 nm/RIU to 1,914 nm/RIU for a 55-nm, gold-coated, tapered multimode optical fiber. The mean FWHM was also increased from 167 to 217 nm with the decrease of the taper-waist diameter. It was also shown that the sensitivity increased from 1,255 to 1,547 nm/RIIU and the mean FWHM decreased from 185 to 138 nm by reducing the taper-waist length from 9 to 3 mm. This results from the diminution of reflections undergone by a ray when the taper waist length is shorter. Besides these two important physical parameters (waist diameter and length) that drive

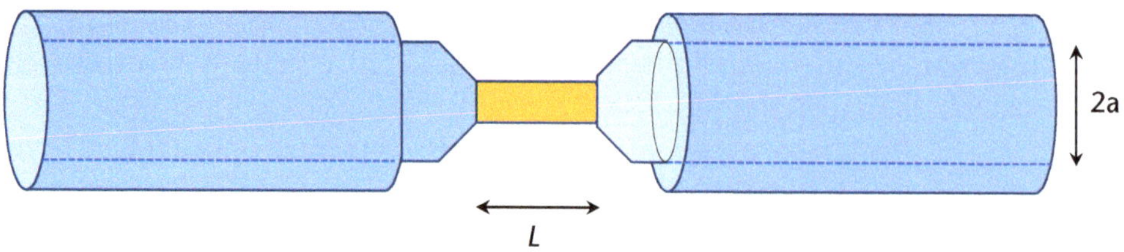

Figure 3.18 Schematic diagram of a tapered, multimode optical fiber configuration.

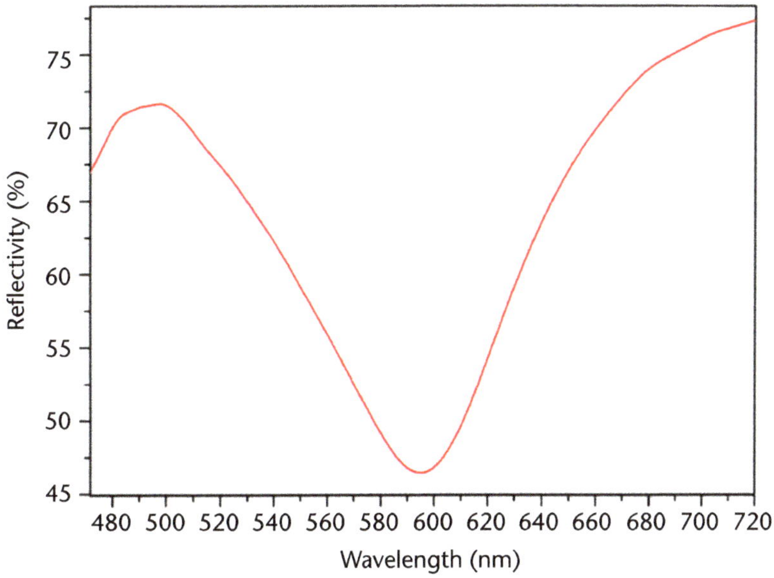

Figure 3.19 Transmitted amplitude spectrum of a gold-coated, tapered, multimode optical fiber (experimental data).

the overall sensing performance, it was shown that sensors with shorter transition regions have higher magnitude of evanescent field emanating from the optical fiber, which leads to an approximately 10% increase in sensitivity to changes of the surrounding refractive index medium.

Looking at the literature, plasmonic, tapered, multimode optical fiber configurations are much less popular than unclad configurations, certainly because of the increased complexity of their fabrication process. We will see in Chapter 4 that single-mode optical fiber counterparts have been reported, with interesting assets, despite a very reduced fiber diameter. Numerous applications have also been reported for such configurations.

Let us finally add that, in [28], a sensitivity reaching 11,885 nm/RIU in the range of 1.404 to 1.434 was demonstrated thanks to the use of an Au/Ti overlay on a tapered fiber with a 40-μm waist diameter. This remains the ultimate refractometric sensitivity reported for such structures.

3.5 D-Shaped Configurations

The first SPR-based, side-polished, multimode fiber sensor configuration was reported in 2007 [29]. It was based on a graded-index multimode fiber with a 62.5-μm core diameter and a 125 μm cladding diameter. The multimode fiber was placed in a V-groove holder and locally polished using polishing diamond films. The length of the polished part was kept very small at 0.5 mm, while the depth was chosen to be 55 μm so that the core is locally exposed to the surrounding medium. Figure 3.20 depicts a sketch of the obtained configuration.

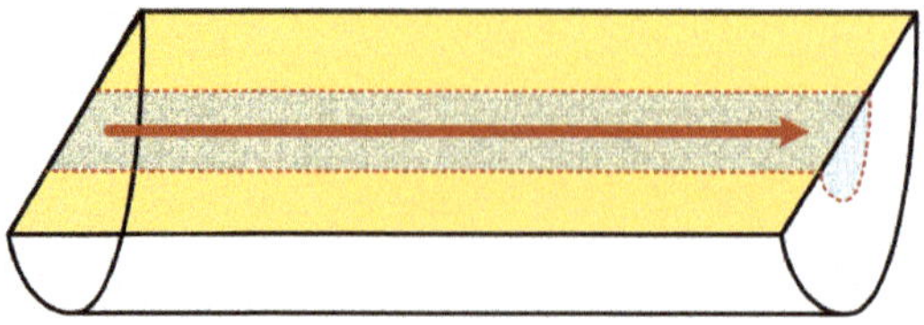

Figure 3.20 Schematic diagram of a side-polished multimode optical fiber configuration.

The authors determined that the optimal gold thickness for this configuration was 37 nm. They also tested thicknesses of 30 and 43 nm and were able to report that the SPR curve was sharper and had a larger peak-to-peak intensity for the 37-nm thickness. The authors used a halogen lamp and an optical spectrum analyzer to record the transmitted amplitude spectrum of their device. They also used a reference path so that the measured spectrum was the subtraction of the intensity measured through the side-polished fiber and the one measured through a reference fiber. Figure 3.21 shows the SPR curve measured when the 37-nm, gold-coated, side-polished fiber is immersed in alcohol (refractive index = 1.361).

The authors showed that the minimum of SPR curve shifted from 745.71 to 796.95 nm when the surrounding refractive index was changed from 1.38 to 1.40. A refractometric sensitivity slightly higher than 3,000 nm/RIU was reported in this pioneering work. An enhancement of the resonance shape was demonstrated for two 5-mm-long, side-polished regions separated by 5 mm [30]. A multi-D-shaped

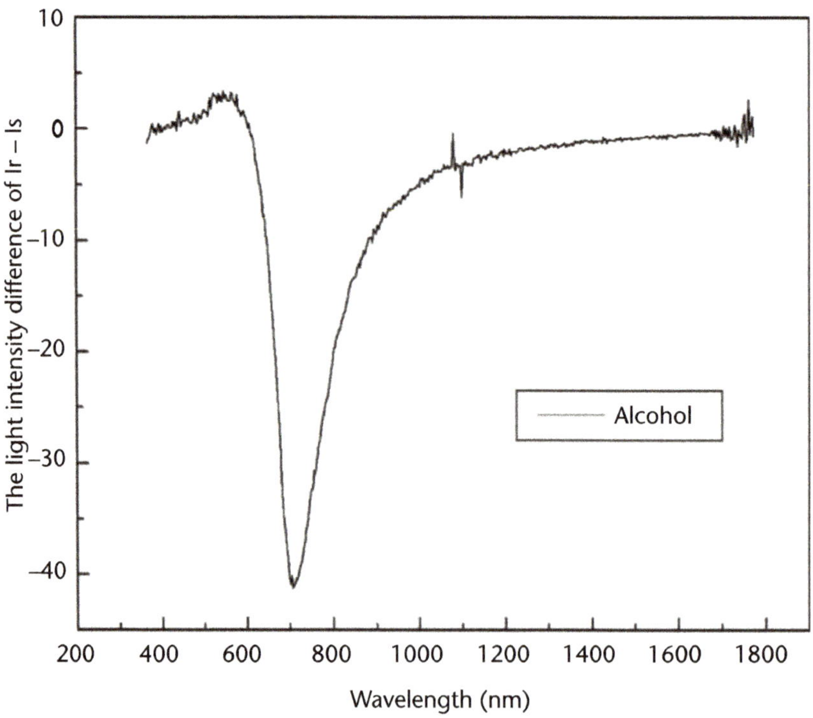

Figure 3.21 SPR curve recorded in alcohol (n = 1.361) for a gold-coated side-polished graded-index multimode optical fiber. (*From:* [29].)

configuration comprising up to 7 cascaded, side-polished regions was even presented in [31]. Interestingly, the D-shaped regions were produced using tightly focused femtosecond laser pulses (800-nm wavelength, approximately 120-fs pulse duration, 1-kHz repetition rate, and approximately 3.5-mJ maximum pulse energy).

In [32], a step-index, 200-μm-core, multimode optical fiber was side-polished over 20 mm down to about half of the core diameter using a diamond polishing film. This region was then coated with a stack of an inner gold layer (18 nm), a ZnS-SiO_2 layer (220 nm), and then an outer gold layer (18 nm). The refractometric sensitivity of the SPR mode was estimated to be 915 nm/RIU.

Side-polished fiber sensors have also been derived from 600-μm-core, plastic-clad, multimode optical fibers with 0.37 numerical aperture [33]. Lapping plate and several kinds of abrasive paper with various degrees of roughness were used to polish the fiber probe. It can be extrapolated from the measured data that the actual refractometric sensitivity is approximately 3,000 nm/RIU.

Few-mode fibers have been studied as well [34–36]. They have a core diameter comprised between two optical fibers: single-mode and multimode. In [36], a 19-μm core and 125-μm cladding silica optical fiber is side-polished over a region of 10 mm. A refractometric sensitivity reaching 4,796 nm/RIU is reported in the surrounding refractive index range (1.394 to 1.404).

It has been recognized that D-shaped or side-polished fiber configurations bring the important practical asset to require metal deposition on only one face of the optical fiber, thereby avoiding the complexity of deposition all around the cylindrical section. In [37], Homola et al. recognized that a modal distribution of light in a multimode optical fiber is very sensitive to mechanical disturbances, so that disturbances occurring close to the sensing area of the fiber may cause intermodal coupling and modal noise. Hence, as for tapered configurations, there are more reports about the use of single-mode fibers for side-polished configurations [38].

3.6 U-Bent Configurations

The most nonintrusive approach to extract light from a fiber core is certainly by bending. It is well known that optical fibers become lossy when bent beyond a certain critical radius, because the evanescent field associated with total internal reflection becomes radiative. Therefore, plasmonic optical fiber devices can be obtained using metal-coated U-shaped optical fibers, as sketched in Figure 3.22. While a reversible bend obtained by flexing the fiber would work, it would not be very reliable because the outer glass surface of the bend would be under very strong tensile stresses. Therefore, it is preferable to fabricate a permanent sharp bend in a fiber. The structures demonstrated so far are obtained from large core fibers by exposing an unclad section to a flame and shaping it to form a U. To that aim, a propane or oxy-butane torch or even a simple wax candle can be used. By controlling the heat, it is possible to produce different bend radii with good repeatability, down to 0.5 mm. Depending on the fiber core diameter, the temperature is adjusted. For 200-μm core-diameter fibers, the temperature of the flame lies in the range of 400°C to 500°C while it can increase up to 700°C for 600-μm-core fibers. Such configurations were first proposed in [39] and theoretically studied in [40] for plasmonic sensing.

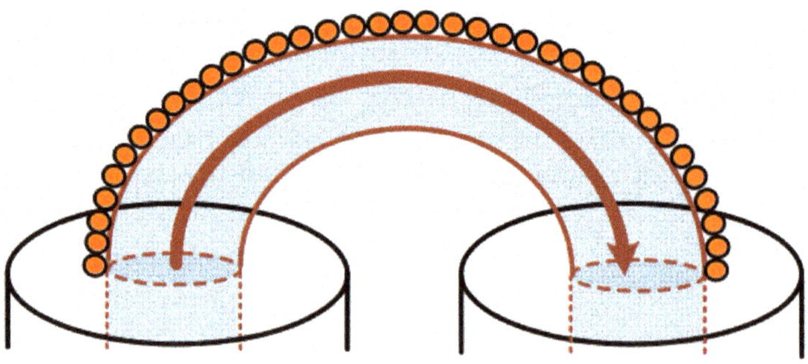

Figure 3.22 Schematic diagram of a U-bent multimode optical fiber configuration.

In practice, U-bent optical fibers can be considered single-ended because the input and output fibers can be colocated in a very small tube, even if they operate in transmission. A very recent and quite complete review about plasmonic U-bent optical fiber configurations can be found in [41]. The interested reader is therefore invited to consult this reference for further details. It can be found that the U-bent configurations in polymer optical fibers are much more popular than those based on silica. They will be reviewed in Chapter 5. Very few configurations were reported with silica fiber and they were not developed with continuous metal films. The example reported in [42] used gold nanoparticles for LSPR generation. In this case, the transmitted amplitude spectrum contains a resonance band whose amplitude can be tightly correlated with the SRI value.

3.7 Interferometers: Hetero-Core Structures

Interferometric systems require the excitation of more than one copropagating mode over a finite distance over which the modes propagate at different phase velocities. Any change in the surrounding refractive index region where the modes propagate will change their relative phase velocities and, hence, their field amplitude superposition. It will thus change the power coupled into the output of the device.

Each mode excited in a waveguide propagates at a different phase velocity. The phase of the *i*th mode (ϕ_i) after propagation in an optical fiber or waveguide of length L is $\phi_i = \beta_i L$. An initial phase (γ_i) of the mode at the optical fiber input can be added to ϕ_i, but, in many practical applications, γ_i can be neglected as it is common to all the modes. When two or more modes propagate in an optical fiber, a beating can be produced between them, thereby inducing a relative phase shift between the modes. This leads to mode interference in an optical fiber, which was an issue (modal dispersion) in the early years of optical telecommunications when multimode fibers were used. However, mode interference can be successfully used for sensing. In these cases, modal interferometers are implemented to extract information from mode interference where the parameter to be sensed or measured is encoded.

The idea to use mode-mode interference for detecting sound was proposed in the late 1970s by Bucaro and Carome [43]. The concept was simple: light at a wavelength below the cutoff wavelength of a step-index, single-mode fiber was used. Under these conditions, the single-mode fiber enables multimode propagation, thereby allowing multimode interference.

Modal interferometers implemented with elliptical-core and birefringent optical fibers were demonstrated in the 1980s [44]. Here the two interfering modes were the fast and slow modes of mutually orthogonal polarizations. In these interferometers, polarizers or mechanisms to preserve the polarization state of the guided light were necessary.

Such modal interferometers bring practical issues due to the need for bulk optic components, such as microscope objectives and polarizers, as well as mechanical positioners to launch light in the fiber core. Therefore, these implementations are nowadays impractical for sensing applications. However, the concepts and ideas introduced in these pioneering works initiated important research and development efforts on modal interferometers and their applications, reporting a great variety of sensing, filtering, and other applications.

In practice, most modal interferometers are interrogated with a broadband optical source and a suitable spectrometer. Thus, the transmitted or reflected amplitude spectrum is analyzed. Changes of $\Delta\phi$ caused by a measurand result in shifts of the transmission or reflection spectrum. In other cases, changes in light intensity at a specific wavelength are monitored and correlated with the parameter being sensed. Shifts of an interference pattern or intensity changes are easy to measure with a spectrometer or photodetector. However, this approach can only be used to measure changes in a property to be sensed, not an absolute static value. Also, it cannot be used to determine changes that occur faster than the refresh rate of the spectrometer unless the wavelength change in question is less than one interfringe distance.

Depending on the launching conditions and the geometrical design, two or more modes can interfere. It is thus possible to devise two-mode or multimode interferometers. The sensor design then consists of choosing modes that will provide the highest selectivity and sensitivity for the target application and the corresponding launch conditions to access the selected modes. There is a great variety of optical fibers and configurations that can be used to produce compact modal interferometers [45], and, in this section, we will just focus on multimode optical fiber interferometers.

The idea of splicing a segment of step-index multimode optical fiber (MMF) between two segments of identical step-index SMFs was proposed in the late 1980s [46]. Such a structure was called a multifiber union, but nowadays it is called single-multi-single (SMS) mode fiber structure. Such structures can be produced with a conventional fusion splicer. It is very relevant for sensing physical parameters but will not be effective to sensing surrounding refractive index changes. To this aim, it is much more interesting to follow the opposite approach: the SMF section is spliced between two identical MMFs. Such a configuration is called a multimode-single-mode-multimode (MSM) configuration. It was investigated in [47], and it is depicted in Figure 3.23. An alternative makes use of a tapered single-mode optical sandwiched between two multimode optical fibers [48]. In both cases, light is outcoupled in the surrounding medium and the analysis of the interference pattern yields a sensitivity to surrounding refractive index changes. For the bare configuration, a sensitivity of 1,880 nm/RIU was reported in [48]. The application of a metal coating on the SMF area enables the excitation of surface plasmons. In [49], a hetero-core configuration was produced by using a 1-cm-long, 3.1-μm core, single-mode fiber

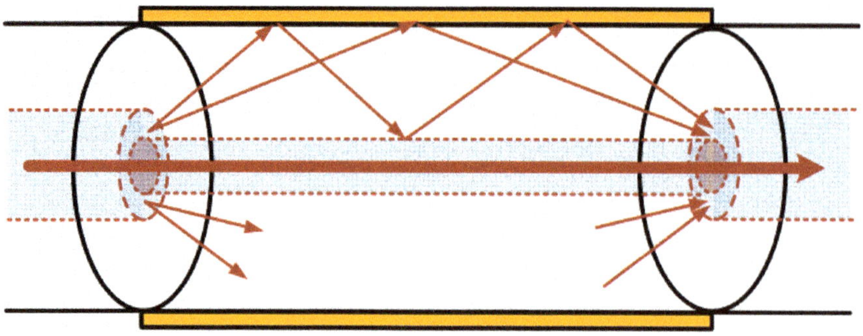

Figure 3.23 Schematic diagram of a plasmonic multimode-single-mode-multimode optical fiber interferometer configuration.

segment spliced between two 50-μm, graded-index, multimode optical fibers. The sensing region was silver-coated using a sputtering process. It exhibits a refractive index sensitivity of approximately 3,200 nm/RIU. Similar configurations were recently demonstrated in [50, 51].

3.8 Fiber End Facets

Finally, there is an important subset of single-ended, intrinsic, fiber-optic sensor probes that has been the object of renewed interest due to advances in nano-patterning technologies. It consists of fiber sensors where the sensing surface is located on the end of a cut fiber [52]. Here we do not consider fiber sensor devices where the fiber is only used to bring pump light to a medium and to recover fluorescence or Raman signals, for instance. However, as illustrated in Figure 3.24, (L)SPR sensor probes have been developed by patterning the flat fiber end face or covering it with an array of nanoparticles. In this case, the core-guided light is directly exposed to the nanoparticles, yielding LSPR generation.

SPR on periodic subwavelength metallic structures has attracted increasing interest since the first observation of extraordinary optical transmission [53]. By extremely concentrating electromagnetic fields beyond the diffraction limit in three dimensions, light-matter interactions are enhanced by several orders of magnitude. Immobilization of nanoparticles on optical fibers has been explored following a self-assembly approach [54, 55]. This brings a poor uniformity and lack of reproducibility. Electron beam lithography (EBL) has also been utilized to make metal nanostructures such as nanorings and nanorods on end facets of optical fibers, either directly or by transfer method [56–58]. In [59], nanohole arrays have been produced by a template transfer method. In [60], focused ion beam milling is used to pattern nanostructures from a metal film deposited on the end face of an optical fiber.

Although the references mentioned in this section are based on the use of multimode optical fibers, such developments are made most of the time on standard optical fibers that reflect light back towards a detector through a splitter at the input end and these configurations will therefore be reviewed in the next chapter.

In conclusion, after a reminder of light propagation mechanisms in optical fibers, this chapter reviewed the operating principle of the main plasmonic configurations

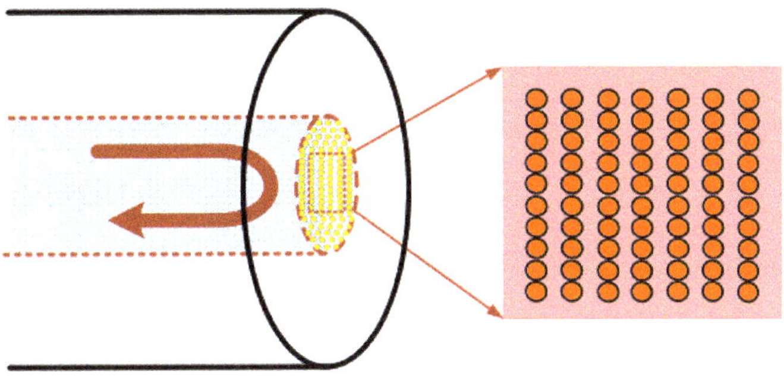

Figure 3.24 Schematic diagram of a plasmonic fiber end facets configuration.

based on the use of multimode optical fibers. The general operating principle consists in exposing the core-guided light to the surroundings by a local mechanical alteration of the fiber cladding, either by removal (totally or in large part) or through bending. In terms of technological maturity, sensors based on unclad multimode optical fiber sections remain the most advanced ones. They can be used as optrodes and present the required sensitivity for use in most biochemical sensing applications, as confirmed in Chapters 6 to 8.

References

[1] Keiser, G., *Optical Fiber Communications*, New York: McGraw-Hill, 2000.

[2] Snyder, A. W., and J. D. Love, *Optical Waveguide Theory*, New York: Springer, 1983.

[3] Taylor, H. F., "Bending Effects in Optical Fibers," *Journal of Lightwave Technology*, Vol. LT-2, 1984, pp. 617–628.

[4] Gupta, B. D., H. Dodeja, and A. K. Tomar, "Fibre-Optic Evanescent Field Absorption Sensor Based on a U-Shaped Probe," *Optical and Quantum in Electronics*, Vol. 28, 1996, pp. 1629–1639.

[5] Dwivedi, Y. S., A. K. Sharma, and B. D. Gupta, "Influence of Design Parameters on the Performance of a Surface Plasmon Sensor Based Fiber Optic Sensor," *Plasmonics*, Vol. 3, 2008, pp. 79–86.

[6] Pollet, J., et al., "Aptamer-Based Surface Plasmon Resonance Probe," *IEEE Sensors*, 2008, pp. 1187–1190.

[7] Gupta, B. D., and C. D. Singh, "Evanescent-Absorption Coefficient for Diffuse Source Illumination: Uniform- and Tapered-Fiber Sensors," *Applied Optics*, Vol. 33, 1994, pp. 2737–2742.

[8] Verma, R. K., A. K. Sharma, and B. D. Gupta, "Modeling of Tapered Fiber-Optic Surface Plasmon Resonance Sensor with Enhanced Sensitivity," *IEEE Photonics Technology Letters*, Vol. 19, 2007, pp. 1786–1788.

[9] Sharma, A. K., and B. D. Gupta Rajan, "Influence of Different Dopants on the Performance of a Fiber Optic SPR Sensor," *Optics Communications*, Vol. 274, 2007, pp. 320–326.

[10] Lahav, A., M. Auslender, and I. Abdulhalim, "Sensitivity Enhancement of Guided-Wave Surface Plasmon Resonance Sensors," *Optics Letters*, Vol. 33, 2008, pp. 2539–2541.

[11] Verma, R. K., A. K. Sharma, and B. D. Gupta, "Surface Plasmon Resonance Based Tapered Fiber Optic Sensor with Different Taper Profiles," *Optics Communications*, Vol. 281, 2008, pp. 1486–1491.

[12] Singh, S., R. K. Verma, and B. D. Gupta, "LED Based Fiber Optic Surface Plasmon Resonance Sensor," *Optical and Quantum Electronics*, Vol. 42, 2010, pp. 15–28.

[13] Verma, R. K., and B. D. Gupta, "Surface Plasmon Resonance Based Fibre Optic Sensor for the IR Region Using a Conducting Metal Oxide Film," *Journal of Optical Society of America A*, Vol. 27, 2010, pp. 846–851.

[14] Bhatia, P., and B. D. Gupta, "Surface-Plasmon-Resonance-Based Fiber-Optic Refractive Index Sensor: Sensitivity Enhancement," *Applied Optics*, Vol. 50, 2011, pp. 2032–2036.

[15] Bhatia, P., and B. D. Gupta, "Surface Plasmon Resonance Based Fiber Optic Refractive Index Sensor Utilizing Silicon Layer: Effect of Doping," *Optics Communications*, Vol. 286, 2013, pp. 171–175.

[16] Singh, S., S. K. Mishra, and B. D. Gupta, "Sensitivity Enhancement of a Surface Plasmon Resonance Based Fibre Optic Refractive Index Sensor Utilizing an Additional Layer of Oxides," *Sensors and Actuators A*, Vol. 193, 2013, pp. 136–140.

[17] Gupta, B. D., S. K. Srivastava, and R. Verma, *Fiber Optic Sensors Based on Plasmonics*, Singapore: World Scientific Publishing, 2015.

[18] Villuendas, F., and J. Pelayo, "Optical Fibre Device for Chemical Sensing Based on Surface Plasmon Excitation," *Sensors and Actuators A*, Vol. 23, 1990, pp. 1142–1145.

[19] Jorgenson, R. C., and S. S. Yee, "A Fiber-Optic Chemical Sensor Based on Surface Plasmon Resonance," *Sensors and Actuators B*, Vol. 12, 1993, pp. 213–220.

[20] Weber, A., and J. S. Schultz, "Fiber-Optic Fluorimetry in Biosensors: Comparison Between Evanescent Wave Generation and Distal-Face Generation of Fluorescent Light," *Biosensors and Bioelectronics*, Vol. 7, 1992, pp. 193–197.

[21] Jorgenson, R. C., and S. S. Yee, "Control of the Dynamic Range and Sensitivity of a Surface Plasmon Resonance Based Fiber Optic Sensor," *Sensors and Actuators A: Physical*, Vol. 43, 1994, pp. 44–48.

[22] Gentleman, D. J., and K. S. Booksh, "Determining Salinity Using a Multimode Fiber Optic Surface Plasmon Resonance Dip-Probe," *Talanta*, Vol. 68, 2006, pp. 504–515.

[23] Kanso, M., S. Cuenot, and G. Louarn, "Sensitivity of Optical Fiber Sensor Based on Surface Plasmon Resonance: Modeling and Experiments," *Plasmonics*, Vol. 3, 2008, pp. 49–57.

[24] Trouillet, A., C. Ronot-Trioli, and H. Gagnaire, "Chemical Sensing by Surface Plasmon Resonance in Multimode Optical Fibre," *Pure Applied Optics*, Vol. 5, 1996, pp. 227–237.

[25] Wu, W., et al., "U-Shaped Fiber Optics Fabricated with a Femtosecond Laser and Integrated into a Localized Plasmon Resonance Biosensor," *Proc. DTIP*, Rome, Italy, 2009.

[26] Kumar, S., G. Sharma, and V. Singh, "Sensitivity of Tapered Optical Fiber Surface Plasmon Resonance Sensors," *Optical Fiber Technology*, Vol. 20, 2014, pp. 333–335.

[27] Al-Qazwini, Y., et al., "Experimental Realization and Performance Evaluation of Refractive Index SPR Sensor Based on Unmasked Short Tapered Multimode-Fiber Operating in Aqueous Environments," *Sensors and Actuators A: Physical*, Vol. 236, 2015, pp. 38–43.

[28] Ju, S., et al., "Experimental Demonstration of Surface Plasmon Resonance Enhancement of the Tapered Optical Fiber Coated with Au/Ti Thin Film," *Journal of Non-Crystalline Solids*, Vol. 383, 2014, pp. 146–152.

[29] Lin, H. Y., et al., "Side-Polished Multimode Fiber Biosensor Based on Surface Plasmon Resonance with Halogen Light," *Applied Optics*, Vol. 46, 2007, pp. 800–806.

[30] Lin, Y. C., et al., "An Enhanced Optical Multimode Fiber Sensor Based on Surface Plasmon Resonance with Cascaded Structure," *IEEE Photonics Technology Letters*, Vol. 20, 2008, pp. 1287–1289.

[31] Chen, C. H., et al., "A Multi-D-Shaped Optical Fiber for Refractive Index Sensing," *Sensors*, Vol. 10, 2010, pp. 4794–4804.

[32] Ahn, J. H., et al., "Fiber-Optic Waveguide Coupled Surface Plasmon Resonance Sensor," *Optics Express*, Vol. 20, 2012, pp. 21729–21738.

[33] Zhao, Y., Z. Q. Deng, and Q. Wang, "Fiber Optic SPR Sensor for Liquid Concentration Measurement," *Sensors and Actuators B: Chemical*, Vol. 192, 2014, pp. 229–233.

[34] Jang, H. S., et al., "Optical Fiber SPR Biosensor with Sandwich Assay for the Detection of Prostate Specific Antigen," *Optics Communications*, Vol. 282, 2009, pp. 2827–2830.

[35] Wang, Y., et al., "Indium Tin Oxide Coated Two-Mode Fiber for Enhanced SPR Sensor in Near-Infrared Region," *IEEE Photonics Journal*, Vol. 9, 2017, pp. 1–9.

[36] Dong, J., et al., "Side-Polished Few-Mode Fiber Based Surface Plasmon Resonance Biosensor," *Optics Express*, Vol. 27, 2009, pp. 11348–11360.

[37] Homola, J., S. S. Yee, and G. Gauglitz, "Surface Plasmon Resonance Sensors: A Review," *Sensors and Actuators B*, Vol. 54, 1999, pp. 3–15.

[38] Homola, J., "Optical Fiber Sensor Based on Surface Plasmon Excitation," *Sensors and Actuators B*, Vol. 29, 1995, pp. 401–405.

[39] Gupta, B. D., H. Dodeja, and A. K. Tomar, "Fibre-Optic Evanescent Field Absorption Sensor Based on a U-Shaped Probe," *Optical and Quantum Electronics*, Vol. 28, 1996, pp. 1629–1639.

[40] Verma, R. K., and B. D. Gupta, "Theoretical Modelling of a Bi-Dimensional U-Shaped Surface Plasmon Resonance Based Fibre Optic Sensor for Sensitivity Enhancement," *Journal of Physics D: Applied Physics*, Vol. 41, 2008, p. 065106.

[41] Danny, C. G., et al., "U-Bent Fiber Optic Plasmonic Sensors: Fundamentals, Applications, Challenges and Future Directions," in *Recent Advances in Plasmonic Probes: Theory and Practice*, R. Biswas and N. Mazumder, (eds.), New York: Springer, 2022.

[42] Sai, V. V. R., T. Kundu, and S. Mukherji, "Novel U-Bent Fiber Optic Probe for Localized Surface Plasmon Resonance Based Biosensor," *Biosensors and Bioelectronics*, Vol. 24, 2009, pp. 2804–2809.

[43] Bucaro, J., and E. Carome, "Single Fiber Interferometric Acoustic Sensor," *Applied Optics*, Vol. 17, 1978, pp. 330–331.

[44] Kim, B. Y., et al., "Use of Highly Elliptical Core Fibers for Two-Mode Fiber Devices," *Optics Letters*, Vol. 12, 1987, pp. 729–731.

[45] Caucheteur, C., et al., "Mode-Division and Spatial-Division Optical Fibers," *Advances in Optics and Photonics*, Vol. 14, 2022, pp. 1–86.

[46] Horche, P. R., et al., "Spectral Behavior of a Low-Cost All-Fiber Component Based on Untapered Multifiber Unions," *IEEE Photonics Technology Letters*, Vol. 1, 1989, pp. 184–187.

[47] Yin, B., et al., "Investigation on a Compact In-Line Multimode-Single-Mode-Multimode Fiber Structure," *Optics & Laser Technology*, Vol. 80, 2016, pp. 16–21.

[48] Zhang, S., et al., "A Compact Refractive Index Sensor with High Sensitivity Based on Multimode Interference," *Sensors and Actuators A: Physical*, Vol. 315, 2020, p. 112360.

[49] Iga, M., A. Seki, and K. Watanabe, "Hetero-Core Structured Fiber Optic Surface Plasmon Resonance Sensor with Silver Film," *Sensors and Actuators B*, Vol. 101, 2004, pp. 368–372.

[50] Akbarpour, Z., V. Ahmadi, and F. A. Roghabadi, "Enhanced Mach-Zehnder Interferometer Multimode–Single-Mode–Multimode Fiber Optic Refractive Index Sensor Based on Surface Plasmon Resonance," *Optical Fiber Technology*, Vol. 73, October 2022.

[51] Wang, H., et al., "Dual-Channel SPR Sensor Based on MSM Fiber for Detection of Glucose Concentration and Temperature," *IEEE Photonics Technology Letters*, Vol. 34, 2022, pp. 919–922.

[52] Tuniz, A., and M. A. Schmidt, "Interfacing Optical Fibers with Plasmonic Nanoconcentrators," *Nanophotonics*, Vol. 7, 2018, pp. 1279–1298.

[53] Ebbesen, T. W., et al., "Extraordinary Optical Transmission Through Sub-Wavelength Hole Arrays," *Nature*, Vol. 391, 1998, pp. 667–669.

[54] Mitsui, K., Y. Handa, and K. Kajikawa, "Optical Fiber Affinity Biosensor Based on Localized Surface Plasmon Resonance," *Applied Physics Letters*, Vol. 85, 2004, p. 4231.

[55] Liang, Y., et al., "A Self-Assembled Plasmonic Optical Fiber Nanoprobe for Label-Free Biosensing," *Scientific Reports*, Vol. 9, 2019, p. 7379.

[56] Feng, S., et al., "A Miniaturized Sensor Consisting of Concentric Metallic Nanorings on the End Facet of an Optical Fiber," *Small*, Vol 8, 2012, pp. 1937–1944.

[57] Smythe, E. J., M. D. Dickey, and F. Capasso, "A Technique to Transfer Metallic Nanoscale Patterns to Small and Non-Planar Surfaces," *ACS Nano*, Vol. 3, 2009, pp. 59–65.

[58] Consales, M., et al., "Lab-on-Fiber Technology: Toward Multifunctional Optical Nanoprobes," *ACS Nano*, Vol. 6, 2012, pp. 3163–3170.

[59] Jia, P., and J. Yang, "Integration of Large-Area Metallic Nanohole Arrays with Multimode Optical Fibers for Surface Plasmon Resonance Sensing," *Applied Physics Letters*, Vol. 102, 2013, p. 243107.

[60] Kim, H. M., et al., "Localized Surface Plasmon Resonance Biosensor Using Nanopatterned Gold Particles on the Surface of an Optical Fiber," *Sensors and Actuators B Chem*, Vol. 280, 2019, pp. 183–191.

CHAPTER 4

Single-Mode Optical Fiber Platforms

We have previously recalled that, to excite SPR from an optical fiber, the core-guided light has to be locally outcoupled and directed towards the surrounding medium. In single-mode optical fibers, this can be achieved either from a geometrical alteration (polishing or etching of the cladding, totally or in part) so as to expose the evanescent wave to the surrounding medium or from a periodic refractive index modulation of the fiber core along the propagation axis. The latter configuration refers to the fiber gratings that are permanently photo-inscribed in the fiber core. Such configurations operate at near-infrared telecommunication wavelengths, which significantly increase the penetration depth of the evanescent wave compared to multimode optical fiber counterparts, as shown next. The main plasmonic configurations derived from single-mode optical fibers will be reviewed, with a focus on their typical performance in terms of surrounding refractive index sensitivity. As in Chapter 3, our objective is to focus on their behavior with respect to surrounding refractive index changes physically rather than describing the coupling mechanisms in such structures (as can be found in numerous textbooks and review papers, for which the references will be provided). For the sake of completeness, the intrinsic temperature and strain sensitivities of standard fiber Bragg gratings will be discussed, as they remain the most prominent features of this technology. The reader may feel that some sections are more elaborate than others: this is just the reflection of the actual literature for which some platforms such as tilted fiber Bragg gratings were much more developed and reported.

4.1 Overview of Single-Mode Optical Fibers

The reminder of light propagation mechanisms in optical fibers made in Chapter 3 has highlighted that single-mode optical fibers are produced in such a way that the core dimension (a) and the difference of refractive indices between the core and the cladding (Δn) enable the normalized frequency $V \leq 2.4$. In this case, only the fundamental mode can propagate, beyond the cutoff wavelength. Solving the Helmholtz equation obtained by combining Maxwell equations and boundary conditions (see Chapter 3), one can show that the fundamental mode is a transverse mode. Telecommunication-grade single-mode optical fibers are usually made of a 8.2-μm core in silica doped with germanium oxide (GeO_2) surrounded by a 125-μm cladding in pure silica. Such fibers are step-index: there is a sharp transition between the core and cladding refractive indices. The standards G.652 and G.657 define the most widely used forms of single-mode optical fiber. They specify the geometrical, mechanical, and transmission attributes of single-mode optical fibers as well as the way to produce cables. Most often, these fibers are surrounded by a polymer jacket of 250 μm diameter in PMMA. This yields the geometry sketched in Figure 4.1. It is worth mentioning that Sir Charles K. Kao received the 2009 Nobel Prize in Physics for his pioneering work on single-mode optical fibers in the 1960s.

In practice, there are two independent and degenerated propagation modes, HE^x_{11} and HE^y_{11}. These modes are very similar but have orthogonal polarization planes. The electric field associated with light propagating in the optical fiber is a linear superposition of these two polarization modes (called the x and y modes). These two modes are degenerated ($\beta_x = \beta_y$) only if the fiber is ideal (i.e., characterized by a perfect circular symmetry). In practice, an optical fiber has always some imperfections such as local variations of the refractive index profile, a slightly elliptical core, or being subject to a slight stress. The perfect circular symmetry is therefore broken and the two modes are no longer degenerated ($\beta_x \neq \beta_y$). Consequently, the two effective refractive indices of these modes are slightly different; this difference is called modal birefringence:

$$\beta_{yx} = k\left(n_{\text{eff},y} - n_{\text{eff},x}\right) \tag{4.1}$$

The intrinsic fiber birefringence is very low, of the order of 10^{-6}.

These fibers present a minimum of lineic attenuation (attenuation as a function of the propagation distance within the optical fiber) at the wavelength of 1,550 nm. This value is close to 0.16 dB/km. With the advent of erbium-doped fiber amplifiers (EDFA), optical telecommunications have been mainly deployed in the C and L-bands. The C-band ranges from 1,530 to 1,565 nm, while the L-band extends from 1,565 to 1,625 nm. Numerous cost-effective fiber-optic devices and components have also been developed around that wavelength. Couplers and circulators that enable working in the reflection mode and isolators that prevent reflected light to go back to the optical source are largely available, as well as optical sources and detectors. For that reason, single-mode optical fiber sensing has also been mainly deployed around 1,550 nm. It is worth mentioning that dispersion is present at that wavelength: optical pulses propagating in optical fiber will be spread because of a difference of velocity between the different wavelength components of the pulse. While this characteristic is an issue for long-haul telecommunications as it limits the bit rate, it is generally not an issue for sensing, especially plasmonic sensing as covered in this book.

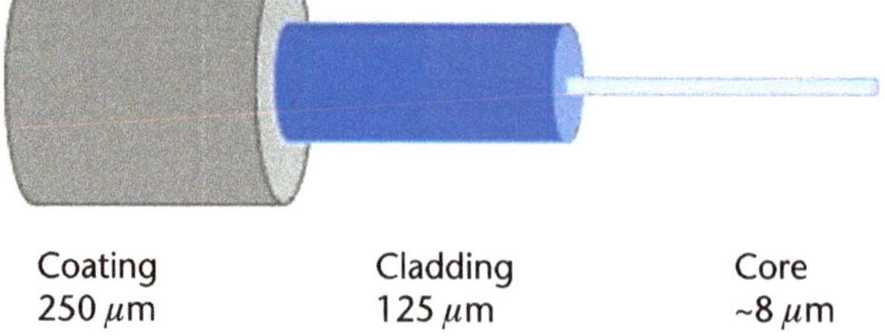

Figure 4.1 Sketch of a step-index, single-mode optical fiber.

Given the importance of light polarization effects in this chapter and in the following chapters, it is relevant to remember some important considerations here. A polarized light wave signal propagating in an optical fiber or in free space is represented by electric and magnetic field vectors perpendicular to each other in a transverse plane, itself perpendicular to the direction of propagation. Commonly, the attention is focused on the electric field because it has the most direct effect during the interaction between wave and matter [1].

The state of polarization is defined as the pattern drawn in the transverse plane by the extremity of the electric field vector as a function of time at a fixed position in space. A monochromatic light source emits a single frequency and is always totally polarized. In some cases, the electric field vector can occupy random orientations in the transverse plane as a function of time. If the orientation changes are fast enough to be beyond observation in the physical context of a particular measurement or application, the light is said to be unpolarized [2]. It is the case of naturally produced light, such as sunlight and firelight, which is characterized by a very large spectral width. In that case, the polarization behavior of each frequency is different and it is therefore impossible to clearly define a fixed polarization state. Between the cases of totally polarized and unpolarized light, we find partially polarized light for which the electric field vector is still characterized by random orientations but stays around a polarization state of maximum probability. It is typically the case when an optical source emits a quasi-monochromatic wave with a small but not null spectral width. Hence, each spectral component of the wave produces a different polarization state at a fixed point of space but all these states remain distributed around the same state.

Partially polarized light can be represented as a superposition of both fully polarized and completely unpolarized light waves. The degree of polarization (DOP) describes partially polarized light and is defined as the ratio of the intensity of the totally polarized component to the total intensity of the wave. The DOP varies from 0 for unpolarized light to 1 for totally polarized light and takes intermediate values for partially polarized light. In free-space propagation, the DOP of a wave is maintained. In a transmission medium, it can evolve depending on the spectral width of the source and the dispersive properties of the transmission path.

For a plane and monochromatic light wave (DOP = 1) propagating along the z-direction of a Cartesian coordinate system (x, y, z), the transverse components of the electric field can be expressed as [1]:

$$E_x(z,t) = E_{0x}\cos\left(\omega t + \delta_x - kz\right) \tag{4.2}$$

$$E_y(z,t) = E_{0y}\cos\left(\omega t + \delta_y - kz\right) \tag{4.3}$$

where E_{0x} and E_{0y} are the amplitudes of the x and y components, δ_x and δ_y are the corresponding phases. The elimination of $(\omega t - kz)$ from (4.1) and (4.2) yields the following relationship:

$$\frac{E_x^2}{E_{0x}^2} + \frac{E_y^2}{E_{0y}^2} + 2\frac{E_x E_y}{E_{0x}E_{0y}}\cos\delta = \sin^2\delta \tag{4.4}$$

where $\delta = \delta_y - \delta_x$ is the phase difference between the two components.

Equation (4.4) is the equation of an ellipse drawn in the transverse plane by the extremity of the electric field vector at a fixed point in space as a function of time. The state of polarization of a fully polarized light wave is therefore elliptical in general. The ellipse defined by (4.4) is illustrated in Figure 4.2 where one can see that parameters other than E_{0x}, E_{0y}, and δ can be used to define a state of polarization. These parameters are the angle ϕ, called the azimuth, between the major axis and the x-axis, the sense of rotation of the electric field vector materialized by an arrow on the ellipse, and the ellipticity degree d_e defined as:

$$d_e = \frac{b}{a} \tag{4.5}$$

where a and b are the half-lengths of the major and minor axes, respectively. The ellipticity e is then defined by:

$$e = \pm d_e = \tan\chi \tag{4.6}$$

It is borne by −1 and +1. The sign of e depends on the sense of rotation. By convention, the ellipse is presented as it would appear through the xy-plane for an observer looking towards the source. Hence, when the electric field vector describes the ellipse in the clockwise (counterclockwise) sense, the state of polarization is right-handed (left-handed) and the sign in (4.6) is positive (negative) [2].

The elliptical polarization can degenerate in two special cases when the ellipticity e takes the values +1, −1, and 0. These degenerated states of polarization can be found from (4.4):

- *Circular polarization state:* Two conditions are required to obtain a circular state: δ has to be an odd multiple of $\pi/2$ and the maximum amplitudes of the two field components must be equal ($E_{0x} = E_{0y}$).

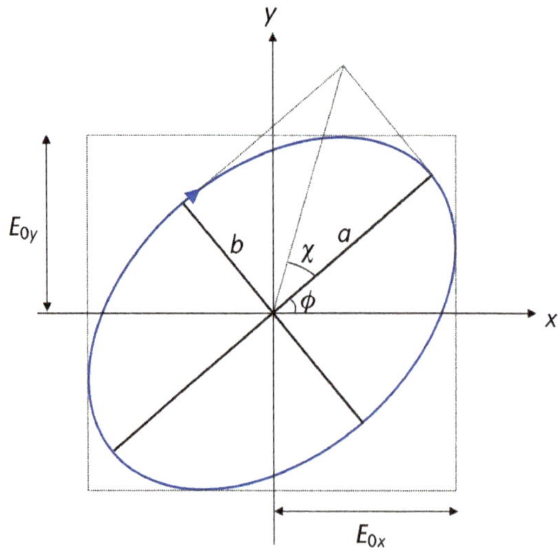

Figure 4.2 Polarization ellipse.

- *Linear polarization state:* One condition is required to obtain a linear state: δ has to be a multiple of π. Equation (4.4) degenerates in a straight line so that the ellipticity is equal to 0 and the angle ϕ is given by

$$\phi = \tan^{-1}\frac{E_{0x}}{E_{0y}} \tag{4.7}$$

The linear state is horizontal when $E_{0y} = 0$, whereas it is vertical when $E_{0x} = 0$.

In the 1940s, Jones introduced a mathematical formalism to describe polarization effects of fully polarized light by means of a matrix representation [3]. In this formalism, a state of polarization is represented by a 2-D vector of complex numbers. This vector **V** contains the amplitude and the phase of the x and y components of the electric field:

$$\mathbf{V} = \begin{pmatrix} E_{0x}e^{j\delta_x} \\ E_{0y}e^{j\delta_y} \end{pmatrix} \tag{4.8}$$

The normalized Jones vector $\mathbf{V}_\mathrm{n}$ defined by (4.9) is more used in practice.

$$\mathbf{V}_n = \frac{\mathbf{V}}{\sqrt{E_{0x}^2 + E_{0y}^2}} \tag{4.9}$$

Next, we will examine normalized Jones vectors for which the notation **V** will be used.

An elliptical state of polarization is described by the following normalized Jones vector [2]:

$$\mathbf{V} = \begin{pmatrix} \cos\phi\cos\chi - j\sin\phi\sin\chi \\ \sin\phi\cos\chi + j\cos\phi\sin\chi \end{pmatrix} \tag{4.10}$$

The normalized Jones vectors associated to the right-handed ($\mathbf{V}_\mathrm{R}$) and left-handed ($\mathbf{V}_\mathrm{L}$) circular polarization states can be written as [2]:

$$\mathbf{V}_\mathrm{R} = \frac{1}{\sqrt{2}}\begin{pmatrix} 1 \\ j \end{pmatrix} \tag{4.11}$$

$$\mathbf{V}_\mathrm{L} = \frac{1}{\sqrt{2}}\begin{pmatrix} 1 \\ -j \end{pmatrix} \tag{4.12}$$

A linear state of polarization making an angle ϕ with the x-axis is represented by:

$$\mathbf{V} = \begin{pmatrix} \cos\phi \\ \sin\phi \end{pmatrix} \tag{4.13}$$

In the Jones formalism, the transmission properties, in terms of polarization, of an optical device can be described by a complex 2-by-2 matrix (called the Jones matrix and noted J), which links the input and output Jones vectors of the optical device as shown in Figure 4.3.

$$\mathbf{V}_{\mathrm{OUT}} = J\mathbf{V}_{\mathrm{IN}} \tag{4.14}$$

In the Jones formalism, the states of polarization are represented by complex numbers and characterized by the amplitudes and phase angles of the electric field components. In practice, it is much easier to measure intensities. Furthermore, the Jones formalism does not allow representing partially polarized light. These two reasons have led to the introduction of other formalisms to express polarization in terms of easily measurable optical powers. One of them is the Stokes formalism in which the involved numbers are real. In the Stokes formalism, a state of polarization is represented by a four-dimensional (4-D) vector **S** called the Stokes vector and is defined by:

$$\mathbf{S} = \begin{pmatrix} S_0 \\ S_1 \\ S_2 \\ S_3 \end{pmatrix} \tag{4.15}$$

where

$$S_0 = \text{Total power of the light} \tag{4.16}$$

$$S_1 = P_0 - P_{\pi/2} \tag{4.17}$$

$$S_2 = P_{\pi/4} - P_{-\pi/4} \tag{4.18}$$

$$S_3 = P_{\mathrm{CR}} - P_{\mathrm{CL}} \tag{4.19}$$

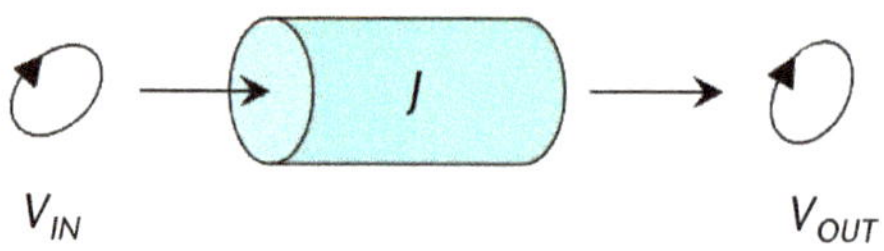

Figure 4.3 Representation of an optical device by a Jones matrix.

where P_θ denotes the power of the light passed through a linear polarizer set at an angle θ (in radians) with respect to the x-axis. P_{CR} and P_{CL} represent the optical powers of the light passed through a right-handed or left-handed circular polarizer, respectively.

In general, it is more convenient to use the normalized Stokes parameters given by:

$$s_1 = \frac{S_1}{S_0} \quad s_2 = \frac{S_2}{S_0} \quad s_3 = \frac{S_3}{S_0} \tag{4.20}$$

The normalized Stokes vector is then expressed by:

$$\mathbf{s} = \begin{pmatrix} 1 \\ s_1 \\ s_2 \\ s_3 \end{pmatrix} \tag{4.21}$$

so that the normalized Stokes parameters vary from −1 to +1.

The correspondence between the Stokes and Jones vectors is obtained by developing (4.16) to (4.19), which yields for fully polarized light:

$$S_0 = E_{0x}^2 + E_{0y}^2 \tag{4.22}$$

$$S_1 = E_{0x}^2 - E_{0y}^2 \tag{4.23}$$

$$S_2 = 2E_{0x}E_{0y}\cos\delta \tag{4.24}$$

$$S_3 = 2E_{0x}E_{0y}\sin\delta \tag{4.25}$$

Unlike the Jones formalism, the Stokes representation enables to describe partially polarized light [4]. Note that in practice, the Stokes parameters can be represented in a graphical way on the Poincaré sphere [2, 4].

The following sections will review the main plasmonic sensing platforms based on modified single-mode optical fibers.

4.2 Etched, Tapered, and D-Shaped Configurations

We will not discuss the operating principle of these configurations, as they are a transposition in single-mode optical fibers of what can be obtained with multimode configurations. We will directly review the literature on the subject and highlight the main reported features.

Etching a single-mode optical fiber is more difficult than for the large multimode optical fibers presented in Chapter 3, as it implies removing totally or in part the fiber cladding that is in silica. For large, multimode optical fibers, the polymer cladding can be quite easily removed with a solvent like acetone. In the case of standard single-mode optical fibers, the etching process is demanding hydrofluoric acid (HF), which is a dangerous chemical compound that requires a lot of experimental precautions. In practice, the fiber is left in the solution placed in a plastic container (as usual glass is damaged by HF) and the acid progressively attacks the silica. Depending on the exact concentration of the acid, the etching rate can be such that a couple of micrometers will be removed per minute of the etching process. Figure 4.4 depicts a standard single-mode optical fiber section whose diameter has been decreased by HF etching.

In [5], a 10-mm section of a standard single-mode optical fiber was etched in a 48% HF solution for 38 minutes to decrease the cladding diameter down to 13 μm. This allows exposing the evanescent field of the core-guided mode to the surroundings. This section was then coated by gold and titanium dioxide to excite a surface plasmon wave. For the interrogation, a supercontinuum light source and an optical spectrum analyzer were used. The surface plasmon resonance appears around 1,550 nm and its refractometric sensitivity has been computed to be 3,800 nm/RIU when the fiber is used in the reflection mode. A combination of etched regions and fiber Bragg gratings (see Section 4.3) has been demonstrated by the same team in [6]. A refractometric sensitivity reaching 5,000 nm/RIU has been reported. An alternative configuration at the fiber tip was reported in [7]. The cleaved end of a single-mode optical fiber was etched in the HF solution and, because of the different composition between the GeO_2-doped core and the pure silica cladding, a difference of etching rate happens between the two materials. This results in the production of a small cone (with an angle of 30°) at the fiber end. The latter was coated by an approximately 10-nm gold sheath to operate at the wavelength of 780 nm. An operation in the refractive index range from 1.33 to 1.40 was reported, with a sensitivity of 0.008 RI units. A similar configuration was studied in [8] with a tip under 100 nm. The dangerousness of the etching process and the fragility of the residual optical fiber section are two important limiting factors of the development of etched SMF platforms. Tapering results in a similar fragility but remains less hazardous to implement, as shown next.

The first report about the use of tapered single-mode optical fibers for surface plasmon excitation dates back to 1997 [9]. In [10], a uniform-waist tapered, single-mode optical fiber with an asymmetric double layer (metal and dielectric) has been proposed and studied. To produce such structures comprising a long waist with a uniform diameter, a heater is displaced in an oscillatory way to heat the complete length of the taper, while the fiber is gently stretched in opposite directions. Such configurations contrast with the abrupt tapers that can be produced with a fiber splicer and that are called biconical tapers. The latter have higher losses than uniform-waist tapers, which can be made almost completely adiabatic (meaning that all the power is transmitted across a taper with no loss). In these devices, light is guided by the cladding in the waist of the tapered region, as the core has been reduced to a negligible diameter in that area. Therefore, the guided field can easily reach the outer medium and any change of its refractive index produces a variation of the signal transmitted through the tapered fiber. In [10, 11], a double layer of Al (8 nm) and TiO_2 (56 nm) was used on a 30-μm waist and 3-cm-long, tapered, single-mode optical fiber. The SPR refractometric sensitivity has been computed to be 2,000 nm/RIU. In [12], a variant based on indium nitride (thickness between 20 and 40 nm) deposited on the 8-nm-thick layer of Al was demonstrated. Depending on the indium nitride thickness, the plasmon resonance can be tuned up to 1,000 nm and the refractometric sensitivity can reach 11,800 nm/RIU in the range of indices between 1.415 and 1.429. Localized SPR generation has been demonstrated in [13] with gold nanoparticles immobilized on the 48-μm waist taper (1.25 mm in length) of a single-mode optical fiber. The refractive index resolution following an interrogation based on the transmission intensity change was calculated to be 3.2 10^{-5} RIU. In [14], the taper was made using a CO_2 laser heating technique. Typically, 20-μm tapered regions were coated with a 30-nm gold film and an Al_2O_3 overlayer to enhance the refractometric sensitivity, thanks to a high-index dielectric layer on the gold film. This configuration is studied theoretically and experimentally based on the Beer-Lambert law of absorption.

The third configuration that also exposes the evanescent wave of the core mode to the surrounding medium is obtained from D-shaped or side-polished optical fibers. In 2015, a graphene-based, indium tin oxide-coated, D-shaped, single-mode optical fiber sensor was reported [15]. The sensor operates in the near-infrared wavelength range, thanks to the conducting metal oxide layer and achieves a maximum refractometry sensitivity of 5,700 nm/RIU in the range of 1.33 to 1.345. An ultrahigh sensitive D-shaped SMF sensor with a silver-graphene layer was reported a couple of years later [16]. In [17], a configuration based on D-shaped (polishing depth 0.8 nm

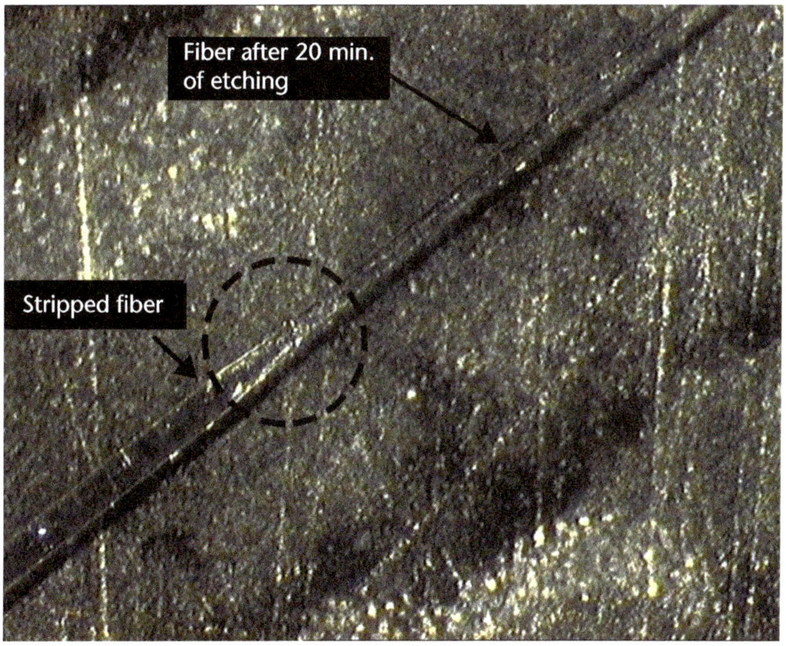

Figure 4.4 Picture of an optical fiber after 20 minutes of etching with a 50% HF solution (optical stereomicroscopy).

above the core) single-mode optical fiber coating with a gold grating (thickness of 45 nm and period of 1 μm) was numerically studied using a finite element method. A numerical refractometric sensitivity of 7,590 nm/RIU was reported. As shown in Chapter 5, D-shaped structures are much more spread with specialty optical fibers rather than more standard single-mode configurations.

Looking at the number of reports on these configurations, it is trivial to say that etched, tapered, or side-polished single-mode optical fiber platforms are much less attractive in practice than their multimode counterparts. We believe that it results from their increased complexity in development and handling, their relative fragility, and the fact that the sensitivity is not improved compared to plasmonic multimode optical fibers.

The next section will be devoted to the use of thinned uniform fiber Bragg grating for refractometric sensing. The operating principle and sensitivities to temperature and strain of standard FBGs will be first presented. For the sake of completeness and as they have made the commercial success of this technology so far, we find it relevant to recall these important properties here, even if they are not directly connected to plasmonic sensing.

4.3 Thinned Uniform Fiber Bragg Grating Configurations

4.3.1 Basics of Uniform Fiber Bragg Gratings

A uniform fiber Bragg grating (FBG) is a periodic and permanent modification of the core refractive index value along the optical fiber axis [18, 19]. This modification is usually obtained by transversally exposing the core of a photosensitive optical fiber to an intense interference pattern of ultraviolet light at a wavelength around 240 nm. Indeed, due to the presence of germanium oxide dopants inside the core of telecommunication-grade, single-mode optical fibers, the latter are photosensitive (i.e., they benefit from the property to permanently change their refractive index when exposed to light) in a wavelength band centered around 240 nm. For this reason, a continuous-wave, frequency-doubled, argon-ion laser emitting at 244 nm or pulsed excimer laser emitting at 248 nm are most often used to manufacture FBGs. Bright interference fringes locally increase the mean refractive index of the core. Different ways can be implemented to photo-inscribe Bragg gratings (see Section 4.4.1). Mass production of identical gratings can be quite straightforwardly achieved with the phase mask technique. A phase mask is a diffractive element made in a pure silica substrate that is optimized to maximize the diffraction in the first (+1 and −1) diffraction orders. Let us note that the fiber photosensitivity can be enhanced by increasing the germanium content in the fiber core or through a hydrogenation process [18].

An interference pattern is therefore produced in the fiber core when the optical fiber is located in close proximity behind the mask. An FBG is defined by several physical parameters, as sketched in Figure 4.5. The grating length L is the optical fiber length along which the refractive index modulation is produced. The periodicity and the amplitude of the refractive index modulation are labeled Λ and δn, respectively.

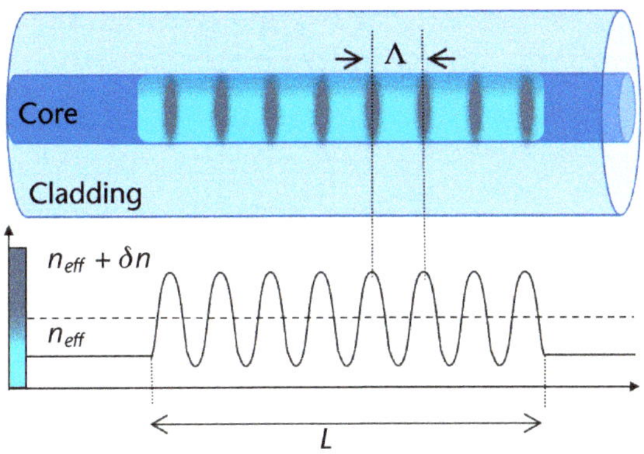

Figure 4.5 Sketch of a uniform FBG and its physical parameters (not in scale).

The order of magnitude of these parameters typically varies between 200 and 1,000 nm for Λ, from a few millimeters to a few tens of centimeters for L and from 10^{-5} to 10^{-3} for δn. Such a perturbation induces light coupling between two counterpropagating core modes. This mode coupling occurs for some wavelengths around the Bragg wavelength defined by:

$$\lambda_B = 2n_{eff}\Lambda \tag{4.26}$$

where n_{eff} is the effective refractive index of the core mode at the Bragg wavelength. A uniform FBG acts as a selective mirror in a wavelength around the Bragg wavelength to yield a passband-reflected amplitude spectrum and a stopband-transmitted amplitude spectrum, as depicted in Figure 4.6 for a 1-cm-long FBG. At each refractive index discontinuity along the fiber axis, a weak Fresnel reflection is generated. They add in phase at the Bragg wavelength, yielding an important reflection band surrounded by side lobes.

In practice, the effective refractive index of the core and the spatial periodicity of the grating are both affected by changes in strain and temperature. In particular, the effective refractive index is modified through the thermo-optic and strain-optic

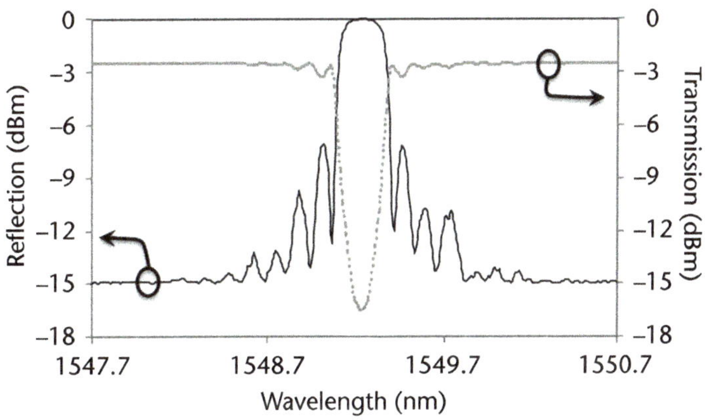

Figure 4.6 Reflected and transmitted amplitude spectra of a 1-cm-long uniform FBG.

effects. Hence, from (4.27), the Bragg wavelength shift $\Delta\lambda_B$ due to strain ($\Delta\varepsilon$) and temperature (ΔT) variations is given by:

$$\Delta\lambda_B = 2\left(\Lambda\frac{dn_{eff}}{dT} + n_{eff}\frac{d\Lambda}{dT}\right)\Delta T + 2\left(\Lambda\frac{dn_{eff}}{d\varepsilon} + n_{eff}\frac{d\Lambda}{d\varepsilon}\right)\Delta\varepsilon \qquad (4.27)$$

4.3.2 Temperature Sensitivity of Uniform FBGs

The first term in (4.4) represents the effect of temperature on the Bragg wavelength. The Bragg wavelength shift due to thermal expansion comes from the modification of the grating spacing and the refractive index. The relative wavelength shift due to a temperature change ΔT can be written as:

$$\frac{\Delta\lambda_B}{\Delta T} = \lambda_B\left(\frac{1}{n_{eff}}\frac{dn_{eff}}{dT} + \frac{1}{\Lambda}\frac{d\Lambda}{dT}\right) \qquad (4.28)$$

where $(1/n_{eff})(dn_{eff}/dT)$ is the thermo-optic coefficient, which is approximately equal to 8.6 10^{-6} K^{-1} for germanium doped silica core optical fiber and $(1/\Lambda)(d\Lambda/dT)$ is the thermal expansion coefficient of the optical fiber, which is approximately equal to 0.55 10^{-6} K^{-1} for silica so that the refractive index change is by far the dominant effect [18]. The order of magnitude of the temperature sensitivity of the Bragg wavelength is 10 pm/°C around 1,550 nm.

Figure 4.7 displays the experimental evolutions of the Bragg wavelength as a function of temperature for a 1-cm-long uniform FBG. The grating was placed inside an oven regulated in temperature with an accuracy of 0.1°C. The evolution

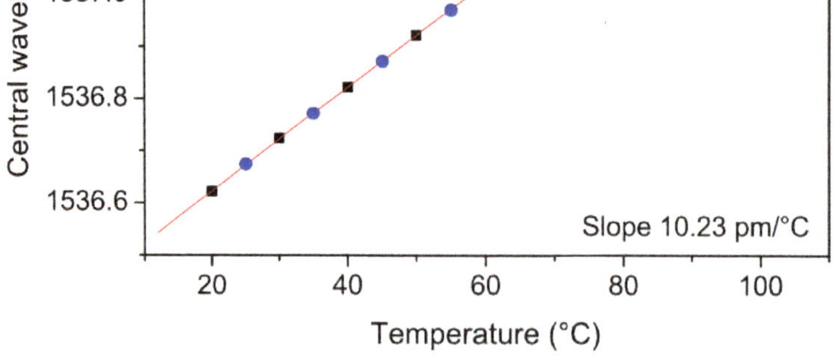

Figure 4.7 Evolution of the Bragg wavelength as a function of temperature changes.

is linear with a slope computed equal to 10.23 pm/°C. Experiments carried out for increasing and decreasing temperature values confirm the absence of hysteresis.

4.3.3 Axial Strain Sensitivity of Uniform FBGs

The second term in (4.27) represents the effect of axial strain on an optical fiber. It corresponds to a change in the grating periodicity and the strain-optic induced change in the refractive index. Defining the strain as $\varepsilon = \Delta\Lambda/\Lambda$, the change of the grating periodicity can be expressed by:

$$\Lambda_S = \Lambda + \Delta\Lambda = \Lambda(1+\varepsilon) \tag{4.29}$$

where Λ_S represents the modified grating period after the application of the perturbation. The elasto-optic effect relates the refractive index change to the applied strain [19]:

$$\Delta\eta_{ij} = \left(\Delta\frac{1}{n_{\text{eff}}^2}\right)_{ij} = \sum_{i=1}^{6}\sum_{j=1}^{6} p_{ij}\varepsilon_{ij} \tag{4.30}$$

where $\Delta\eta_{ij}$ is the change in the electric impermeability tensor, ε_{ij} are the strain components, and p_{ij} are the elements of the elasto-optic tensor. The shape of the elasto-optic tensor, but not the magnitude of the coefficients p_{ij}, can be derived from the symmetry of the material under consideration. In the most general case, there are 36 different coefficients p_{ij}. For the class of isotropic materials to which fused silica glass belongs, this number is reduced to two independent coefficients p_{11} and p_{12} and the elasto-optic tensor becomes [19]:

$$\begin{pmatrix} p_{11} & p_{12} & p_{12} & 0 & 0 & 0 \\ p_{12} & p_{11} & p_{12} & 0 & 0 & 0 \\ p_{12} & p_{12} & p_{11} & 0 & 0 & 0 \\ 0 & 0 & 0 & \frac{1}{2}(p_{11}-p_{12}) & 0 & 0 \\ 0 & 0 & 0 & 0 & \frac{1}{2}(p_{11}-p_{12}) & 0 \\ 0 & 0 & 0 & 0 & 0 & \frac{1}{2}(p_{11}-p_{12}) \end{pmatrix} \tag{4.31}$$

The magnitudes of the individual coefficients p_{ij} are dependent on the material considered. For pure bulk silica, it was found that typical values measured at 628 nm are $p_{11} = 0.121$ and $p_{12} = 0.270$ [15]. These values are often used when computing the influence of mechanical perturbations such as elongation and lateral compression of optical fibers. However, due to the presence of doping elements in the core, the effective values for fibers may be different from those for bulk silica. Measurements on single-mode optical fibers at 628 nm yielded values of $p_{11} = 0.113$ and p_{12}

= 0.252, respectively [20]. An error on these values of approximately 5% is possible because of the uncertainty on the value of the Poisson's ratio of the optical fiber.

Assuming that the grating is strained in the z-direction only (along the fiber axis) and that the fiber material follows Hooke's law, we obtain that:

$$\varepsilon_x = \varepsilon_y = -\nu\varepsilon_z = -\nu\varepsilon \tag{4.32}$$

where ν is the Poisson ratio.

As all other strain components are null, (4.30), combined with (4.31) and (4.32), gives:

$$\Delta n_x = -\frac{n_{\text{eff}}^3}{2}\left[-\upsilon p_{11} + (1-\upsilon)p_{12}\right]\varepsilon \tag{4.33}$$

$$\Delta n_y = -\frac{n_{\text{eff}}^3}{2}\left[-\upsilon p_{11} + (1-\upsilon)p_{12}\right]\varepsilon \tag{4.34}$$

so that the effect of axial strain is the same in the x and y directions (perpendicular to the optical fiber axis).

Taking into account (4.26), the second term of (4.27) can be rewritten as:

$$\frac{\Delta\lambda_{\text{B}}}{\Delta\varepsilon} = \lambda_{\text{B}}\left(\frac{1}{n_{\text{eff}}}\frac{\text{d}n_{\text{eff}}}{\text{d}\varepsilon} + \frac{1}{\Lambda}\frac{\text{d}\Lambda}{\text{d}\varepsilon}\right) \tag{4.35}$$

This allows defining the effective strain-optic constant p_e:

$$\text{p}_e = -\frac{1}{n_{\text{eff}}}\frac{\text{d}n_{\text{eff}}}{\text{d}\varepsilon} \tag{4.36}$$

such that under the assumption of small strain variations, the Bragg wavelength shift in response to axial changes is finally given by:

$$\Delta\lambda_{\text{B}} = \lambda_{\text{B}}(1-p_e)\Delta\varepsilon \tag{4.37}$$

$$p_e = \frac{n_{\text{eff}}^2}{2}\left[p_{12} - \upsilon(p_{11}+p_{12})\right] \tag{4.38}$$

Substitution of parameters ($p_{11} = 0.113$, $p_{12} = 0.252$, $\nu = 0.16$, and $n_{\text{eff}} = 1.482$) in (4.30) and (4.31) gives a strain-optic constant $p_e = 0.21$ and an axial strain sensitivity of the Bragg wavelength of 1.2 pm/$\mu\varepsilon$ around 1,550 nm.

Figure 4.8 displays the evolution of the Bragg wavelength as a function of an axial strain change for a 1-cm-long uniform FBG. To apply controlled axial strain

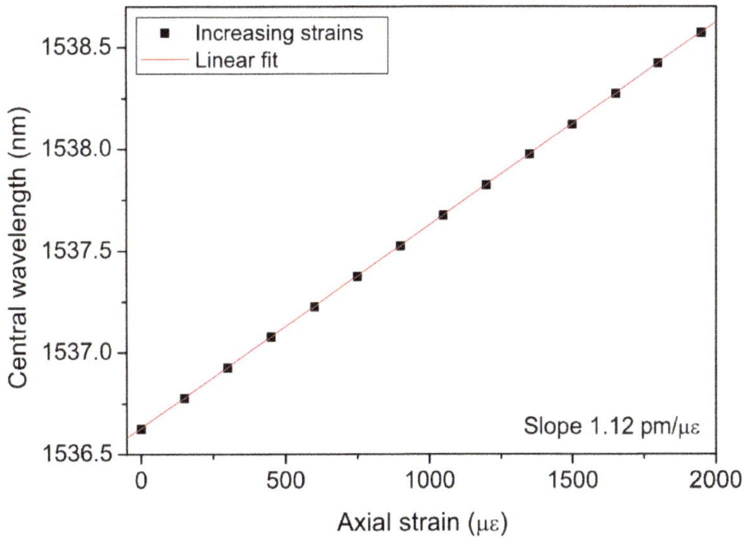

Figure 4.8 Evolution of the Bragg wavelength as a function of axial strain changes.

values, the fiber containing the grating was fixed on two clamps whose one was moved in the direction parallel to the fiber axis. As in the case of temperature, the evolution is linear and without hysteresis. The slope of the linear evolution was computed equal to 1.12 pm/$\mu\varepsilon$.

4.3.4 Pressure Sensitivity of Uniform FBGs

Starting from (4.26), it can be derived that a pressure p leads to a corresponding shift of the Bragg wavelength given by:

$$\frac{\Delta\lambda_B}{\lambda_B} = \left(\frac{1}{n_{\text{eff}}}\frac{dn_{\text{eff}}}{dp} + \frac{1}{\Lambda}\frac{d\Lambda}{dp}\right)p \tag{4.39}$$

The strain components resulting from the pressure load are:

$$\varepsilon_x = \varepsilon_y = \left(\frac{1-\nu}{E}\right)p \tag{4.40}$$

$$\varepsilon_z = \left(\frac{-2\nu}{E}\right)p \tag{4.41}$$

where E is the Young's modulus of the optical fiber. Substituting these values into (4.33) gives:

$$\Delta n_x = \Delta n_y = -\frac{n_{\text{eff}}^3}{2}\left[(1-\upsilon)p_{11} + (1-3\upsilon)p_{12}\right]\frac{\text{p}}{E} \tag{4.42}$$

Given that $\varepsilon_z = \Delta\Lambda/\Lambda$, the components of (4.39) are given by:

$$\frac{1}{\Lambda}\frac{d\Lambda}{dp} = \frac{-2v}{E} \tag{4.43}$$

$$\frac{1}{n_{eff}}\frac{dn_{eff}}{dp} = -\frac{n_{eff}^2}{2E}\left[(1-\upsilon)p_{11} + (1-3\upsilon)p_{12}\right] \tag{4.44}$$

The wavelength-pressure sensitivity can be formulated as:

$$\frac{\Delta\lambda_B}{\lambda_B} = \left[\frac{-2v}{E} - \frac{n_{eff}^2}{2E}\left[(1-\upsilon)p_{11} + (1-3\upsilon)p_{12}\right]\right]p \tag{4.45}$$

After substitution of the numerical values given above, we obtain:

$$\frac{\Delta\lambda_B}{\lambda_B} = -0.57\frac{\mathrm{p}}{E} \tag{4.46}$$

The typical hydrostatic pressure sensitivity of a uniform FBG around 1,550 nm is about –4 pm/MPa.

4.3.5 Transverse Strain Sensitivity of Uniform FBGs

Let us consider now that the applied stress σ is parallel to the y-axis (perpendicular to the optical fiber axis) so that the strain components related to the stress field are, respectively, given by:

$$\varepsilon_y = \frac{\sigma}{E} \tag{4.47}$$

$$\varepsilon_x = \varepsilon_z = -v\varepsilon_y \tag{4.48}$$

The change of the Bragg wavelength due to a transverse strain is given by the following relationship:

$$\frac{\Delta\lambda_B}{\lambda_B} = \left(\frac{1}{n_{eff}}\frac{dn_{eff}}{d\sigma} + \frac{1}{\Lambda}\frac{d\Lambda}{d\sigma}\right)\sigma \tag{4.49}$$

Substituting the values of strain components into (4.33) gives:

$$\Delta n_x = -\frac{n_{eff}^3}{2}\left[-\upsilon p_{11} + (1-\upsilon)p_{12}\right]\frac{\sigma}{E} \tag{4.50}$$

$$\Delta n_y = -\frac{n_{\text{eff}}^3}{2}\left[p_{11} - 2\upsilon p_{12}\right]\frac{\sigma}{E} \tag{4.51}$$

Hence, due to the asymmetry of the load with respect to the (x, y)-plane, dissimilar refractive index changes are obtained for the x and y-directions. Consequently, the shift of the Bragg wavelength will be dependent on the polarization state of the light coupled into the optical fiber.

$$\left(\frac{\Delta\lambda_B}{\lambda_B}\right)_x = \left(-v - \frac{n_{\text{eff}}^3}{2}\left[-\upsilon p_{11} + (1-\upsilon)p_{12}\right]\right)\frac{\sigma}{E} \tag{4.52}$$

$$\left(\frac{\Delta\lambda_B}{\lambda_B}\right)_y = \left(-v - \frac{n_{\text{eff}}^3}{2}\left[p_{11} - 2\upsilon p_{12}\right]\right)\frac{\sigma}{E} \tag{4.53}$$

In other words, transverse strain leads to birefringence, which is spectrally manifested by a broadening of the main reflection mode prior to is splitting in two different bands corresponding to both polarization modes. It is the reason why uniform FBGs written into polarization maintaining fibers (highly birefringent fibers) are most often preferred in practice to measure transverse force effects because the resonance bands corresponding to the x and y polarization modes are well resolved.

It now becomes apparent that any change in wavelength, associated with the action of an external perturbation to the grating, is the sum of strain and temperature terms. Therefore, in sensing applications for which only one perturbation is of interest or when the two perturbations have to be simultaneously and unambiguously measured, it is required to separate the effects of temperature and strain. This is not possible with a single grating written into a standard single-mode optical fiber as it exhibits a strong sensitivity to both perturbations. Several techniques have thus been developed to achieve temperature compensation by measuring simultaneously the strain and temperature effects.

As already mentioned, the main advantage of FBG sensors is that the information about the perturbation is wavelength-encoded. This property makes the sensor self-referencing and independent of fluctuating light levels. The system is therefore immune to source power and connector losses that affect many other types of optical fiber sensors. The very low insertion loss and narrowband wavelength reflection of FBGs provide convenient serial multiplexing along a single-mode optical fiber.

4.3.6 Refractometric Sensitivity of Uniform FBGs

As light remains guided in the fiber core, uniform FBGs are not intrinsically sensitive to changes of the surrounding refractive index value. This sensitivity can only happen when the core mode is exposed to the surrounding medium, when the cladding diameter is sufficiently decreased. In [21], the refractive index sensitivity of thinned FBGs has been numerically and experimentally studied. To that aim, a single-mode optical fiber comprising a 6-mm-long uniform FBG was etched over 1 cm in a buffered HF solution. Residual cladding diameters ranging between 20.0 μm and 8.5 μm (corresponding to an almost full etching of the cladding) have been analyzed. Figure 4.9 presents the measured wavelength shift of the reflected spectrum of a thinned FBG (residual diameter of 8.5 μm) as a function of the surrounding value. A comparison is made with a numerical analysis obtained from the resolution of the dispersion equation in the case of a doubly cladding fiber model [22]. Both evolutions are consistent and it turns out, as expected, that the sensitivity is not linear with the refractive index value. The authors concluded that resolutions of 10^{-5} and 10^{-4} are possible for the outer refractive index around 1.45 and 1.333, respectively, when devices with a minimum detectable wavelength shift of 1 pm are used. The detection limit of etched FBG sensors was studied in [23] for the *in situ* measurement of electrostatic layer-by-layer (LbL) assembly of polyelectrolytes on their silica surface. A good match between experimental evolutions and simulations made using a finite element model was reported in [23]. A theoretical and experimental study of the effect of the etching process on the properties of FBGs was also reported in [24]. In particular, it was shown that the residual fiber radius at the FBG location can be estimated from the measurement of the Bragg wavelength shift depending on the etching process. In [25], a Bragg grating written in standard single-mode optical fiber at 1,550 nm was etched in an HF solution at 48% to obtain a residual cladding thickness of approximately 25 μm in diameter. A refractometric sensitivity of 17.4 nm/RIU was reported. This configuration, operating in reflection mode thanks to a mirror deposited on the cleaved fiber end after the grating location, was then functionalized and used to detect thrombin.

A nonuniform etching along the grating length or the use of side polishing were both demonstrated to enable simultaneous temperature and refractive index measurements [26, 27].

In the past few years, microfibers have also attracted increasing interest due to their intrinsic advantages such as large evanescent field, small effective mode field

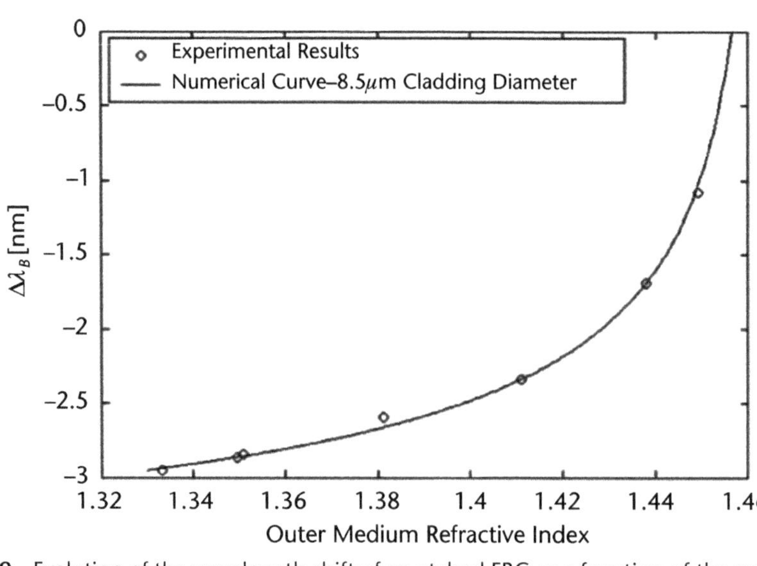

Figure 4.9 Evolution of the wavelength shift of an etched FBG as a function of the surrounding medium refractive index. (*From:* [21].)

diameter, and low-loss interconnection to single-mode fibers. Microfibers can be produced with the standard flame brushing technique. In [28], FBGs were produced in microfibers with diameters ranging from 2 to 10 μm by using femtosecond pulse laser irradiation. A maximum sensitivity of 231.4 nm/RIU was demonstrated for refractive index values near 1.44 for a microfiber with a 2-μm diameter. In [29], the microfiber diameter was approximately 6 μm and the gratings were written using a KrF excimer laser. A maximum refractometric sensitivity of approximately s102 nm/RIU was reported around the refractive index of 1.378.

The possibility to use an etched FBG for surface plasmon generation at a predetermined wavelength of interest was first theoretically studied in [30]. In [31], a plasmonic-based refractometer based on a nanoshell-coated etched FBG has been demonstrated. The sensor works in reflection mode and the reflected light intensity shows a high refractive index sensitivity of −4,400%/RIU. The sensitive length of the used fiber samples was approximately 2.7 mm with a residual diameter of 6 μm. A gold-coated configuration was reported in [32], where the grating was used for light scattering in the cladding. It was used to sense the concentration of aqueous ethanol.

As for the sensing platforms reported in Section 4.2, because of the very reduced fiber diameter that considerably weakens the mechanical resistance of the fiber, etched FBGs remain a marginal approach in the context of plasmon generation. Tilted FBGs have been much more studied and developed to that aim with several tens of publications to date. They are the focus of the next section.

4.4 Weakly and Highly Tilted Fiber Bragg Gratings

4.4.1 Weakly Tilted Fiber Bragg Gratings

Weakly tilted fiber Bragg gratings (TFBGs) belong to the short period gratings family as the periodicity of the refractive index modulation is of the order of 500 nm [33, 34]. They are characterized by grating planes blazed by a certain angle θ with respect to the fiber axis, as sketched in Figure 4.10. Tilting the grating planes results in light (that is otherwise guided in the fiber core) being coupled into cladding or radiation modes. The tilt angle of the grating planes and the strength of the refractive index modulation determine the coupling efficiency and the bandwidth of the light that is outcoupled from the fiber core.

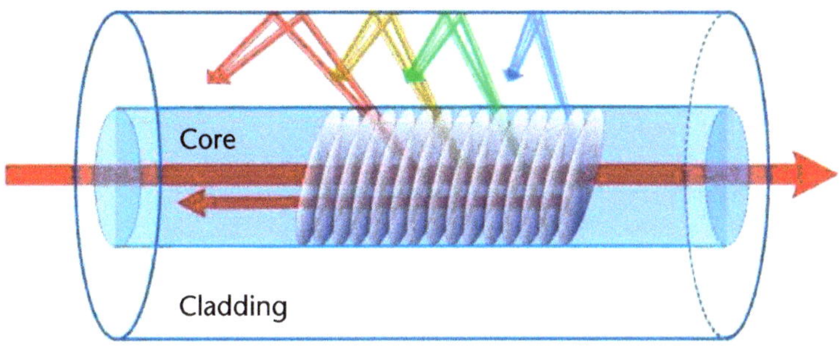

Figure 4.10 Sketch of a weakly tilted FBG and its physical parameters (not in scale).

In the frame of sensing, the most interesting feature of TFBGs is their transmitted amplitude spectrum. It is indeed composed of cladding mode resonances below the Bragg wavelength. Each cladding mode resonance corresponds to the coupling between the forward-going core mode and a backward-going cladding mode. TFBGs are governed by phase-matching conditions that give the wavelength positions of the resonance bands corresponding to the couplings between two modes. The resonance wavelength λ_B and the wavelength at which the discrete coupling to the *j*th cladding mode occurs $\lambda_{c,j}$ (characterized by an effective refractive index $n_{c,j}$) are given by the following set of equations:

$$\lambda_{\text{Bragg}} = 2n_{\text{eff,core}} \frac{\Lambda}{\cos\theta} \tag{4.54}$$

$$\lambda_{\text{clad},j} = \left(n_{\text{eff,core}} + n_{\text{eff,clad},j}\right) \frac{\Lambda}{\cos\theta} \tag{4.55}$$

Figure 4.11 depicts the transmitted amplitude spectrum of a 1 cm-long 10° TFBG. Each resonance of the spectral comb corresponds to the coupling from the core mode to a group of backward propagating cladding modes. As a result of phase matching, the spectral position of a resonance now depends on the effective index of its associated cladding mode, which, in turn, depends on the optical properties of the medium over or near the cladding surface.

Obviously, TFBGs are sensitive to temperature and strain. It was shown in [35] that the cladding mode resonances present a similar temperature sensitivity to the Bragg resonance. However, the axial strain sensitivity tends to increase with the cladding mode order so that a temperature-insensitive axial strain sensor can be obtained with such gratings.

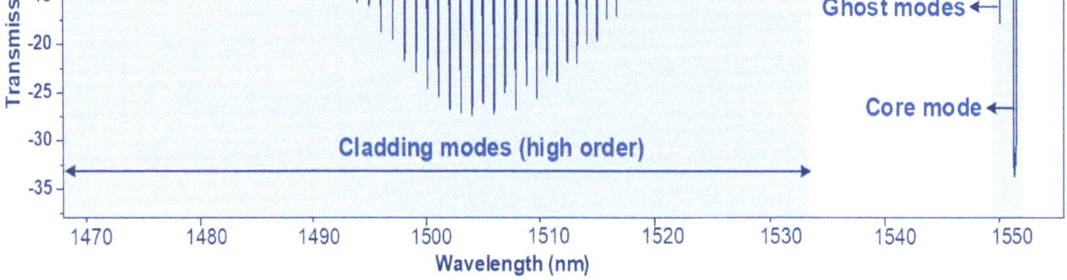

Figure 4.11 Transmitted amplitude spectrum of a 1-cm long 10° TFBG. This graph corresponds to a bare grating measured in air using linearly polarized input light, as described in Section 4.3.

More interestingly, TFBGs are sensitive to the surrounding refractive index. Therefore, spectral shifts of individual resonances can be used to measure changes in fiber coatings and surroundings. Laffont et al. were the first to study the behavior of TFBGs for refractive in sensing in 2001. They reported that, when the surrounding refractive index increases in the range of 1.30 to 1.45, a progressive smoothing of the transmitted spectrum starting from the shortest wavelengths is obtained [36]. This smoothing can be explained by the fact that one can assign to any resonance $\lambda_{c,j}$ a discrete cladding mode with an effective refractive index denoted $n_{c,j}$, which decreases while $\lambda_{c,j}$ decreases. So when the external refractive index rises and reaches the value $n_{c,j}$, this mode becomes weakly guided due to the decrease of the overlapping integral between the fundamental guided mode and the jth cladding mode, thereby reducing the amplitude of the coupling coefficient and consequently the amplitude of the cladding mode resonance. When the surrounding refractive index matches the $n_{c,j}$ value, the cladding mode is no longer guided and the coupling occurs with a continuum of radiation modes. This corresponds to the mode cutoff. The latter corresponds to the wavelength at which the effective refractive index of a cladding mode resonance matches the one of the surrounding medium.

Several demodulation techniques can be used to quantitatively correlate the spectral content with the SRI value, either based on a global spectral evolution or a local spectral feature change. This surely results from the richness of the TFBG cladding mode resonances' spectrum. The first method involves monitoring the area delimited by the cladding mode resonance spectrum, through a computation of the upper and lower envelopes as resonances gradually disappear when the SRI reaches the cutoff points of each cladding mode [36, 37]. Another technique tracks the wavelength shift and amplitude variation of individual cladding mode resonances as they approach the cutoff [38]. Both techniques present minimum detectable SRI changes of approximately 10^{-4} RIU when applied to bare gratings. In terms of wavelength shift, refractive sensitivities close to 25 nm/RIU were reported for cladding mode resonances near cutoff. In all cases, the Bragg wavelength provides an absolute power and wavelength reference, which can therefore be used to remove uncertainties related to unwanted optical power fluctuations or temperature changes. Thus, weakly tilted FBGs inherently provide temperature-insensitive SRI measurements and large SNR. For biosensing purposes, it is usually necessary to improve the LOD levels to at least 10^{-5} RIU, by increasing the wavelength shift sensitivity while keeping noise level down and spectral features narrow [39]. It has been demonstrated that the addition of a nanometric-scale gold coating overlay on the TFBG outer surface considerably enhances the refractometric sensitivity via SPR excitation [40–42]. The main achievements obtained in this growing research field will be summarized hereafter.

As demonstrated in [33], the tilt angle (we consider here values below 45°; higher values yield to the development of highly (or excessively) tilted FBGs) influences the cladding mode resonances' distribution in the transmitted amplitude spectrum. Although the evolution is not monotonic, the global trend results in an increase of the coupling to high-order modes (therefore at shorter wavelengths) for increasing tilt angle values. For tilt angle values less than 10°, gratings couple to cladding mode resonances with effective refractive indices ranging between approximately 1.30 and approximately 1.45, as shown in Figure 4.11. Hence, such angles are preferred to operate when the SRI lies near the refractive index of water and aqueous solutions, which is often the case in biochemical research, because then the strongest cladding mode resonances are located near 1,550 nm (when the Bragg wavelength is near 1,610 nm), where they are easier to measure considering available equipment at that wavelength. For refractometric sensing purposes in gaseous media [43, 44], higher tilt angles can be envisaged to couple light to cladding mode resonances with an effective refractive index close to 1.00. As depicted in Figure 4.12, this is possible for tilt angles higher than 30°. In this case, the cladding mode resonances on the short wavelength side of the Bragg resonance can be divided into two main subsets:

1. In the wavelength range (1,480 to 1,550 nm), cladding mode resonances have effective refractive indices ranging between 1.30 and 1.44 and are therefore suited for measurement in aqueous solutions, similar to the case of weakly tilted FBGs with tilt angles limited to 10°.
2. In the wavelength range (1,270 to 1,410 nm), cladding mode resonances present effective refractive indices ranging between 0.92 and 1.18, according to the aforementioned phase-matching condition.

It was also shown that cascading TFBG sections with different tilt angles can easily provide a continuous comb of cladding mode resonances with effective refractive indices ranging between 1.45 and less than 1.00 [41].

TFBGs are fabricated using the same tools and techniques as standard FBGs (i.e., from a permanent refractive index change induced in doped glasses from the interference between two ultraviolet laser beams). Hence, all the technological improvements and reliability studies that have accompanied the development of a worldwide FBG industry for telecommunications and sensing applications over the last 25 years are directly applicable to TFBGs, including low-cost mass production.

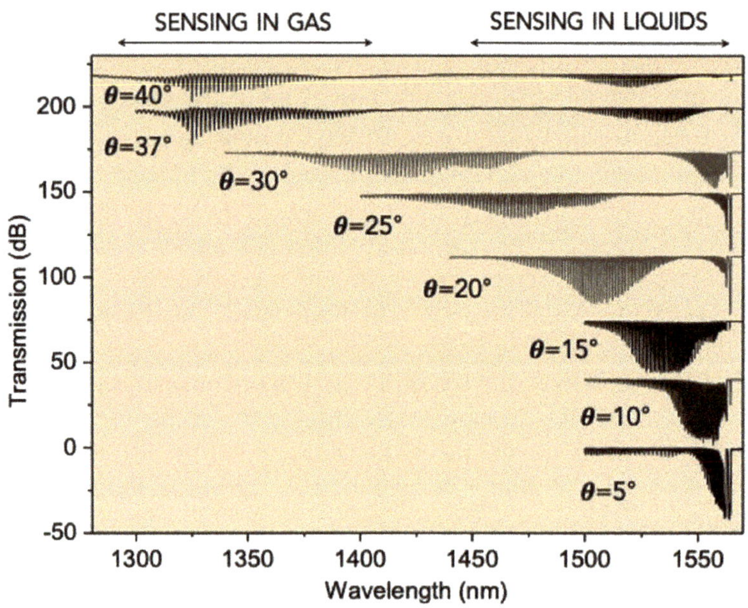

Figure 4.12 Effect of the tilt angle on the cladding mode distribution in TFBGs.

We only describe here the modifications required to make tilted FBGs, as shown in Figure 4.13.

With an interferometric setup (Figure 4.13(a)), it is sufficient to tilt the fiber relative to the fringe pattern. This was the approach used by Meltz and coworkers in their pioneering work of the early 1990s and it is still used [46, 47]. The main advantage of this approach is its flexibility to change periods and tilt angles. However, when large numbers of identical gratings are needed, another approach is usually preferred, with the interference pattern generated by a phase mask located close to the fiber, as sketched in Figure 4.13(b). The period of the grating is fixed by the phase mask and because of the proximity of the fiber, low coherence ultraviolet sources can be used, such as high-energy pulsed excimer lasers. In this case, tilting can be done in two ways: rotating the phase mask and fiber as in the previous case (Figure 4.13(b)) or keeping the fiber and phase mask perpendicular to the incident writing beam but rotating the phase mask around the axis of the writing beam (Figure 4.13(c)). Note that in the first case this also requires to tilt the cylindrical lens that is used to focus the writing light intensity along the fiber axis.

In all cases if a moderate power continuous-wave laser beam is used to write the grating (typically a continuous wave frequency-doubled Ar ion laser emitting at 244 nm), the laser beam must be scanned to obtain typical grating lengths of the order of 1 to 2 cm. Scanning is not necessary with high-energy pulsed excimer lasers (at 248 or 193 nm) that have energies per pulse ranging from 100 to 400 mJ and beam size of 10×40 mm^2 because the beam can be expanded in one dimension (along the fiber axis) and thus used to write long gratings (up to 10 cm and

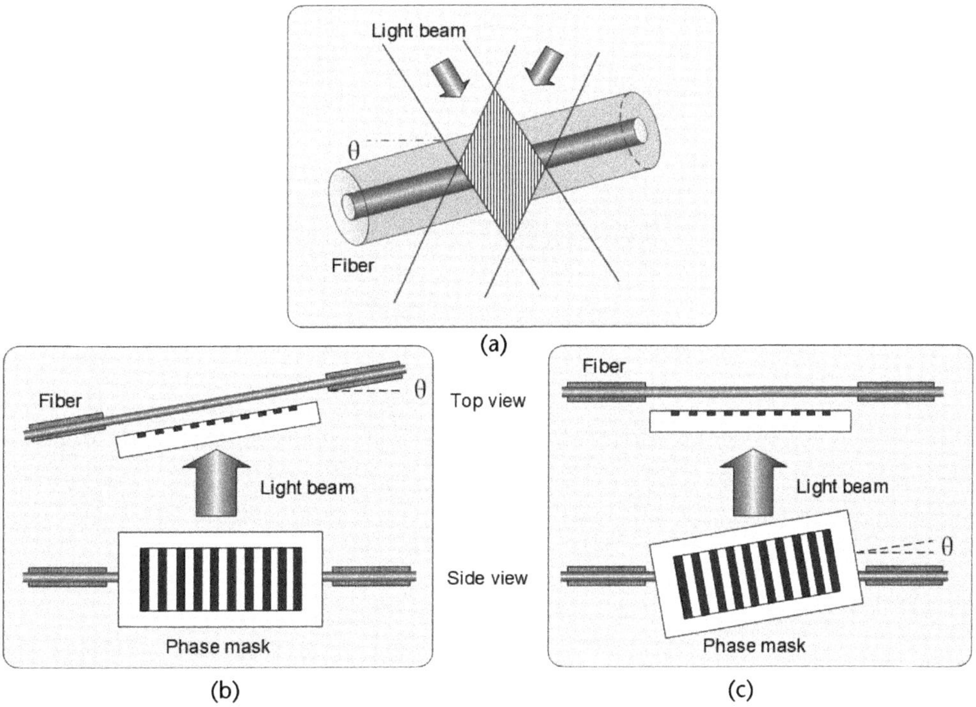

Figure 4.13 Sketch of the main techniques used to photo-inscribe TFBGs in photosensitive optical fibers.

more) in a single exposure. We have found by experience that the fiber-phase mask assembly rotation technique provides the best spectral responses for strong TFBGs with tilt angles between 4° and 10°.

The first experimental demonstration that a gold-coated TFBG can excite a surface plasmon dates back to 2007 and was reported by the team of Professor J. Albert at Carleton University of Ottawa in Canada [40]. (Professor Albert also did us the honor of writing the foreword to this book.) Since then, numerous studies have been conducted and practical applications have been demonstrated, especially for use as biosensors. Most often, the interrogation of gold-coated TFBGs is based on transmitted amplitude spectra measurements. A polarization controller is usually placed behind the optical source to control and orient the state of polarization (SOP) of the light launched into the TFBG. In such a configuration, care is taken to avoid polarization instabilities (use of short fiber lengths, strong curvatures avoided, and ambient temperature kept constant to within 1°C). It is worth mentioning that these gratings can operate in reflection mode, provided that a gold or silver mirror is deposited on the cleaved fiber end face, after the TFBG section.

Figure 4.14 displays the typical transmitted amplitude spectrum of a gold-coated TFBG immersed in salted water (refractive index measured close to 1.356 at 589 nm). The latter was recorded with a linear input SOP optimized to maximize coupling to the SPR, corresponding to the radial polarization, as explained further. This spectrum displays the typical SPR signature around 1,551 nm, which is due to the maximum phase matching of the cladding mode to the surface plasmon mode of the gold water interface, according to [40]. The core mode resonance appears at the right end side, corresponding to a Bragg wavelength of 1,602 nm. In practice, the Bragg wavelength is used to remove any effect from surrounding temperature changes by monitoring its shift. This intrinsic feature is very interesting in practice as a change of 0.1°C induces an SRI change of 10^{-5}, thus prone to generate erroneous spectral modifications.

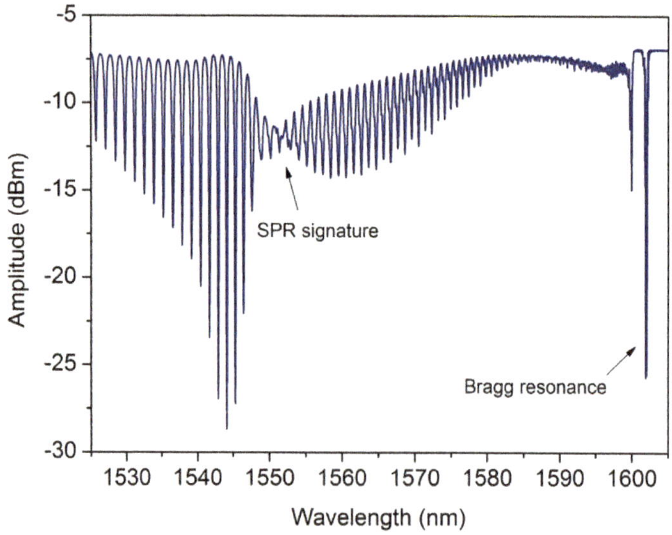

Figure 4.14 Transmitted amplitude spectrum of an SPR-TFBG immersed in salted water (radial polarization – SRI = 1.358).

Figure 4.15 displays a zoom around the SPR signature for radially (EH and TM modes) and azimuthally polarized (HE and TE modes) amplitude spectra that yield antagonist behaviors in liquids, as explained in [42]. It is obvious from this figure that EH and HE modes come in pairs and, in order to facilitate the explanations of the cladding mode behavior, the mode resonances are labeled as a function of their position with respect to the most important one (strongest peak-to-peak amplitude) in the EH spectrum. This mode is the cutoff of the bare grating immersed in the same refractive index solution. Let's label this mode as the number 0. Modes to the left of the 0 mode are labeled with a negative integer (growing from right to left), while modes above the 0 mode are labeled with a positive integer (growing from left to right).

As for other optical fiber configurations, the refractometric sensitivity of gold-coated TFBGs is usually determined by immersing them in different calibrated refractive index liquids. Figure 4.16(a) shows the transmitted amplitude spectra of a 50-nm gold-coated TFBG measured for 3 different refractive index values. In the case of large SRI changes, the SPR location can be unambiguously located by following the strongest attenuation in each spectrum. Therefore, the interrogation relies on the tracking of the wavelength shift of the center of the envelope of the most attenuated resonances. Using this technique, a linear response is obtained, as depicted in Figure 4.16(b) and the SRI sensitivity is approximately 550 nm/RIU in the range between 1.32 and 1.42.

For high-resolution refractometric sensing over an SRI range limited to 10^{-3} typically, which corresponds to variations obtained in the case of biochemical sensing, accurate measurements of the SPR mode are not possible from the radially polarized spectrum. Indeed, the SPP is only revealed by its absence from the spectrum and it cannot be reliably measured for wavelength shifts limited to a few picometers, as this region of the spectrum appears quite noisy. Hence, different methods have

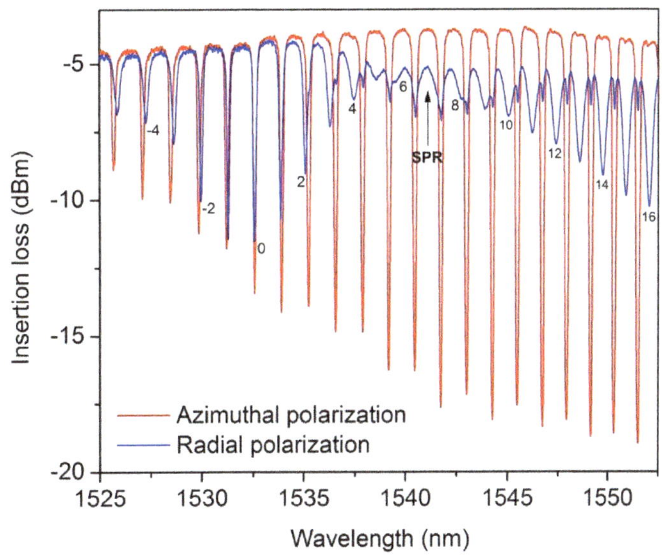

Figure 4.15 Transmitted amplitude spectra for two orthogonal SOPs (radial and azimuthal polarizations) of a TFBG immersed in salted water (SRI = 1.338).

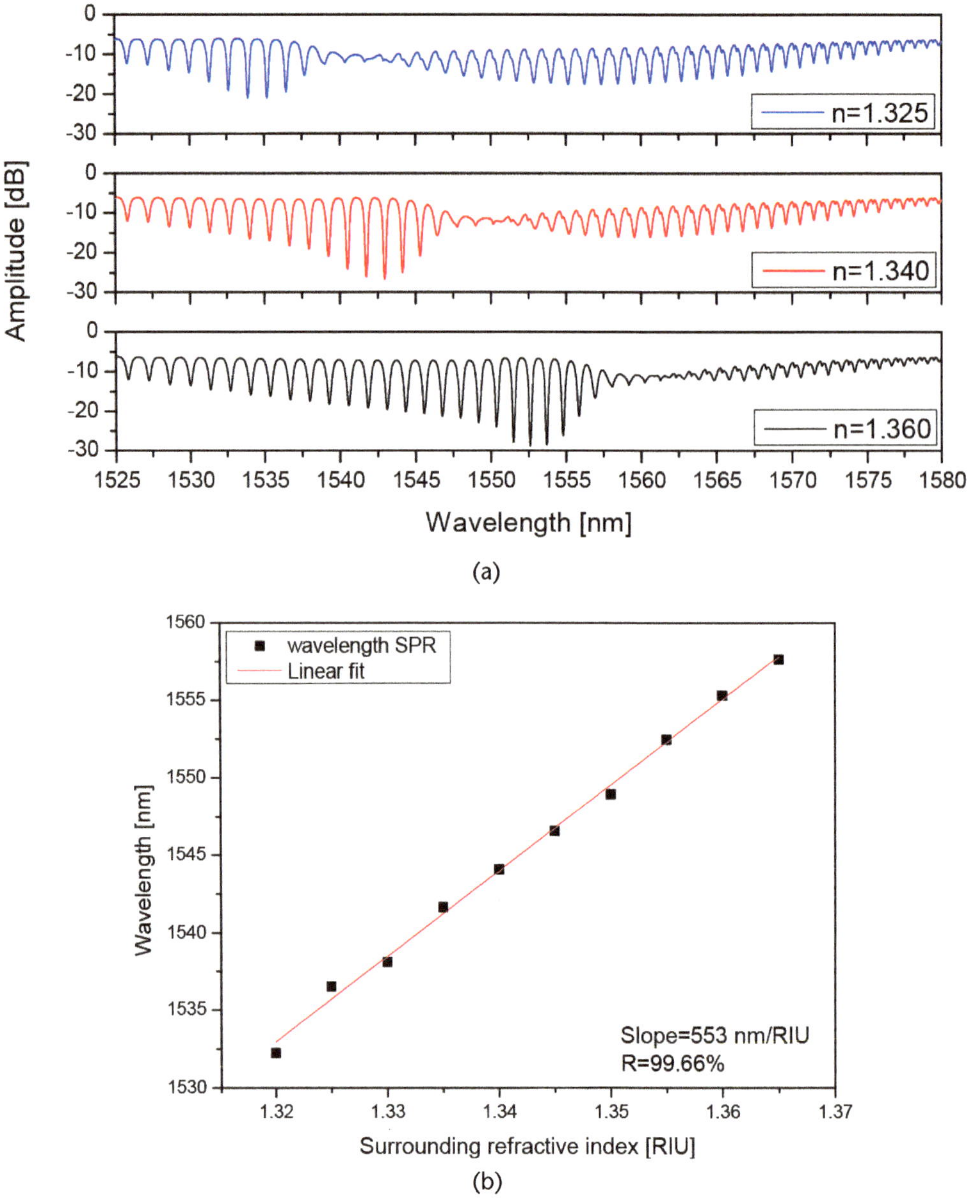

Figure 4.16 (a) SPR signature in the transmitted spectrum of gold-coated TFBG for coarse changes in SRI and (b) SPR wavelength shift as a function of the SRI value.

been developed and used to track the SPR shift [41, 42]. In practice, modes slightly off the SPR are used (usually +2 to +4 modes depending on the metal thickness), because they combine relatively high sensitivity with a narrow spectral width and they can be followed by a combination of their changes in amplitude and wavelength. Indeed, as they stand on the shoulder of the SPR envelope, a slight change of the SPP location yields a modification of the peak-to-peak amplitude of these modes, as well as a wavelength shift. A direct tracking of the +2 mode wavelength shift yields a refractometric sensitivity that ranges between 30 and 100 nm/RIU, depending on the actual gold thickness [42, 48, 49].

To gain in sensitivity, an alternative demodulation process based on the computation of the spectral envelopes has been developed [50, 51]. In particular in [51], a demodulation process based on the computation of the crossing point between the spectral envelopes (themselves computed from the upper and lower spectral envelopes) has been reported. The corresponding point and its evolution as a function of the surrounding refractive index are depicted in Figure 4.17. A sensitivity reaching 530 nm/RIU has been reported.

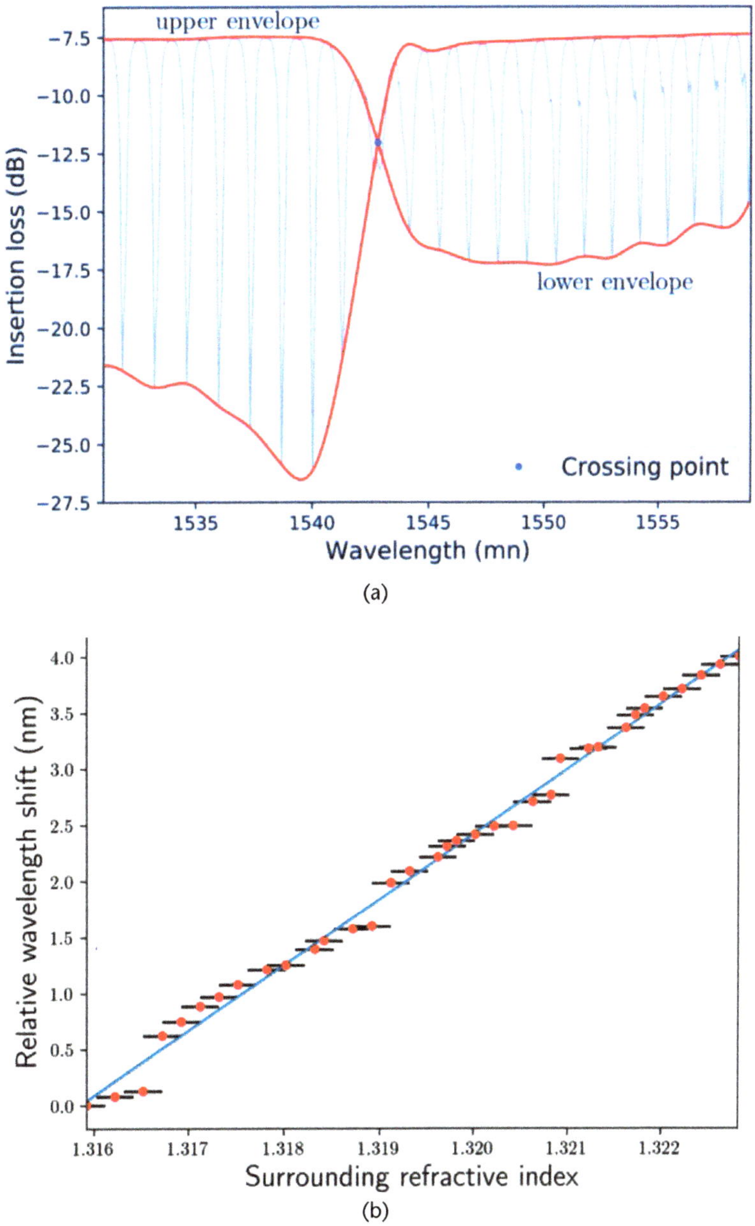

Figure 4.17 (a) Operating principle of the demodulation process based on the computation of the intersection point between the spectral envelopes of a gold-coated TFBG and (b) corresponding wavelength shift as a function of the surrounding refractive index value.

In terms of experimentally demonstrated FOM, TFBGs exhibit a value reaching 5,000, which outperforms all other optical fiber configurations by more than one order of magnitude. This results from the fact that they exhibit narrow resonance bands (FWHM ~0.2 nm) compared to even the best possible theoretical value (~5 nm obtained by calculating the reflection from the base of a prism in the Kretschmann-Raether configuration), also keeping in mind that the experimental SPR FWHM from all other fiber configurations exceeds 20 nm and more.

4.4.2 Excessively Tilted Fiber Bragg Gratings

Excessively tilted fiber gratings (ex-TFGs) were first proposed in 2006 at Aston University in the United Kingdom and have been used extensively since then [52, 53]. The technology was recently comprehensively reviewed in [54] and, although it has not been yet used to excite surface plasmon wave, we find it relevant to discuss this configuration here, as its sensing potential is manifest. Typical ex-TFGs have periodically slanted refractive index planes inclined more than 66.9° with respect to the perpendicular to the optical fiber axis, as sketched in Figure 4.18. In practice, the grating period of such structures lies between the ones of uniform Bragg gratings and long period fiber gratings presented in Section 4.6.

Hence, the grating period (perpendicular to the grating planes) is short relatively to long period fiber gratings, but the large angle lengthens the axial period by $1/\cos\theta$. As a result, such gratings do not operate in reflection but enable a codirectional mode coupling. The usual resulting period is typically around tens of microns and therefore supports the excitation of higher order cladding modes. The phase matching conditions for ex-TFGs can be expressed as follows:

$$\lambda_{\text{clad},i} = \left(n_{\text{eff,core}} - n_{\text{eff,clad},i}\right)\Lambda_z \tag{4.56}$$

where $\Lambda_z = \Lambda/\cos\theta$ denotes the grating period along the fiber axis. Because of their slanted nature and similarly to weakly TFBGs, ex-TFGs introduce an intrinsic birefringence in the core. The orthogonal polarizations corresponding to TE and TM modes are therefore well resolved spectrally, as will be shown.

Prior to discussing the typical transmitted amplitude spectrum of such structures, let us discuss their manufacturing process. To produce ex-TFGs, the amplitude mask method is the one most used in practice. An amplitude mask (mask usually made

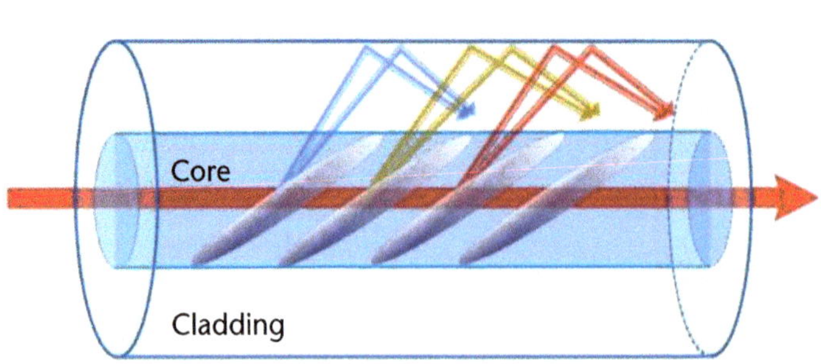

Figure 4.18 Sketch of an ex-TFBG and its physical parameters (not in scale).

in metal that contains periodical apertures along its main length) is tilted by the required angle in front of the fiber and the grating structure is produced with the direct laser beam, as sketched in Figure 4.19. There is a one-to-one correspondence between the tilt angle and period of the mask and those of the inscribed grating [54]. Generally, a custom amplitude mask containing a period of a few microns is used.

Yan et al. investigated the transmitted amplitude spectrum of ex-TFGs in detail, both theoretically and experimentally [55]. A typical spectrum of an ex-TFG measured with unpolarized light is reproduced in Figure 4.20(a). A series of resonances can be seen in the wavelength range 1,300 to 1,700 nm, with bandwidths and wavelength separations similar to those of LPGs, except for the fact that each resonance appears to be split in two. Additional measurements with linearly polarized core guided input light in Figure 4.20(b) further show that the splitting corresponds to the separate excitation of the EH/TM group and the HE/TE group by input light polarized in the plane and out of the plane of the tilt, respectively [55]. The observed splitting is due to the break in cylindrical symmetry of the fiber associated with the formation of the ex-TFBG.

The refractive index sensitivity of ex-TFGs has been first investigated in [56]. As for thinned uniform FBGs and weakly TFBGs, it arises from the interaction between the grating evanescent field and the surrounding medium, which changes the effective refractive index of the corresponding resonance peak. It was shown that the sensitivity of TM mode is 291.5 nm/RIU while the one of TE mode is 252.5 nm/RIU. More recently, Yan et al. conducted a more comprehensive study (both theoretically and experimentally) and have reported that the refractive index sensitivity of ex-TFGs depends on the order number of the cladding mode. The sensitivity is the highest when the surrounding refractive index value is close to the one of the cladding mode. Reducing the cladding size, the sensitivity can be further enhanced. Two ways can be followed to that aim: direct inscription of an ex-TFG in a thin cladding fiber [57] or reducing the fiber cladding by etching [58]. It was shown that the sensitivity can reach 1,180 nm/RIU (1,150 nm/RIU) for the TM (TE) mode for a grating inscribed in an 80-μm optical fiber. In a fiber with a reduced cladding diameter of approximately 15 μm after etching, a refractometric sensitivity as high as 1,600 nm/RIU was reported. Obviously, the reduction of the cladding diameter at the grating location decreases the mechanical strength of the sensor. The use of a TiO_2 coating (high refractive index material) deposited on the cladding surface at the ex-TFG location was demonstrated to enhance the refractometric sensitivity as well, while keeping the fiber integrity. A sensitivity peaking at 32,261.2 nm/

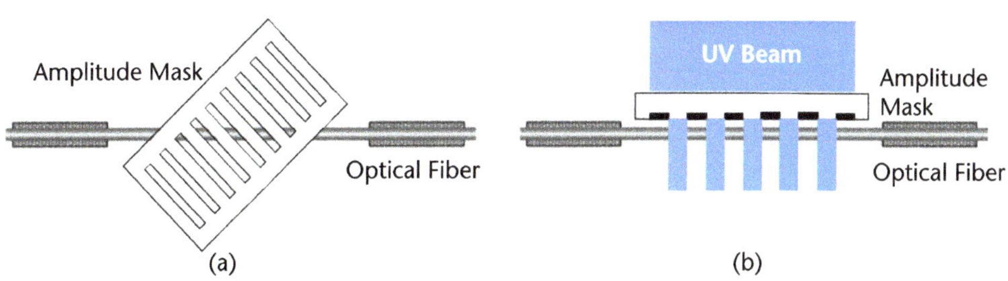

Figure 4.19 Schematic of the ex-TFG inscription process: (a) front view in the plane perpendicular to the incident laser beam and (b) top view.

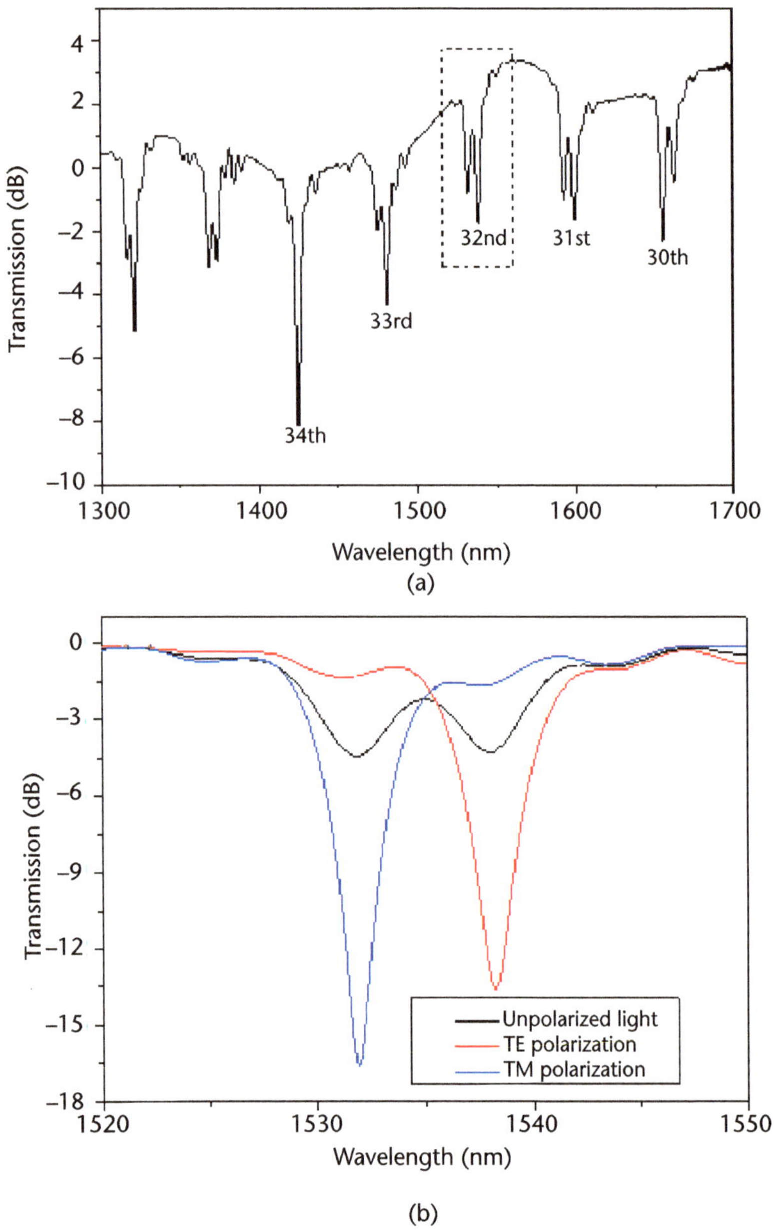

Figure 4.20 Transmitted amplitude spectrum of a 81° TFG: (a) dual-peak resonances observed with unpolarized light launched in the core, and (b) zooming on one dual resonance and its identification with one EH(TM) and HE(TE) mode pair. (*Adapted from:* [55].)

RIU was reported [59]. The use of graphene as a coating material was investigated as well in [60]. In [61], gold nanoshells here immobilized on ex-TFGs increase the polarization dependency between TE and TM modes.

Hence, ex-TFGs remain today an emerging technology for which the number of applications is growing, with an important part devoted to refractive index sensing. There is almost no doubt that this platform will be soon associated with continuous metal sheaths for surface plasmon generation. The following section will briefly present an alternative to weak TFBGs that can be produced thanks to tightly focused laser pulses.

4.5 Eccentric Fiber Bragg Gratings

Aside from the aforementioned classical writing methods, the advent of femtosecond pulses lasers has allowed alternative gratings production through the development of the point-by-point, line-by-line, and plane-by-plane techniques [62, 63]. There the laser beam is focused through a microscope objective (either in index-matching oil or in air) towards the fiber location where the inscription (a controlled glass damage process depending on the pulse energy) can take place, as sketched in Figure 4.21. The laser beam waist is generally smaller than the grating period. Because the laser is pulsed, the stepwise translation of the fiber according to the pulse rate allows patterning the glass modification and thereby creating a controlled refractive index modulation. First-order gratings (with a physical period of approximately 530 nm, like in uniform FBGs) can be produced with the technique, but, on many occasions, high-order gratings are produced. Indeed, (4.26) corresponds to a first-order grating, but more generally this equation can be rewritten as:

$$m\lambda_B = 2n_{eff}\Lambda \tag{4.57}$$

where m is the grating order. Hence, tenth-order gratings can be produced at approximately 1,550 nm when their physical period is of the order of 5 μm.

Gratings can also be located away from the core or partly offset from the fiber axis in a 3-D fashion because traditional photosensitivity mechanisms are no longer required to produce permanent, subwavelength-scale refractive index modifications in just about all kinds of transparent materials. This technique is certainly the most flexible to date and allows producing many more structures in optical fibers than standard fiber grating technologies.

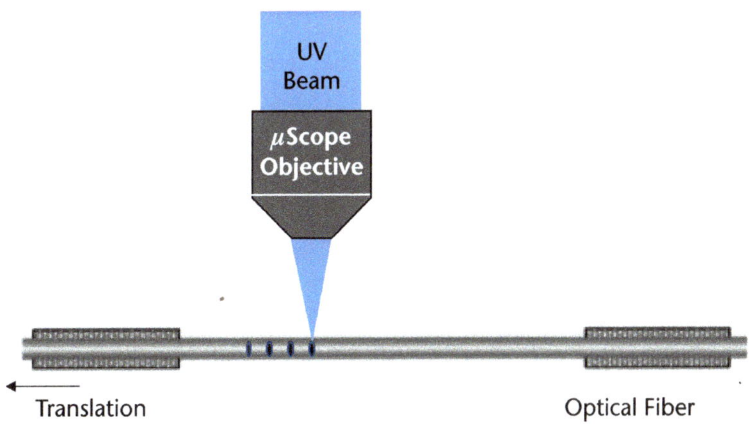

Figure 4.21 Sketch of the direct writing technique with focused femtosecond laser pulses.

An alternative to TFBGs that shares many of their characteristics has been reported using highly localized grating inscription techniques relying on point-by-point femtosecond pulsed laser irradiation [63, 64]. Instead of grating planes extending across the fiber core, smaller dimension periodic structures along the fiber axis can be localized anywhere in the fiber cross-section and, in this particular case, away from the fiber core, as sketched in Figure 4.22.

As in tilted gratings, this localized refractive index modulation breaks the cylindrical symmetry of the fiber. It couples a forward propagating core mode to a very large number of backward propagating cladding modes. Unlike TFBGs, however, where the amplitudes of the cladding mode resonances strongly depend on the tilt angle, peaking at a specific distance from the Bragg resonance, localized FBGs generate an apparently infinite number of cladding mode resonances extending for hundreds of nanometers from the Bragg wavelength with quite similar peak-to-peak amplitudes. Figure 4.23 shows the measured transmitted amplitude spectrum of such grating with air as the surrounding medium. Cladding mode resonances around the wavelength of 1,300 nm present an effective refractive index close to 1. The decrease in peak-to-peak amplitude of these modes therefore represents the cutoff because the grating is surrounded by air. The 1-cm-long grating was written using 120-fs third-harmonic (266-nm) laser pulses focused through a 40× long working distance (4-mm) 0.5 numerical aperture microscope objective in air (no refractive index-matching oil). A two-axis, air-bearing, translation stage was used for precise alignment and displacement during the inscription process. The strength of the mode coupling is linked to the position of the refractive index modulation created in the fiber core during the photo-inscription [64, 65].

In [66], an approximately 30-nm gold coating was deposited on both sides of eccentric FBGs thanks to a standard sputtering process. Measurements conducted with polarized light have confirmed the possibility to excite a surface plasmon wave for radially polarized (EH modes) light but not for azimuthally polarized ones (HE modes), as depicted in Figure 4.24 for a gold-coated EFBG immersed in an aqueous solution. The refractometric sensitivity of such structures was experimentally studied in calibrated refractive index liquids. As for TFBGs, the demodulation was based on tracking the wavelength shift of the most sensitive cladding mode resonance located on the left shoulder of the SPR signature (identified by an arrow in Figure 4.24). A refractometric sensitivity equal to 50 nm/RIU has been reported, which is very similar to the one of TFBGs. Other demodulation techniques based on spectral envelope computation like in TFBGs could be applied to these structures as well.

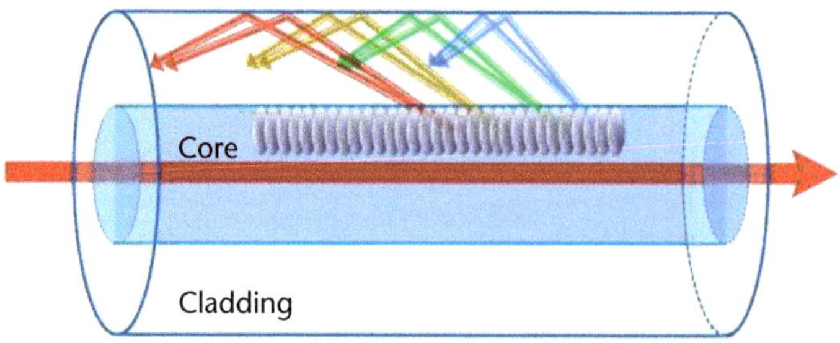

Figure 4.22 Sketch of an eccentric FBG and its physical parameters (not in scale).

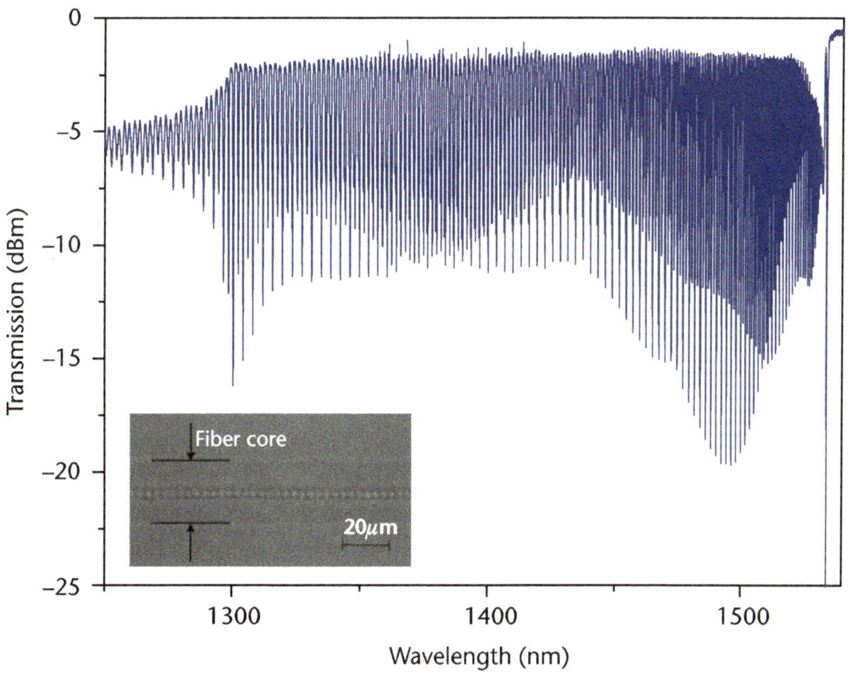

Figure 4.23 Transmitted amplitude spectrum of a localized FBG written by the point-by-point technique with a femtosecond pulse laser. Inset: microscope picture of the refractive index modulation localized in the core.

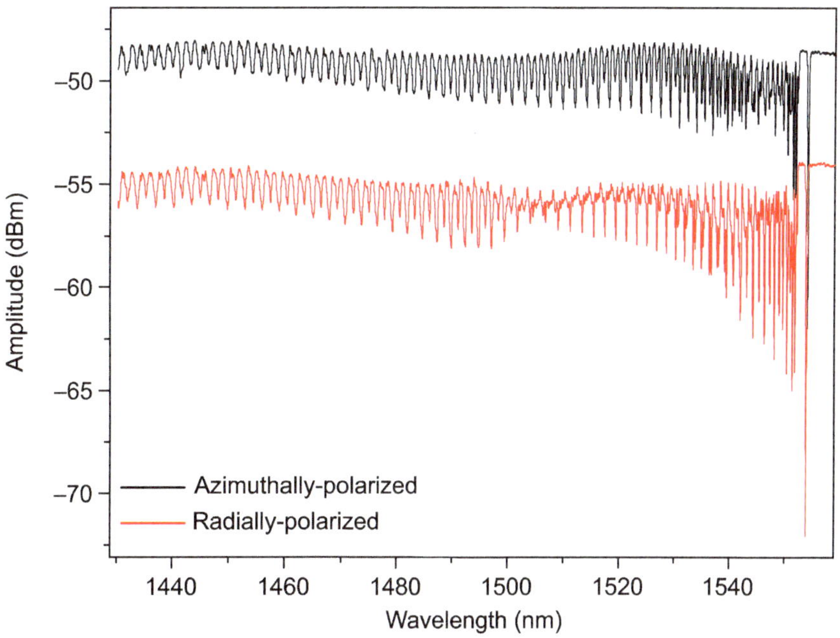

Figure 4.24 Transmitted amplitude spectra (azimuthal and radial polarizations) of a gold-coated eccentric Bragg grating immersed in a calibrated refractive index liquid.

Compared to TFBGs photo-inscribed by the phase mask technique, eccentric FBGs bring relevant practical assets, such as enhanced design flexibility due to their point-by-point nature and resistance to high temperatures, up to 1,000°C [67]. Also, they can be used for plasmonic sensing in gaseous media, as they enable coupling to cladding modes with an effective refractive index close to 1.

4.6 Long-Period Fiber Gratings

As described above, subwavelength grating periods lead to contradirectional coupling. Oppositely, gratings with periods much higher than the light operational wavelength couple modes in the forward direction. Because of their longer period (typically between 100 and 500 μm), long-period fiber gratings (LPFGs) can be fabricated much more easily than standard FBGs, even using periodic heating with fiber fusion tools.

LPFGs were popularized after an initial report by Vengsarkar in 1996 [68] about periodic structures able to couple the guided fundamental core mode of a single-mode fiber into forward-propagating cladding modes [69]. Because cladding modes decay rapidly as they propagate along the fiber axis due to scattering losses at the cladding-surrounding medium interface and they are strongly perturbed by bending, they can only be used over short distances, which is not a problem in sensing. Similar to FBGs, the coupling in LPGs is wavelength-selective and cladding mode resonances appear in the transmitted amplitude spectrum of an LPFG at the following central wavelengths:

$$\lambda_{\text{clad},i} = \left(n_{\text{eff,core}} - n_{\text{eff,clad},i}\right)\Lambda \tag{4.58}$$

where $n_{\text{eff,clad},i}$ is the effective index of the ith cladding mode. The structure of an LPFG is sketched in Figure 4.25.

The transmitted amplitude spectrum of an LPFG is composed of several well-separated broadband (FWHM approximately 10 to 50 nm) resonances, as depicted in Figure 4.26. The wavelength position of these bands depends on the grating period and the order of the resonances increases from left to right, in agreement with (4.58).

While the sensitivities of LPG resonances are based on the same principle as those of FBGs, there are two important differences: the cladding mode resonances

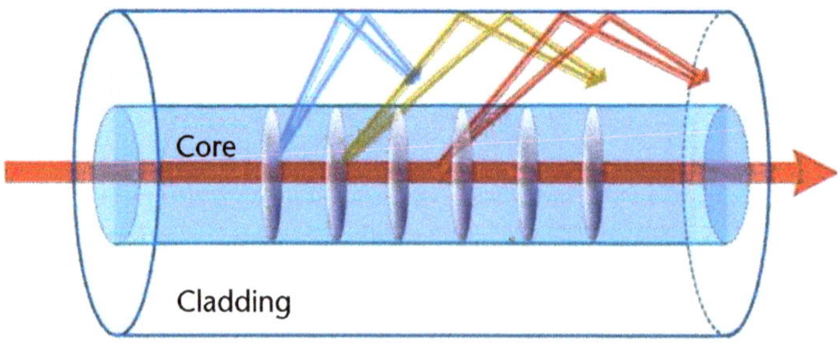

Figure 4.25 Sketch of an LPFG and its physical parameters (not to scale).

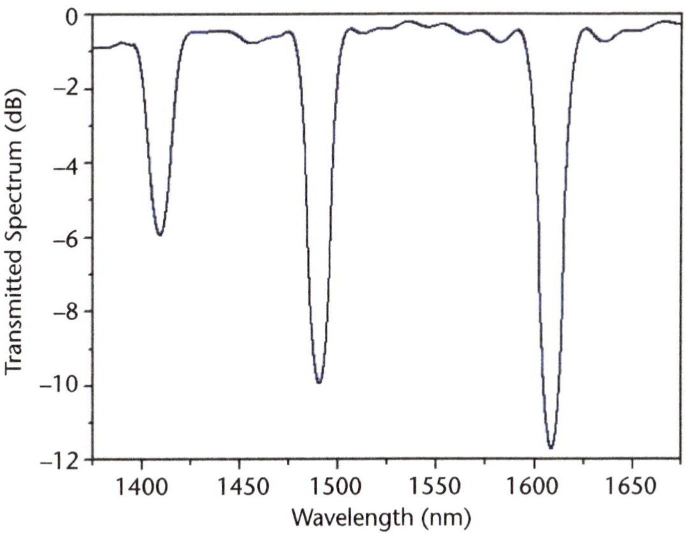

Figure 4.26 Transmitted amplitude spectrum of an LPFG.

are sensitive to bending and to the surrounding refractive index, and they are significantly more sensitive in general because the wavelength shifts depend on the difference of the mode dispersions (instead of their sums in FBGs).

The sensitivity of these gratings greatly depends on the grating period value. Dispersion curves drawn in Figure 4.27 indicate the presence of a turnaround point (TAP) for higher-order cladding modes where the slope of the curve changes signs from positive to negative [70]. LPFGs featuring resonances at these TAPs become highly sensitive to perturbations, including surrounding refractive index change, as the slope $d\lambda_{res}/d\Lambda$ of the dispersion curve reaches infinite values. For lower-order

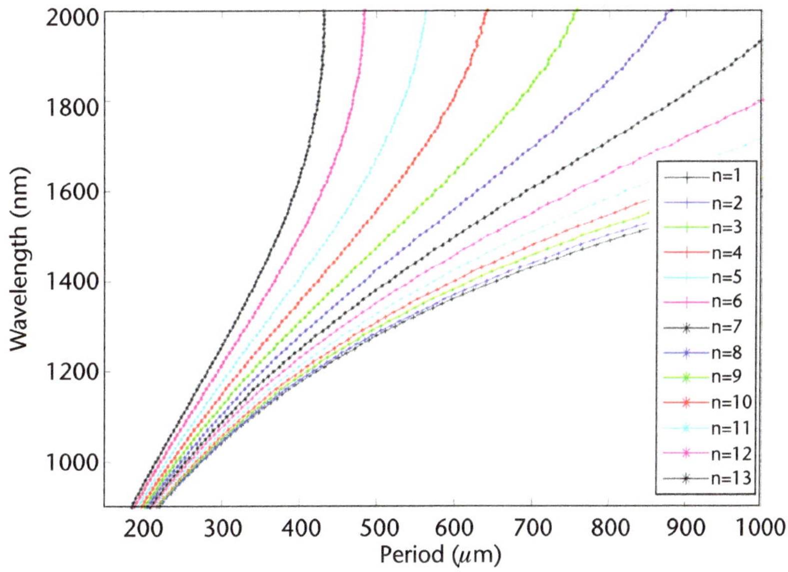

Figure 4.27 Phase matching curves for an LPFG in an SMF-28 (n = 1 to 13).

modes, the TAPs occur at longer wavelengths, beyond the optical telecommunication window.

LPFGs can be manufactured following different ways. Making use of an amplitude mask (usually made of metal) containing periodic apertures corresponding to half of the grating period as well as point-by-point UV or CO_2 laser irradiation are popular techniques for making LPGs, as sketched in Figure 4.28. LPFGs can also be produced mechanically by pressing fibers between grooved plates or by using longitudinally distributed electric arc discharges (from a fusion splicer, for instance).

The spectral evolution of LPFGs as a function of coatings and thin films was analyzed in several works [71–73]. The possibility of generating SPR was first theoretically studied in [74] based on the coupled-mode equations. Plating the fiber cladding surface with a metal layer, as done in [75], enables the coupling of the cladding mode to an SPP and the measurement of changes at the metal surface by monitoring the LPFG resonance. In this work, LPFGs were produced using the point-by-point technique with a frequency-doubled argon ion laser. The grating period was 140 μm to get a good coupling of light to the $HE_{1,20}$ cladding mode at a wavelength around 660 nm. A 30-μm gold coating was then plated on the LPFG and the refractometric sensitivity was measured in calibrated liquids. In [76], a similar experimental study was conducted on an LPFG with a period of 80 μm surrounded by gold nanoparticles to enable localized SPR excitation. At the turning point of the cladding mode resonance and for a refractive index of 1.4058, a 118% increase in sensitivity was observed compared to a bare grating. Material dispersion was taken into account in a design study based on the coupled mode theory made in [77]. Another numerical study was conducted in [78] using the finite element method and eigenmode expansion method. A theoretical refractometric sensitivity of 27,000 nm/RIU was reported. In [79], an LPFG (grating length of 1.5 cm and grating period of 600 μm) was coated with an Ag film (thickness of 50 nm) and then with a monolayer of graphene for methane detection.

A variant fiber configuration was used in [80]. The used fiber contains a photosensitive cladding so that the LPFG is produced in the cladding instead of the core. The structure was coated with silver and a refractometric sensitivity of 4,900 nm/RIU was reported.

The scientific literature on plasmonic sensors based on LPFGs is much less important than the one on TFBGs. This may be due to the quite important sensitivity of LPFGs to bending, which implies securing the fibers during experiments, in order to avoid unwanted fluctuations.

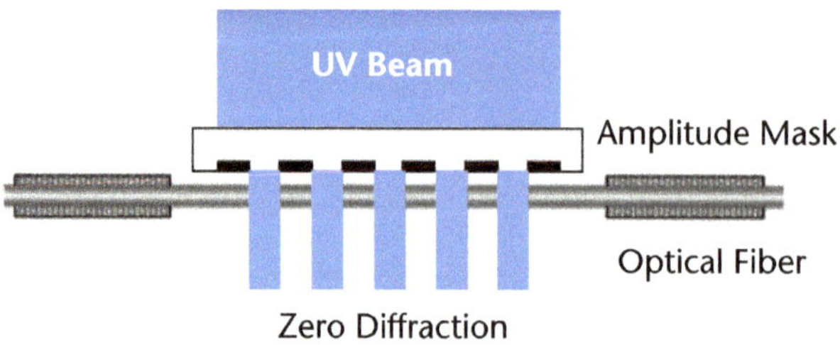

Figure 4.28 Sketch of the amplitude mask method usually used to produce LPFGs in single-mode optical fibers.

In conclusion, this chapter reviewed the operating principle of the main plasmonic configurations based on the use of single-mode optical fibers. Gratings (i.e., periodic and permanent refractive index modulations of the fiber core along the propagation axis) constitute the main platforms derived from single-mode optical fibers. Depending on the grating type, these devices intrinsically operate in transmission or in reflection mode. For configurations active in transmission, an operation in reflection is possible when the fiber is cleaved after the grating location and a mirror is deposited on the fiber end. These grating configurations can operate at telecommunication wavelengths (around 1,550 nm) and can therefore be interrogated with cost-effective equipment, including the FBG interrogators.

References

[1] Rogers, A., *Essential of Optoelectronics*, London, U.K.: Chapman and Hall, 1993.

[2] Azzam, R., and N. Bashara, *Ellipsometry and Polarized Light*, North-Holland, 1977.

[3] Jones, R., "A New Calculus for the Treatment of Optical Systems: I. Description and Discussion of the Calculus," *Journal of the Optical Society of America A*, Vol. 31, 1941, pp. 488–493.

[4] Derickson, D., *Fiber Optic Test and Measurement*, Upper Saddle River, NJ: Prentice Hall, 1998.

[5] Coelho, L., et al., "Sensing Structure Based on Surface Plasmon Resonance in Chemically Etched Single Mode Optical Fibers," *Plasmonics*, 2014, s11468-014-9811-3.

[6] Coelho, L., et al., "Multiplexing of Surface Plasmon Resonance Sensing Devices on Etched Single Mode Fiber," *Journal of Lightwave Technology*, Vol. 33, 2015, pp. 432–438.

[7] Kurihara, K. et al., "Fiber-Optic Conical Microsensors for Surface Plasmon Resonance Using Chemically Etched Single-Mode Fiber," *Analytica Chimica Acta*, Vol. 523, 2004, pp. 165–170.

[8] Tubb, A. J. C., et al., "Single-Mode Optical Fibre Surface Plasma Wave Chemical Sensor," *Sensors and Actuators B*, Vol. 41, 1997, pp. 71–79.

[9] Chang, Y., et al., "Nanofiber Optic Sensor Based on the Excitation of Surface Plasmon Wave Near Fiber Tip," *Journal of Biomedical Optics*, Vol. 11, 2014, p. 014032.

[10] González-Cano, A., et al., "Multiple Surface-Plasmon Resonance in Uniform-Waist Tapered Optical Fibers with an Asymmetric Double Layer Deposition," *Applied Optics*, Vol. 44, 2005, pp. 519–526.

[11] Navarrete, M., et al., "Surface Plasmon Resonance in the Visible Region in Sensors Based on Tapered Optical Fibers," *Sensors and Actuators B: Chemical*, Vol. 190, 2014, pp. 881–885.

[12] Esteban, O., et al., "High-Sensitive SPR Sensing with Indium Nitride as a Dielectric Overlay of Optical Fibers," *Sensors and Actuator B: Chemical*, Vol. 158, 2011, pp. 372–376.

[13] Lin, H., et al., "Tapered Optical Fiber Sensor Based on Localized Surface Plasmon Resonance," *Optics Express*, Vol. 20, 2012, p. 21693.

[14] Wieduwilt, T., et al., "Optical Fiber Micro-Taper with Circular Symmetric Gold Coating for Sensor Applications Based on Surface Plasmon Resonance," *Plasmonics*, Vol. 8, 2013, pp. 545–554.

[15] Patnaik, A., K. Senthilnathan, and R. Jha, "Graphene-Based Conducting Metal Oxide Coated D-Shaped Optical Fiber SPR Sensor," *IEEE Photonics Technology Letters*, Vol. 27, 2015, pp. 2437–2440.

[16] Liang, H., B. Liu, and J. Hu, "An Ultra-Highly Sensitive Surface Plasmon Resonance Sensor Based on D-Shaped Optical Fiber with a Silver-Graphene Layer," *Optik*, Vol. 149, 2017, pp. 149–154.

[17] Khanikar, T., and V. K. Singh, "Gold Grating Assisted SPR Based D-Shaped Single-Mode Fiber for Detection of Liquid Refractive Index," *Optical and Quantum Electronics*, Vol. 51, 2019, p. 296.

[18] Othonos, A., and K. Kalli, *Fiber Bragg Gratings: Fundamentals and Applications in Telecommunications and Sensing*, Norwood, MA: Artech House, 1999.

[19] Kashyap, R., *Fiber Bragg Gratings*, San Diego, CA: Academic Press, 1999.

[20] Barlow, A., and D. Payne, "The Stress-Optic Effect in Optical Fibers," *Journal of Quantum Electronics*, Vol. 19, 1983, pp. 834–839.

[21] Iadicicco, A., et al., "Thinned Fiber Bragg Gratings as High Sensitivity Refractive Index Sensor," *IEEE Photonics Technology Letters*, Vol. 16, 2004, pp. 1149–1151.

[22] Monerie, M., "Propagation in Doubly Clad Single-Mode Fibers," *IEEE Journal of Quantum Electronics*, Vol. 18, 1982, pp. 535–542.

[23] Nanjunda Shivananju, B., et al., "Detection Limit of Etched Fiber Bragg Grating Sensors," *Journal of Lightwave Technology*, Vol. 31, 2013, pp. 2441–2447.

[24] Tsigaridas, G., et al., "Theoretical and Experimental Study of Refractive Index Sensors Based on Etched Fiber Bragg Gratings," *Sensors and Actuators A: Physical*, Vol. 209, 2014, pp. 9–15.

[25] Bekmurzayeva, A., et al., "Etched Fiber Bragg Grating Biosensor Functionalized with Aptamers for Detection of Thrombin," *MDPI Sensors*, Vol. 18, 2018, p. 4298.

[26] Iadicicco, A., et al., "Nonuniform Thinned Fiber Bragg Gratings for Simultaneous Refractive Index and Temperature Measurements," *IEEE Photonics Technology Letters*, Vol. 17, 2005, pp. 1495–1497.

[27] Schroeder, K., et al., "A Fibre Bragg Grating Refractometer," *Measurement Science and Technology*, Vol. 12, 2001, pp. 757–764.

[28] Fang, X., C. R. Liao, and D. N. Wang, "Femtosecond Laser Fabricated Fiber Bragg Grating in Microfiber for Refractive Index Sensing," *Optics Letters*, Vol. 35, 2010, pp. 1007–1009.

[29] Zhang, Y., et al., "Refractive Index Sensing Based on Higher-Order Mode Reflection of a Microfiber Bragg Grating," *Optics Express*, Vol. 18, 2010, pp. 26345–26350.

[30] Nemova, G., and R. Kashyap, "Fiber Bragg Grating Assisted Surface Plasmon Polariton Sensor," *Optics Letters*, Vol. 31, 2006, pp. 2118–2120.

[31] Burgmeier, J., et al., "Plasmonic Nanoshell Functionalized Etched Fiber Bragg Gratings for Highly Sensitive Refractive Index Measurements," *Optics Letters*, Vol. 40, 2015, pp. 546–549.

[32] Arasu, P., et al., "Absorbance Properties of Gold-Coated Fiber Bragg Grating Sensor for Aqueous Ethanol," *Journal of the European Optical Society—Rapid Publications*, Vol. 9, 2014, p. 14018.

[33] Erdogan, T., and J. E. Sipe, "Tilted Fiber Phase Gratings," *Journal of the Optical Society of America A*, Vol. 13, 1996, pp. 296–313.

[34] Albert, J., L. Y. Shao, and C. Caucheteur, "Tilted Fiber Bragg Grating Sensors," *Laser & Photonics Reviews*, Vol. 7, 2013, pp. 83–108.

[35] Chen, C., and J. Albert, "Strain-Optic Coefficients of Individual Cladding Modes of Single-Mode Fibre: Theory and Experiment," *Electronics Letters*, Vol. 42, 2006, pp. 1027–1028.

[36] Laffont, G., and P. Ferdinand, "Tilted Short-Period Fibre-Bragg-Grating Induced Coupling to Cladding Modes for Accurate Refractometry," *Measurement Science and Technology*, Vol. 12, 2001, pp. 765–770.

[37] Caucheteur, C., and P. Mégret, "Demodulation Technique for Weakly Tilted Fiber Bragg Grating Refractometer," *Photonics Technology Letters*, Vol. 17, 2005, pp. 2703–2705.

[38] Chan, C. F., et al., "Optical Fiber Refractometer Using Narrowband Cladding-Mode Resonance Shifts," *Applied Optics*, Vol. 46, 2007, pp. 1142–1149.

[39] White, I. M., and X. D. Fan, "On the Performance Quantification of Resonant Refractive Index Sensors," *Optics Express*, Vol. 16, 2008, pp. 1020–1028.

[40] Shevchenko, Y. Y., and J. Albert, "Plasmon Resonances in Gold-Coated Tilted Fiber Bragg Gratings," *Optics Letters*, Vol. 32, 2007, pp. 211–213.

[41] Caucheteur, C., et al., "High Resolution Interrogation of Tilted Fiber Grating SPR Sensors from Polarization Properties Measurement," *Optics Express*, Vol. 19, 2011, pp. 1656–1664.

[42] Caucheteur, C., V. Voisin, and J. Albert, "Polarized Spectral Combs Probe Optical Fiber Surface Plasmons," *Optics Express*, Vol. 21, 2013, pp. 3055–3066.

[43] Liedberg, B., C. Nylander, and I. Lungstrom, "Surface Plasmon Resonance for Gas Detection and Biosensing," *Sensors and Actuators*, Vol. 4, 1984, pp. 299–304.

[44] Caucheteur, C., et al., "Ultrasensitive Plasmonic Sensing in Air Using Optical Fibre Spectral Combs," *Nature Communications*, Vol. 7, 2016, p. 13371.

[45] Chen, X., et al., "Wide Range Refractive Index Measurement Using a Multi-Angle Tilted Fiber Bragg Grating," *IEEE Photonics Technology Letters*, Vol. 29, 2017, pp. 719–722.

[46] Zhou, K., et al., "Side Detection of Strong Radiation-Mode Out-Coupling from Blazed FBGs in Single-Mode and Multimode Fibers," *IEEE Photonics Technology Letters*, Vol. 15, 2003, pp. 936–938.

[47] Zhou, K., et al., "Optic Sensors of High Refractive-Index Responsivity and Low Thermal Cross Sensitivity That Use Fiber Bragg Gratings of >80° Tilted Structures," *Optics Letters*, Vol. 31, 2006, pp. 1193–1195.

[48] Voisin, V., et al., "Interrogation Technique for TFBG-SPR Refractometers Based on Differential Orthogonal Light States," *Applied Optics*, Vol. 50, 2011, pp. 4257–4261.

[49] Caucheteur, C., et al., "A Thin Metal Sheath Lifts the EH to HE Degeneracy in the Cladding Mode Refractometric Sensitivity of Optical Fiber Sensors," *Applied Physics Letters*, Vol. 99, 2011, p. 041118.

[50] Manuylovich, E., K. Tomyshev, and O. V. Butov, "Method for Determining the Plasmon Resonance Wavelength in Fiber Sensors Based on Tilted Fiber Bragg Gratings," *Sensors*, Vol. 19, 2019, p. 4245.

[51] Lobry, M., et al., "Plasmonic Fiber Grating Biosensors Demodulated Through Spectral Envelopes Intersection," *Journal of Lightwave Technology*, Vol. 39, 2021, pp. 7288–7295.

[52] Zhou, K., et al., "Low Thermal Sensitivity Grating Devices Based on Ex-45° Tilting Structure Capable of Forward-Propagating Cladding Modes Coupling," *Journal of Lightwave Technology*, Vol. 24, 2006, pp. 5087–5094.

[53] Yan, Z., et al., "Refractive Index and Temperature Sensitivity Characterization of Excessively Tilted Fiber Grating," *Optics Express*, Vol. 25, 2017, pp. 3336–3346.

[54] Yuezhen, S., et al., "Excessively Tilted Fiber Grating Sensors," *Journal of Lightwave Technology*, Vol. 39, 2021, pp. 3761–3770.

[55] Yan, Z., et al., "Theoretical and Experimental Analysis of Excessively Tilted Fiber Gratings," *Optics Express*, Vol. 24, 2016, pp. 12107–12115.

[56] Mou, C., et al., "Liquid Level Sensor Based on an Excessively Tilted Fibre Grating," *Optics Communications*, Vol. 305, 2013, pp. 271–275.

[57] Yan, Z., et al., "Numerical and Experimental Analysis of Sensitivity-Enhanced RI Sensor Based on Ex-TFG in Thin Cladding Fiber," *Journal of Lightwave Technology*, Vol. 33, 2015, pp. 3023–3027.

[58] Lu, H., et al., "Study on Spectral and Refractive Index Sensing Characteristics of Etched Excessively Tilted Fiber Gratings," *Applied Optics*, Vol. 57, 2018, pp. 2590–2596.

[59] Li, Z., et al., "Theoretical Analysis of Tuning Property of the Graphene Integrated Excessively Tilted Fiber Grating for Sensitivity Enhancement," *Journal of the Optical Society of America B*, Vol. 36, 2019, pp. 108–118.

[60] Jiang, B., et al., "Graphene-Induced Unique Polarization Tuning Properties of Excessively Tilted Fiber Grating," *Optics Letters*, Vol. 41, 2016, pp. 5450–5453.

[61] Luo, B., et al., "Plasmonic Gold Nanoshell Induced Spectral Effects and Refractive Index Sensing Properties of Excessively Tilted Fiber Grating," *Chinese Optics Letters*, Vol. 16, 2018, p. 100603.

[62] Martinez, A., et al., "Direct Writing of Fibre Bragg Gratings by Femtosecond Laser," *Electronics Letters*, Vol. 40, 2004, pp. 1170–1172.

[63] Dragomir, A., et al., "Inscription of Fiber Bragg Gratings by Ultraviolet Femtosecond Radiation," *Optics Letters*, Vol. 28, 2003, pp. 2171–2173.

[64] Thomas, J., et al., "Cladding Mode Coupling in Highly Localized Fiber Bragg Gratings: Modal Properties and Transmission Spectra," *Optics Express*, Vol. 19, 2011, pp. 325–341.

[65] Thomas, J. U., et al., "Cladding Mode Coupling in Highly Localized Fiber Bragg Gratings II: Complete Vectorial Analysis," *Optics Express*, Vol. 20, 2012, pp. 21434–21449.

[66] Chah, K., et al., "Surface Plasmon Resonance in Eccentric Femtosecond-Laser-Induced Fiber Bragg Gratings," *Optics Letters*, Vol. 39, 2014, pp. 6887–6890.

[67] Chikh-Bled, H., et al., "Behavior of Femtosecond Laser-Induced Eccentric Fiber Bragg Gratings at Very High Temperatures," *Optics Letters*, Vol. 41, 2016, pp. 4048–4051.

[68] Vengsarkar, A. M., et al., "Long-Period Fiber Gratings as Band-Rejection Filters," *Journal of Lightwave Technology*, Vol. 14, 1996, pp. 58–65.

[69] Bhatia, V., "Applications of Long-Period Gratings to Single and Multi-Parameter Sensing," *Optics Express*, Vol. 4, 1999, pp. 457–466.

[70] Shu, X., L. Zhang, and I. Bennion, "Sensitivity Characteristics of Long-Period Fiber Gratings," *Journal of Lightwave Technology*, Vol. 20, 2022, pp. 255–266.

[71] Wang, Z., et al., "Analysis of Optical Response of Long Period Fiber Gratings to nm-Thick Thin-Film Coatings," *Optics Express*, Vol. 13, 2005, pp. 2808–2813.

[72] Del Villar, I., I. R. Matias, and F. J. Arregui, "Deposition of Coatings on Long-Period Fiber Gratings: Tunnel Effect Analogy," *Optics and Quantum Electronics*, Vol. 38, 2006, pp. 655–665.

[73] Cusano, A., et al., "Sensitivity Characteristics in Nanosized Coated Long Period Gratings," *Applied Physics Letters*, Vol. 89, 2006, p. 201116.

[74] He, Y. J., Y. L. Lo, and J. F. Huang, "Optical-Fiber Surface-Plasmon-Resonance Sensor Employing Long-Period Fiber Gratings in Multiplexing," *Journal of the Optical Society of America B*, Vol. 23, 2006, pp. 801–811.

[75] Schuster, T., et al., "Miniaturized Long-Period Fiber Grating Assisted Surface Plasmon Resonance Sensor," *Journal of Lightwave Technology*, Vol. 30, 2011, pp. 1003–1008.

[76] Heidemann, B. R., et al., "Matching Long-Period Grating Modes and Localized Plasmon Resonances: Effect on the Sensitivity of the Grating to the Surrounding Refractive Index," *Applied Optics*, Vol. 55, 2016, pp. 8979–8985.

[77] Shi, Y. J., and Z. T. Gu, "Design of High-Sensitivity Metal-Coated LPFG Sensor Based on Material Dispersion," *Optoelectronics Letters*, Vol. 8, 2012, pp. 269–272.

[78] He, Y. J., "Investigation of LPG-SPR Sensors Using the Finite Element Method and Eigenmode Expansion Method," *Optics Express*, Vol. 21, 2013, pp. 13875–13895.

[79] Wei, W., et al., "Graphene-Based Long-Period Fiber Grating Surface Plasmon Resonance Sensor for High-Sensitivity Gas Sensing," *MDPI Sensors*, Vol. 17, 2017, p. 2.

[80] Li, Z., et al., "Highly Sensitive Surface Plasmon Resonance Sensor Utilizing a Long Period Grating with Photosensitive Cladding," *Applied Optics*, Vol. 55, 2016, pp. 1470–1480.

CHAPTER 5

Specialty Optical Fiber Platforms

Chapters 3 and 4 were devoted to plasmonic platforms based on multimode and single-mode optical fibers, respectively. This chapter focuses on some other optical fiber types that have already been successfully implemented for plasmons excitation and that bring singular assets compared to more classical telecommunication-grade optical fibers, despite a somewhat increased complexity. Some generalities about each fiber type will first be given in each corresponding section, followed by a review of their use for plasmonic refractive index sensing built from the abundant literature on the subject. Under the umbrella of specialty optical fibers, this chapter will notably cover the case of polarization-maintaining (or highly birefringent) optical fibers, microstructured (or photonic crystal) fibers, and polymer (or plastic) optical fibers. To provide evidence of the difference between polarization-maintaining optical fibers and standard ones for non-experts in this field, we wanted to bring additional information about birefringence effects, in addition to the concepts on light polarization reviewed in Chapter 4.

5.1 Polarization-Maintaining Optical Fibers

5.1.1 Introduction to the Concept of Polarization-Maintaining Optical Fibers

As mentioned in Chapter 4, an ideal single-mode optical fiber allows the propagation of two degenerate core modes (HE^{x}_{11} and HE^{y}_{11}) with orthogonal polarizations [1]. This degenerate nature is only perfectly respected in ideal optical fibers characterized by a perfect circular symmetry. In practice, real optical fibers present appreciable variation in the shape of their core along their length so that at any location of the optical fiber, the core is slightly elliptical. Optical fibers may also contain internal stress induced either by the manufacturing process or by the presence of a nonuniform dopant concentration in its cross section. Degeneracy is consequently removed, which introduces a small difference in the refractive index value for a particular pair of orthogonal polarization states called eigenmodes or polarization modes. This property is called birefringence.

The difference in the refractive index value between the two polarization modes leads to a difference in their velocities and therefore in their propagation times. This leads to the definition of slow and fast eigenmodes. The effect of birefringence can be expressed by the difference β_{xy} between the eigenmodes propagation constants:

$$\beta_{xy} = \frac{2\pi}{\lambda}\left(n_{\text{eff},x} - n_{\text{eff},y}\right) = \frac{\omega}{c}\Delta n \tag{5.1}$$

where λ is the wavelength, $n_{\text{eff},x}$ and $n_{\text{eff},y}$ are the effective refractive indices of the fast and slow eigenmodes, ω is the angular frequency, and c is the speed of light in the vacuum. Δn is the refractive index difference also called birefringence.

If we consider that the optical fiber is characterized by a uniform birefringence along its length L, the phase delay between the two polarization modes can be written as:

$$\phi_d = \beta_{xy} L \tag{5.2}$$

The birefringence between the fast and slow eigenmodes leads to the broadening of an optical pulse during its propagation in the optical fiber. The difference in group velocities yields a differential group delay (DGD) $\Delta\tau$, as illustrated in Figure 5.1 for a linear birefringent element for which the two eigenmodes are linear. $\Delta\tau$ is obtained by the derivative of (5.2) with respect to ω:

$$\Delta\tau = \frac{d\beta_{xy}}{d\omega} L = \left(\frac{\Delta n}{c} + \frac{\omega}{c} \frac{d\Delta n}{d\omega} \right) L \tag{5.3}$$

The birefringence also modifies the state of polarization of light traveling inside the optical fiber. Any state of polarization launched at the fiber input can be decomposed into the two eigenmodes and their phase difference will be modified during the propagation according to (5.3), assuming that the birefringence is uniform along the optical fiber. This varying phase difference alters the state of polarization as light propagates down the optical fiber. Moreover, there is a length of fiber through which polarized light undergoes a complete revolution of polarization and goes back to its initial state of polarization. This length is called the beat length L_B and is obtained when ϕ_d is equal to 2π. It is given by the following relationship:

$$L_B = \frac{\lambda}{\Delta n} = \frac{2\pi}{\beta_{xy}} \tag{5.4}$$

For an optical element characterized by a uniform birefringence (a short enough fiber sample, for example), the birefringence properties can be divided into linear and circular birefringences.

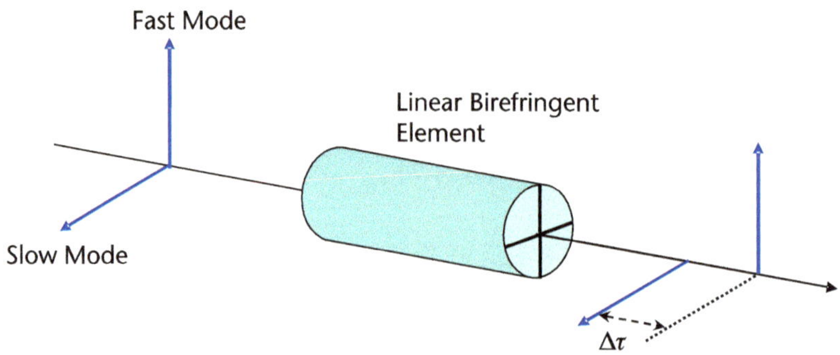

Figure 5.1 Differential group delay introduced by a linear birefringent element.

- In a medium exhibiting only linear birefringence, the eigenmodes are linearly polarized. The phase difference between the polarization modes per unit of length at a given wavelength is called the linear birefringence δ and is expressed in rad/m [2]. The orientation of the eigenmodes is defined by the parameter q, the azimuth of the fast polarization mode. Two parameters (δ and q) are consequently required to represent the linear birefringence. In the case of linear birefringence, the polarization modes can also be called eigenaxes.
- In a medium exhibiting only circular birefringence, the eigenmodes are circularly polarized, one right-handed and the other left-handed. A velocity difference exists between them, the value being positive if the right-handed mode is faster. The circular birefringence, corresponding to the phase difference per unit of length, is defined as 2ρ and is expressed in rad/m [2]. The factor 2 is introduced as the effect of the circular birefringence is to rotate a state of polarization by an angle ρ. A pure circular birefringent element is indeed a rotator.

In the most general case, linear and circular birefringences are simultaneously present. As a result, the eigenmodes are elliptically polarized. The phase delay (Δ_d) per unit of length between the two orthogonal ellipses can be written in terms of its linear and circular components as:

$$\Delta_d = \sqrt{\rho^2 + \frac{\delta^2}{4}} \tag{5.5}$$

The polarization properties are thus characterized by means of three parameters, δ and q, related to the linear birefringence, and ρ, related to the circular birefringence. The resulting Jones matrix (see Chapter 4 for the definition of the Jones matrix) for such an elliptical birefringent element can be written as [2, 3]:

$$J = \begin{pmatrix} \alpha + j\beta\cos 2q & -\gamma + j\beta\sin 2q \\ \gamma + j\beta\sin 2q & \alpha - j\beta\cos 2q \end{pmatrix} \tag{5.6}$$

where

$$\alpha = \cos\Delta_d L_{el} \tag{5.7}$$

$$\beta = \frac{\delta}{2}\frac{\sin\Delta_d L_{el}}{\Delta_d} \tag{5.8}$$

$$\gamma = \rho\frac{\sin\Delta_d L_{el}}{\Delta_d} \tag{5.9}$$

where L_{el} is the length of the optical element. Moreover, one can write:

$$\alpha^2 + \beta^2 + \gamma^2 = 1 \tag{5.10}$$

Different causes lead to birefringence in optical fibers. Among others, we can cite:

1. *Noncircular core:* The geometrical anisotropy of a noncircular core introduces a linear birefringence in the optical fiber. The light travels fastest when it is polarized along the smallest transverse dimension of the fiber core.
2. *Transverse stress:* Any transverse stress introduces a linear birefringence via elasto-optic index changes. The stress may be frozen internally in the optical fiber or may result from external lateral pressure. The linear birefringence value proportionally varies with the applied stress.
3. *Bending:* Through the photo-elastic effect, the bending produces an asymmetry of the refractive index and leads to birefringence. A bent optical fiber is characterized by the outer part of the cross-section being under tension and the inner part being in compression.
4. *Twist:* Twisting an optical fiber results in circular birefringence that is proportional to the twist rate. The coefficient of proportionality is 0.146 for silica fibers [4].
5. *Transverse electric field:* A transverse electric field introduces linear birefringence through the electrooptic Kerr effect. The Kerr effect in silica is such that the linear birefringence phase retardation is proportional to the square of the transverse electric field component. The linear birefringence induced by a transverse electric field E is given by [4]:

$$\delta = 2\pi \mathrm{K} E^2 \tag{5.11}$$

   where K is the Kerr electrooptic constant of the fiber material.
6. *Longitudinal magnetic field:* A longitudinal magnetic field introduces circular birefringence via the Faraday magneto-optic effect. The birefringence is related to the magnetic field H component along the optical fiber axis:

$$\rho = \mathrm{V}H \tag{5.12}$$

   where V is the Verdet constant of the material.

The birefringence parameters (δ, ρ, and q) vary along the length of an optical fiber. As the birefringence is not uniform in real optical fibers, the polarization eigenmodes are not constant and they randomly change their orientations along the optical fiber. An optical fiber can be represented as a concatenation of uniform elements. These elements are characterized by different eigenmodes and different phase delays. Hence, when a light pulse is launched in an optical fiber, the electric field components coming out an element are projected onto the two eigenmodes of the next one while the local birefringence introduces a phase delay between them during the propagation. This disruptive phenomenon is called polarization mode coupling. In practice, for a standard single-mode optical fiber, the birefringence is relatively weak and the mode coupling is randomly distributed along the fiber length such that a pulse will be broadened rather than divided into several pulses. Therefore, this phenomenon results in a dispersion called a polarization mode dispersion (PMD). Its effect limits the capacity of transmission of digital optical links.

For sensing over short distances, such as what usually happens with plasmonic optical fiber sensors, birefringence effects are so small that they can be neglected. When relying on polarized light waves as in the case of fiber gratings, the state of polarization will naturally vary along the connecting optical fibers. Hence, to make sure that the P-polarized is injected into the sensor, the spectrum of the latter is monitored during the optimization of the polarization state (usually thanks to a polarization controller or a linear polarizer) and the connecting fibers are positioned to avoid any unwanted power fluctuations. This is perfectly mastered in practice and does not affect at all the sensing performance.

There also exists a special type of optical fibers called polarization maintaining fibers (PMFs) for which a strong, stress-induced birefringence is introduced during the manufacturing process, as sketched in Figure 5.2 for the most often encountered configurations. This birefringence is ensured by a strong asymmetry in the fiber section. For a PMF, the induced birefringence is uniform along the fiber length so that there is no polarization mode coupling. Moreover, there exist two constant orthogonal polarization modes that propagate without deformation. If the input state of polarization corresponds to one of these eigenmodes, the same state of polarization will be obtained at the fiber output.

Hence, birefringence is usually induced by introduced twofold (or more) rotational symmetry, either in the refractive index profile or in the stress distribution. The first option is called form birefringence, while the second option is the stress birefringence. PMFs have emerged as very important devices to guide polarized light and several types of PMFs are often used in communications, sensing, and fiber lasers, among other applications [5, 6].

Form birefringence originates from a vector electromagnetic effect in optical fibers possessing twofold (or less) rotational symmetry [7, 8]. The simplest way to do so is by making the shape of the core elliptical.

Stress birefringence has a thermal origin and is implemented in PANDA and bow-tie fibers [9, 10]. These names result from the drawing made by stress-applying parts located each side of the core, which have a thermal expansion coefficient different

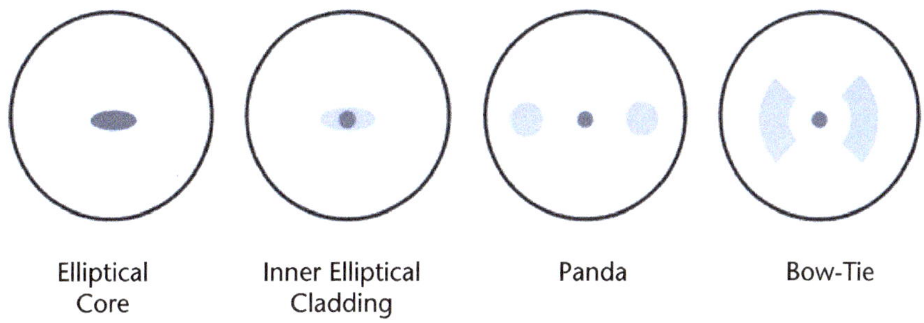

Figure 5.2 Sketch of the cross-section of some polarization-maintaining optical fibers.

from the rest of the fiber. Note that, in addition to its similarity with the panda bear, PANDA also stands for polarization-maintaining and absorption reducing.

Large modal birefringence reduces polarization crosstalk, thereby improving the ability to maintain polarization modes. PMFs typically exhibit modal birefringence of the order or larger than 10^{-4}. In practice, the larger the birefringence, the shorter the beat length.

Two polarization modes exist in a PMF: one polarized along the x-axis and the other polarized along the y-axis. Their propagation constants are different. The polarization direction of the mode characterized by the larger propagation constant is called the slow axis, as its phase velocity is slower than the other. The other mode corresponds to the fast axis. Slow and fast axes are depicted in Figure 5.2. The mode polarized along the slow axis is usually more confined and robust against external perturbations and is thus more often used in telecommunication applications.

5.1.2 Use of Polarization-Maintaining Optical Fibers for Plasmonic Excitation

Plasmonic unclad multimode optical fiber configurations exhibit a limited resolution, as they suffer from inherent modal noise that causes the strength of the interaction between the core-guided light and the surface plasmon to fluctuate. Intensity-modulated SPR sensing was developed on single-mode optical fibers to circumvent this practical issue [11]. As stated by the authors, light polarization in this configuration needs to be precisely controlled to ensure a stable sensor output. Otherwise, power fluctuations in the sensor output may prevent its proper operation. For this reason, the same authors have proposed a plasmonic sensor based on PMF [12] containing a gold-coated, side-polished section. When one of the birefringence axes of the fiber is precisely aligned with the gold film, one polarization (e.g., slow polarization) excites the surface plasmon. If the birefringence axis is not perfectly perpendicular to the metal layer surface and light of both orthogonal polarizations is guided, only the TM-polarized light (i.e., light with its electric field perpendicular to the side-polished surface) will excite the surface plasmon. As the light polarization in the interaction region depends on the amplitudes of the two main polarizations and their relative phase shift, variations in the phase difference between the two fiber modes induced by optical fiber deformations influence the strength of interaction between the outcoupled light and the surface plasmon. Hence, the magnitude of this effect depends on the precision of alignment of birefringent axes. With the elasto-optic method [13], alignment errors within 1° may be achieved.

The PMF used in [12] has the following characteristics: numerical aperture of 0.13, cutoff wavelength equal to 749 nm, beat length of 1.9 mm at the wavelength of 800 nm, birefringence of $5\ 10^{-4}$, cladding diameter of 125 μm, and mode field diameter of 4.2 μm. The fiber jacket was stripped over 2 cm and the bare fiber was glued into a curved slot made within a silica block so that the polarization axis of the fiber was perpendicular to the block surface. The bare fiber was then side-polished and the residual cladding thickness was measured using the drop liquid method [14]. The side-polished region was coated with a 2-nm-thick chromium layer to promote adhesion of a 55-nm gold layer and another 17-nm tantalum pentoxide tuning layer. All depositions were made using electron beam evaporation in a vacuum. The SPR appears around the wavelength of 810 nm when the surrounding refractive index value is close to 1.33, as depicted in Figure 5.3. An experimental refractive index sensitivity of 3,200 nm/RIU was reported.

In [15], tilted fiber Bragg gratings (TFBGs) were produced in elliptical-core optical fibers and coated with an approximately 40-nm gold layer. A study of their robustness against polarization instabilities was made compared to a grating written in standard single-mode optical fiber. To this aim, connecting fibers were bent in the experimental setup and the position of the SPR resonance in the transmitted amplitude spectrum of the TFBG was recorded and measured as a function of the external mechanical perturbation. Figure 5.4 confirms that the TFBG in PMF is very stable as a function of the mechanical perturbation while a standard TFBG is much more affected. This practical asset comes at the expense of a limited spectral quality: peak-to-peak amplitudes of cladding mode resonances are limited to 3 dB (half of the signal) because of the use of a broadband optical source and the inherent nature of the coupling in the PMF. The latter is also much more expensive than a standard single-mode optical fiber.

More recently, a hetero-core structure comprising a section of single-mode PANDA optical fiber spliced between two multimode graded-index optical fibers (core of 50 μm, cladding of 125 μm) was demonstrated in [16]. The structure was coated with gold and the SPR was measured in transmission around the wavelength of 600 nm using a white light source and a spectrometer. In this single experimental report about such a configuration, the authors claimed a relatively easy manufacturing process as it relies on fusion splicing. This approach seems marginal compared to the use of a single-mode optical fiber sandwiched between two multimode optical fibers.

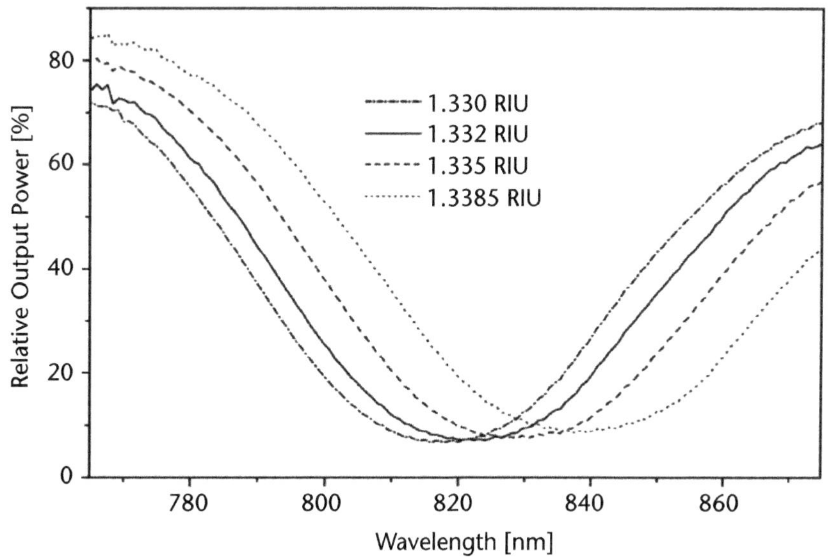

Figure 5.3 SPR for various surrounding refractive index values of a side-polished PMF section. Curves have been normalized to the transmitted amplitude spectrum in the absence of a sample. (*From:* [12].)

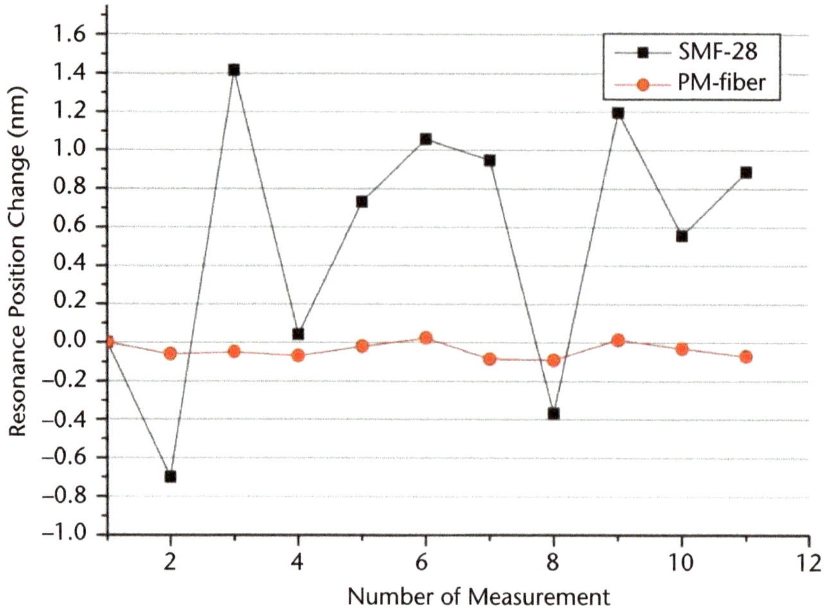

Figure 5.4 Evolution of the SPR wavelength position as a function of the number of measurements (made under mechanical solicitations of the connecting fiber) for gold-coated TFBGs in SMF and PMF. (*From:* [15].)

5.2 Microstructured Optical Fibers

A microstructured optical fiber (MOF) possesses a transverse wavelength-scale pattern that consists of air holes running along the length of the waveguide. MOFs have emerged in the 1990s as a new family of optical fibers featuring unprecedented properties in nonlinear optics (including supercontinuum generation), lasers and amplifiers, optical sensing, and imaging [17, 18]. Compared to standard optical fibers, they bring much more flexibility in terms of tailoring their optical waveguiding properties and adapting these properties for different applications. Their success has also arisen from the fact that MOFs can be manufactured with a single material, oppositely to standard fibers that are made of two materials. The vast majority of these fibers are fabricated in pure silica. MOFs fabricated from polymer materials have been introduced more recently [19]. They are usually called microstructured polymer optical fibers (mPOFs). They will be addressed in Section 5.3.

MOFs comprising a periodic microstructure of air holes are generally called photonic crystal fibers (PCFs). Different types of MOFs can be obtained depending on the light guidance mechanisms. When the latter result from a modified form of total internal reflections, the corresponding fibers are referred to as index-guiding PCFs. When light is confined in a low-index core because of the presence of a 2-D photonic bandgap, the fibers are called photonic bandgap fibers. The photonic bandgap guidance can be achieved in PCFs with a solid core or with a hollow one.

In this classification based on the core design (as sketched in Figure 5.5 for hexagonal lattice configurations), solid-core PCFs are those that guide light based on total internal reflections. For hollow-core PCFs, the guiding mechanism strongly depends on the materials injected in the hollow core. Total internal reflections are

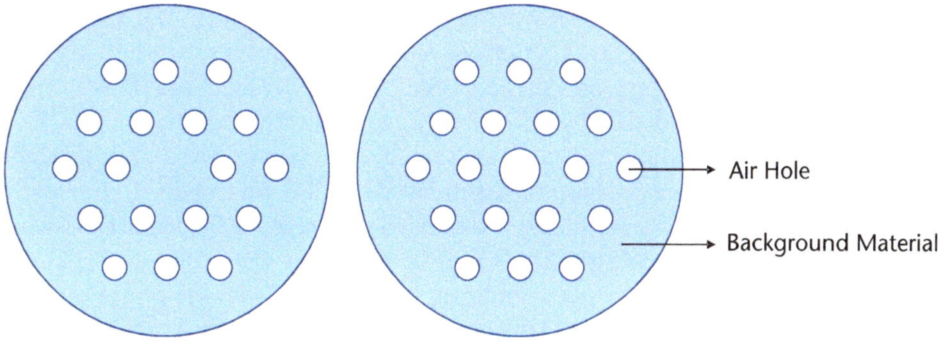

Figure 5.5 Sketch of the cross-section of hollow-core (left) and solid-core (right) PCFs. Note that several design parameters (e.g., core size, air-holes dimension, number of air holes and their inter-distance, host material) can be adjusted to obtain the target optical properties.

only possible when the core is filled with a relatively high index liquid. Otherwise, light guidance happens via the photonic bandgap mechanism. PCFs can also host Bragg gratings. A review on the challenges of their fabrication in such fibers can be found in [20].

For plasmonic sensing applications, both solid-core and hollow-core PCFs were used. The flexibility of PCF designs is such that numerous SPR-PCF sensors have been reported to date. They can be categorized according to the following types: D-shaped structures, grating-based devices, microfluidic slot-based devices, and external metal-coated and internal metal-coated configurations [21]. Internal-metal coated configurations are very promising, but are almost exclusively conceptually studied because of practical challenges. These configurations will be reviewed next. Although these are purely conceptual works that will be very hard to obtain experimentally, we have chosen to provide a quite detailed overview as the publication trend on this thematic is particularly intriguing. Among all possible plasmonic optical fiber sensor configurations, those based on PCF were the most studied conceptually (with hundreds of papers on purely theoretical concepts) but by far (up to now at least) the less developed experimentally.

Preliminary studies have reported numerical studies of some SPR-PCF configurations using the loss spectrum analysis based on the coupled mode theory or finite element method (FEM)-based commercial software [22–25]. In [26–28], two different internal metal-coated PCF structures were proposed and numerically studied for possible biosensing applications, as sketched in Figure 5.6 (left and middle configurations). "Internal metal-coated" means that the microstructure contains holes that are internally coated with a thin metal film. In such structures, SPR generation can be achieved from the fundamental mode of the microstructured fiber. Phase matching between plasmon and a core mode can be enforced by introducing air-filled microstructure into the fiber core, thus allowing tuning of the modal refractive index and its matching with that of a plasmon. Surrounding air-filled holes enable guiding in the fiber core while controlling coupling strengths between the core mode and a plasmon. Holes of the second layer are filled with analytes and metallized for plasmons excitation. A theoretical refractive index resolution of 10^{-4} RIU was demonstrated. A structure similar to one sketched in Figure 5.6 with selectively coated air holes was numerically studied in [29]. A theoretical sensitivity of 5,500 nm/RIU was reported. Another variant was studied in [30] by means of an FEM software and a theoretical sensitivity reaching 11,000 nm/RIU was reported. Multicore PCFs were also numerically studied. In [31], a PCF (hexagonal lattice configuration; see the right sketch of Figure 5.6) was proposed with 6 cores surrounding a central hole that is metal-coated. It was again studied with the FEM and an average refractometric sensitivity close to 3,000 nm/RIU was reported in the range of 1.33 to 1.42. A two-core variant was studied in [32]. The central hole was coated with silver and graphene and an average refractometric sensitivity of 4,350 nm/RIU was determined through FEM simulations. Following a first numerical study based on a three-hole PCF design [33], a large-size, square-lattice configuration was modeled in [34]. A sensitivity of approximately 2,300 nm/RIU was obtained through FEM simulations.

All these conceptual works are very interesting to catalyze progress on the technology, but they come with numerous practical complexities such as enabling an efficient coupling between the fundamental light mode and the SPR, coating the inner holes of the MOFs with a thin metal film over a sufficiently long distance or ensuring a proper fluid flow in the selected holes.

D-shaped PCF configurations were also mainly conceptually studied, for both solid-core and hollow-core types [35–38]. In particular, a sensitivity of 7,000 nm/RIU was reported in [38] for a very thin conceptual fiber configuration (air holes' diameter less than 1 μm). In 2018, a D-shaped PCF-SPR biosensor based on gold gratings (i.e., the gold layer deposited on the flat surface of the fiber cross-section contains a grating) was numerically reported [39]. A maximum resolution of 5.98 10^{-6} RIU in the range of 1.36 to 1.38 and a refractometric sensitivity of 3,340 nm/RIU were achieved. High refractive index sensing (between 1.45 and 1.6) was numerically studied in [40]. A refractometric sensitivity up to 9,300 nm/RIU in the range of 1.45 to 1.525 was numerically demonstrated. Other numerical explorations have focused on very specific configurations such as the use of two symmetric D-shape PCFs or solid-core offset PCF [41, 42]. Refractometric sensitivities up to approximately 15,000 nm/RIU were numerically obtained. All these numerical analyses confirm the flexibility offered by PCF configurations. Depending on the actual geometry, the used metal layer and its thickness, it is an understatement to say that an important

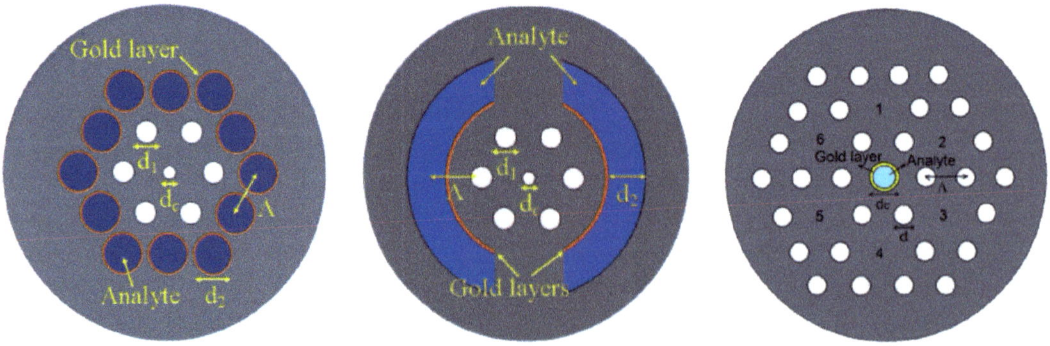

Figure 5.6 Schematic of conceptual internal metal-coated, MOF-based, SPR sensors. (*Adapted from:* [25, 31].)

variation of relative performance was numerically obtained. Based on such results, one can therefore wonder what happens in real practical implementations.

The first practical achievement related to the use of D-shaped PCF for SPR excitation was quite recent at the time of this writing and dated back to 2016 [43]. In their paper, the authors specified that they chose a strong enough solid-core PCF to sustain the polishing process. As shown in Figure 5.7, the selected fiber has an hexagonal lattice with 6 rings of holes (hole diameter of 3.2 μm and interdistance between the holes of 8 μm). A PCF section of 50 cm was spliced to two SMFs and used in transmission mode. It was side-polished over a length of 15 mm and a depth of 60 μm thanks to the use of an abrasive wheel. The side-polished PCF was coated with a 45-nm gold layer. For measurements, a white light source and a spectrometer were used and the SPR sensor was placed in a flow cell in which different calibrated refractive index liquids were flowed. The left part of Figure 5.7 shows the corresponding SPR signal measured for different surrounding refractive index values. The mean refractive index sensitivity has been computed to be approximately 2,340 nm/RIU, which coincides well with FEM simulations.

It has to be noted that H-shaped PCF geometries (i.e., their cross-section has been numerically defined in the form of an H) were also conceptually studied in a mostly recent body of works [44–46]. Refractometric sensitivities comparable to those in D-shaped PCFs were reported.

Another relevant practical implementation of SPR-PCF devices is in the form of an interferometer [47]. A short length (1 cm) of PCF (Nufern, LMA10) was spliced between two multimode graded-index optical fibers, all with an outer diameter of 125 μm. An approximately 50-nm gold coating was applied on the PCF and the sensing platform was connected to a white light source and a spectrometer for measurements in transmission. For surrounding refractive index values close to 1.33, the average refractometric sensitivity was computed to be 2,018 nm/RIU.

The group of Monro has also largely contributed to advance SPR-MOF based sensing. The originality of their works relies on the design, fabrication, and use of

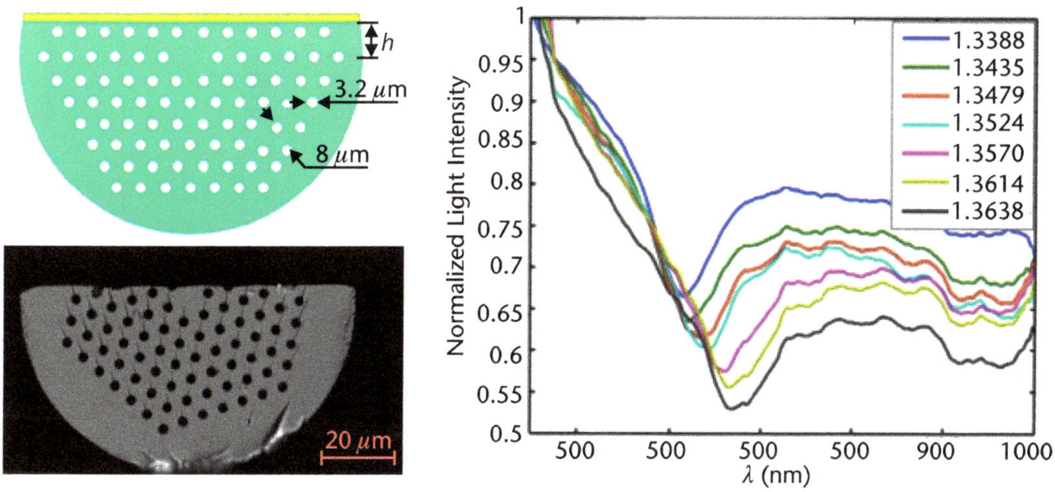

Figure 5.7 Side-polished solid-core hexagonal-lattice PCF used for SPR excitation (left) and normalized SPR resonance for some calibrated surrounding refractive index values (right). (*Adapted from:* [43].)

silica MOFs with their core exposed along its length to the surrounding medium [48]. Such fibers are modified suspended-core fibers and are therefore nonsymmetric in their cross-section, as depicted in Figure 5.8. The relatively small core of 10-μm diameter is supported by an outer silica structure providing robustness to the fiber. A silver film of approximately 50-nm thickness was deposited on a 1-cm-long section using an electroless plating method [49]. The fiber was connected in transmission to a broadband light source and a spectrometer and an SPR curve (FWHM of 75 nm) was monitored around the wavelength of 610 nm in a 1.33 refractive index liquid. A refractometric sensitivity of 1,753 nm/RIU was obtained, which is very close to the one obtained for a 140-μm bare core multimode optical fiber based on experimental measurements conducted in the same work. The MOF brings a twofold improvement in terms of factor of merit compared to the tested multimode optical fiber configuration.

To end this section on PCF-based plasmonic sensing, let's focus now on grating-based devices. A few relevant experimental works can be quoted so far but there is almost no doubt that this field will grow in the near future, supported by the strong progress in femtosecond pulses, laser-based grating fabrication.

Fiber Bragg grating (FBG) in hexagonal lattice PCFs excited for cladding modes excitation have been reported by Eggleton et al. [50] and Chen et al. [51]. Experimental reports confirmed that a nearly zero sensitivity was obtained for an RI of 1.46 [50] and a maximal sensitivity of 12 nm/RIU for the resonances with the effective index matching the SRI in the range between 1.36 and 1.42 [51]. In both works, the absence of cladding modes with an effective index matching the refractive index of water can explain the low sensitivity in that range of surrounding refractive index values. In [52], straight FBGs were written in two types of hydrogen-loaded hexagonal lattice PCFs with six rings of air holes and cladding diameters of 125 and 86 μm, respectively. A 193-nm ArF excimer laser was used and gratings were produced without taking care of the microstructure orientation with respect to the incident laser beam. A gold coating was deposited around the grating location using a sputtering process and the transmitted amplitude spectrum of the gratings was analyzed as a function of surrounding refractive index changes, as depicted in Figure 5.9. Two

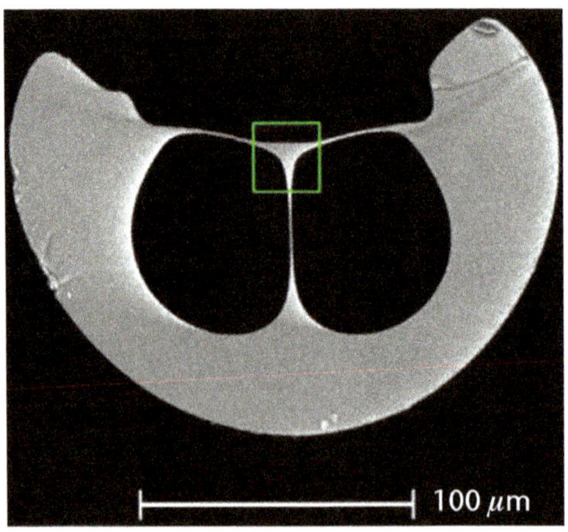

Figure 5.8 Scanning electron microscopy (SEM) picture of the modified suspended core MOF with its core exposed to the surrounding medium used for SPR excitation. (*From:* [46].)

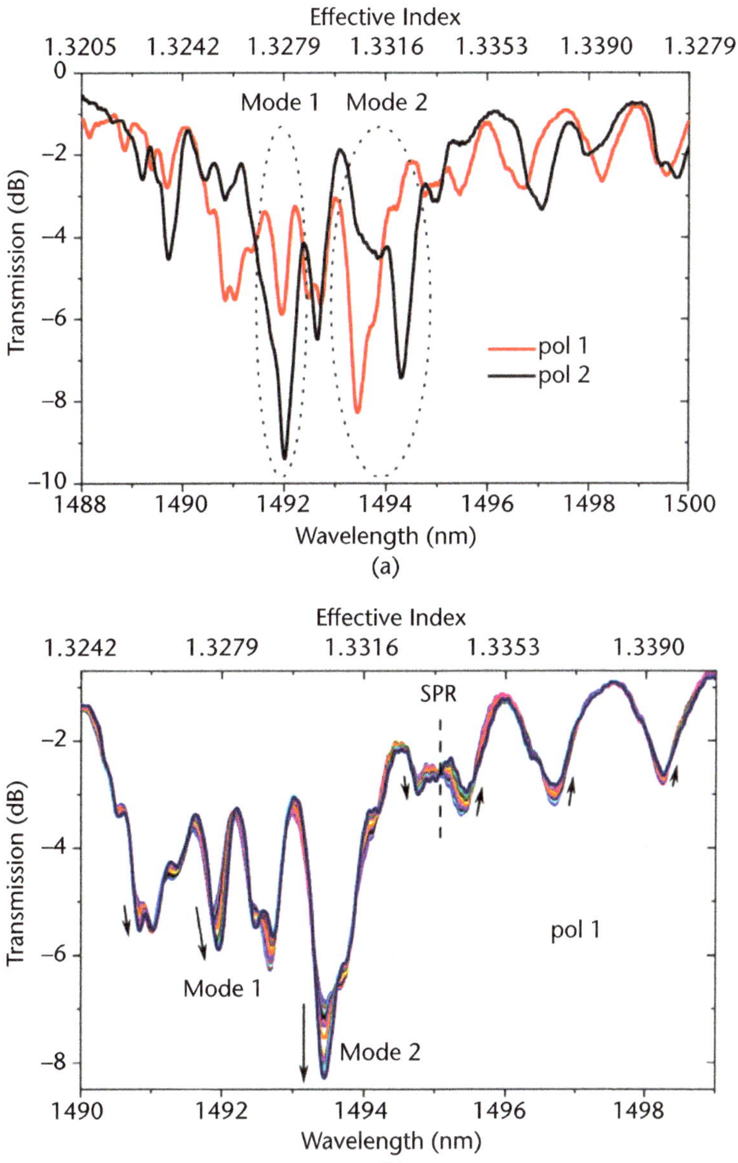

Figure 5.9 (a) Transmitted amplitude spectrum of a gold-coated straight FBG in hexagonal lattice PCF for two orthogonal polarization states (1 and 2) and (b) evolution of the transmitted amplitude spectrum corresponding to pol 1 as a function of the surrounding refractive index value. The most sensitive cladding mode resonances are 1 and 2. (*Adapted from:* [52].)

orthogonal polarizations were first recorded (pol 1 and pol 2), with pol 1 enabling SPR excitation. It was then tracked to compute the refractometric sensitivity. The two most sensitive modes are those labeled modes 1 and 2. They display a sensitivity of 40.3 nm/RIU and 35.1 nm/RIU, respectively. The main practical asset of this experimental achievement is the possibility to excite SPR with narrowband optical resonances in a limited bandwidth. The spectral content is limited to a few nanometers so that several gratings can be wavelength-division-multiplexed in the typical available bandwidth (approximately 80 nm) of standard FBG interrogators.

Last but not least, readers interested in receiving additional information about this growing field of plasmonic PCF-based sensing are invited to consult [53, 54]. It is worth ending this section in emphasizing that, despite 16 years of work since the seminal conceptual work [16], there has been no decisive practical improvement to using MCF for plasmonic sensing compared to standard silica fibers. The main claim from the original publications indicated a potential increase in sensitivity of about one order of magnitude (based on mode dispersion control in such fibers), but that has never been materialized in the very few articles reporting experimental results. An important room for improvement is therefore left for experimentalists on the subject.

5.3 Polymer Optical Fibers

In plasmonic sensing as well as in other applications, polymer optical fibers (POFs), also referred to as plastic optical fibers, are particularly advantageous due to their excellent flexibility, ease of manipulation, great numerical aperture, large diameter, and, last but not least, the fact that plastic can withstand smaller bend radii than glass. Polymers also have excellent compatibility with organic materials, giving them great potential for biomedical applications. In general, POFs provide a lower-cost alternative to silica optical fibers, at the expense of much higher transmission losses. POFs have thus been mainly deployed for data transmission over short distances.

POFs have the same geometry as silica optical fibers, with a core, cladding, and sometimes a jacket. A historical review of the development of POFs can be found in [55]. A variety of optical polymers are used for the fabrication of POFs, including the most spread polymethyl-methecrylate (PMMA), amorphous fluorinated polymer (CYTOP), cycloolefin polymer (TOPAS and ZEONEX), polystyrene (PS), and polycarbonate (PC), among others. An extensive overview of common POF materials and their corresponding properties can be found in [56].

Unlike silica optical fibers, POFs are primarily available as multimode fibers, which have larger diameters and propagate multiple, interacting modes. Because the larger diameter of multimode POFs makes them easier to handle, cleave, and connect, multimode POF sensors are often promoted as less expensive and easier to install than their silica counterparts. Hence, configurations outlined in Chapter 3 can be produced in POFs as well, with one that is particularly easy to achieve given the nature of constitutive materials of POFs: U-bent POFs. Such configurations are practically relevant compared to other designs, essentially because of their ease of fabrication and handling, compact design, and increased depth of evanescent wave, resulting in high sensitivity.

In [57], POFs of 750-μm diameter made of a PMMA core and fluorinated polymer cladding were used. The refractive indices of the core and cladding are 1.49 and 1.41 at 590 nm, respectively, yielding a numerical aperture of 0.5. A 1-cm length of the fiber was uncladded by dipping in ethyl acetate solution for 1 minute. The fiber was bent manually at the uncladded portion by placing it inside a glass capillary (inner diameter of 3 mm) and kept in a hot air oven at 80°C for 10 minutes to obtain a bend diameter of 2.25 mm. The U-bent section was coated with gold (thickness of 60 nm) in a sputter chamber. The ends of the fiber section (22 cm in length in total) were then polished for optical connections. A white light source and a spectrometer were used to record the transmission amplitude spectrum. The SPR was measured around the wavelength of 680 nm for a surrounding refractive index value close to 1.33. The refractometric sensitivity was computed to be 1,040 nm/RIU. Figure 5.10 depicts the actual U-bent POF fiber probe.

In [58], similar fibers were used by another research team and molded in custom-made devices to obtain U-shaped POF in a curve with a diameter of 8 mm. The fiber was heated by a hot-air blower gun for 15 seconds and the temperature was maintained below 70°C during the whole molding process so that the fiber melting temperature was not reached. Gold was sputtered over the U-bent section and different thicknesses up to 100 nm were studied. The experimental work was conducted towards the detection of *Escherichia coli* bacteria. The demodulation was focused on changes of relative amplitude of the SPR spectrum and the refractometric sensitivity for a 70-nm, gold-coated, U-bent POF section was computed to be approximately 338 nm/RIU.

A variant of U-bent fiber probes was produced in [59] where the shaping was made by rapidly moving (10 cycles) the uncladded POF section over a flame. A film of polyvinyl alcohol, graphene, and silver nanoparticles was used on top of the U-bent section to obtain localized SPR excitation. A refractometric sensitivity of approximately 700 nm/RIU was experimentally reported.

Recent experimental achievements [57–60] confirm the efficiency and relative ease of production of plasmonic U-bent POF probes. Compared to silica fibers, less stringent equipment can be used for production. This is also true for side-polished POF sections that were reported a couple of years before U-bent geometries. The pioneering works date back to the early 2010s [61, 62]. In both works conducted by two different teams, "low cost" is highlighted in the titles, to further evidence the ease of production. In [62], a 980-μm-core PMMA fiber with a 20-μm fluorinated polymer cladding was used as a substrate for plasmonic sensing. A POF section without a jacket was first embedded in a resin block, with the purpose of facilitating the polishing process. The latter was carried out with polishing papers so as to laterally remove the cladding and part of the core. After 20 complete strokes following a figure-8 pattern with a 5-μm grain-polishing paper in order to completely expose the core, a 1-μm grain-polishing paper was used for another 20 complete strokes. The achieved sensing region with a D-shape was about 10 mm in length. Prior the deposition of an approximately 40-nm gold film, a buffer photoresist layer was deposited by spin coating on the flat fiber surface. Its thickness was

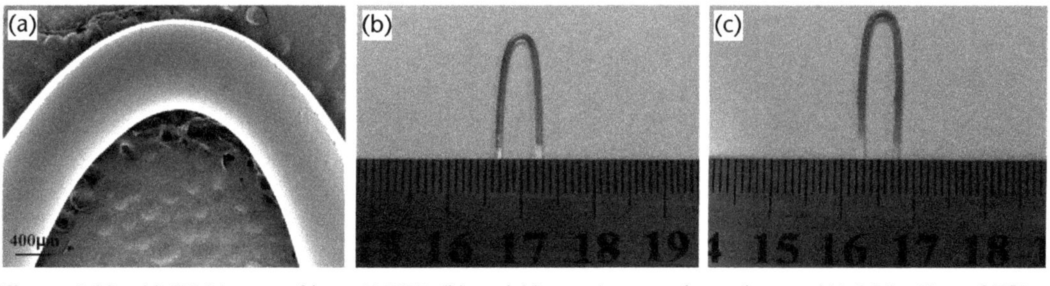

Figure 5.10 (a) SEM image of bare U-POF; (b) and (c) are pictures of metal-coated U-POFs. (*From:* [63].)

about 1.5 μm. The minimum of the SPR curve was located around 610 nm when the D-shaped optical fiber was immersed in a 1.332 refractive index liquid and the FWHM was close to 150 nm. The benefit of the photoresist buffer was to improve the SNR (more than twofold compared to a configuration without buffer) while decreasing the FWHM by approximately 15% (which, in turn, improves the figure of merit). The refractometric sensitivity is close to 2,500 nm/RIU. The same team has further integrated this biosensing platform in [64]. It was successfully applied to the detection of therapeutic antibodies in human serum. Figure 5.11 displays a picture of the experimental setup as well as the corresponding SPR signal. The overall quality of the response can be appreciated.

A double side-polished configuration was proposed in [65]. Such a structure was obtained by polishing two sides of a multimode PMMA POF section symmetrically along the fiber axis. The depth of each D-shape was equal to 200 μm and the

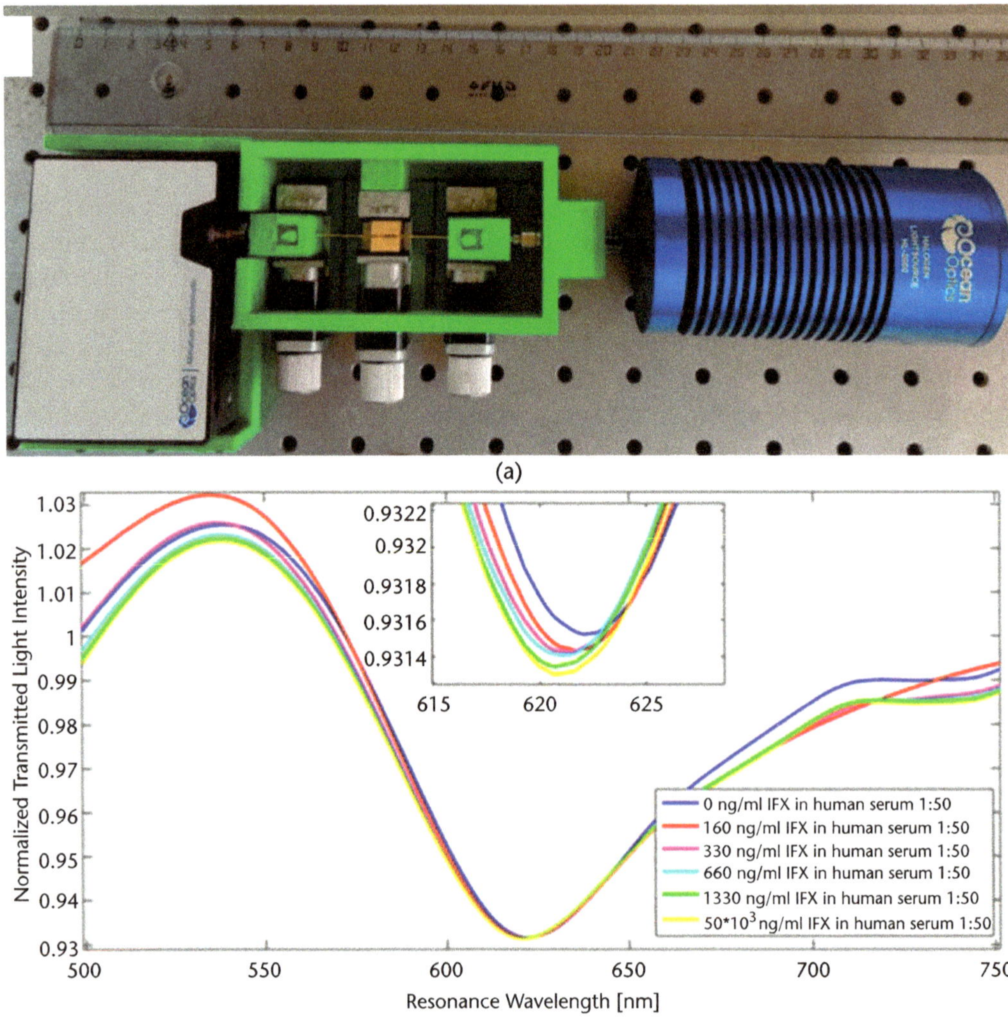

Figure 5.11 Picture of a relatively integrated plasmonic side-polished POF configuration (a) and SPR signal evolution for the biosensing of infliximab (b). (*Adapted from:* [62].)

polished length was 5 mm. An approximately 50-nm gold coating was applied on both sections. A sensitivity of 4,284.8 nm/RIU was demonstrated in liquids with a refractive index close to 1.42.

Gold-coated, side-polished refractometric sensing was also reported based on multimode CYTOP optical fibers [66]. The graded-index fiber had core, cladding, and additional reinforcement layer (overcladding) diameters of 62.5, 102.5, and 490 μm, respectively. The refractive index of the core at 589 nm decreases from 1.357 at the fiber center to 1.342 at the core/cladding interface gradually and remains unchanged in the cladding. The numerical aperture of the fiber is 0.185. The relatively low refractive index of the material is interesting as it is closer to the refractive index value of aqueous solutions. This, in turn, leads to a very high refractometric sensitivity. The D-shaped section obtained with a wheeled polishing machine (different abrasive papers were used depending on the fiber constitutive layers) was coated with an approximately 55-nm gold layer. A white light source and a spectrometer were used to collect the SPR signal in transmission mode. The SPR minimum happens around 750 nm for a surrounding refractive index of 1.3. The sensitivity reaches 22,779 nm/RIU, which is one order of magnitude higher than that measured in silica fibers using similar geometries.

To end this review on plasmonic side-polished POF configurations, let us quote the work performed in [67] where a single-mode birefringent optical fiber was used. The core was produced in PMMA/PS copolymer. The refractometric sensitivity reached 2,765 nm/RIU.

Switching now to single-mode POFs, we would like to figure out that many of them were specifically made for Bragg grating production in their core, since the first report of FBG production in POF in 1999 [68]. Dopants such as diphenyl sulfide, trans-4-stilbenemethanol, and benzyl dimethyl ketal were introduced to enhance the fiber photosensitivity and allow the production of photo-chemically induced gratings with specific laser (including a helium-cadmium laser emitting at 325 nm) in addition to more physically induced gratings obtained with femtosecond pulses laser and point-by-point processing. The goal of this section is certainly not to review the important progress in polymer optical fiber Bragg gratings (POFBGs) writing but rather focus on their use for plasmonic sensing. The reader interested in this progress is invited to consult [69–71].

To date, the only experimental report on the use of POFBG for plasmonic sensing can be found in [72]. Weakly tilted FBGs were produced in step-index PMMA POF (core and cladding diameters of 8.2 and 150 μm, respectively) using the phase mask technique and a HeCd laser. They were coated with a gold film and interrogated using polarized light. Figure 5.12 displays the transmitted amplitude spectra recorded for two orthogonally polarized input light states. Tracking the most sensitive cladding mode resonance as done for TFBGs in silica fibers (see Chapter 4), a refractometric sensitivity of approximately 550 nm/RIU was obtained for surrounding refractive index values between 1.40 and 1.43. While the transmitted amplitude spectrum of the plastic TFBGs are not as clean as their equivalent in standard glass fiber, they are sufficient for an effective readout and the polymer materials might be important in some biochemical applications, as stated above.

Finally, microstructured POFs have also been largely investigated these last few decades [73, 74] and, as a result of this progress, some configurations were

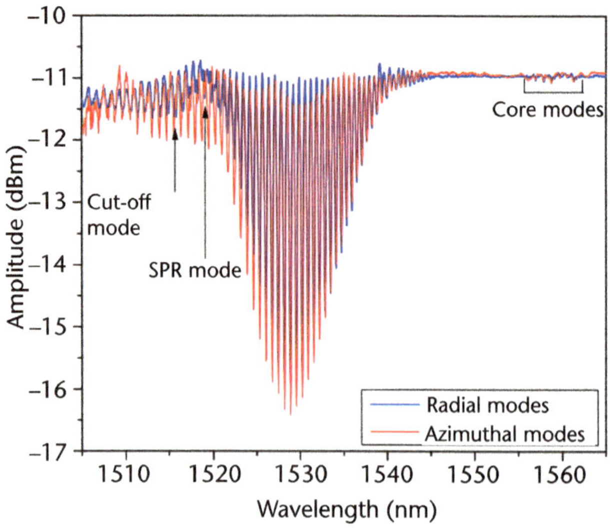

Figure 5.12 Transmitted amplitude spectrum of a gold-coated TFBG inscribed in a step-index PMMA POF.

studied for plasmonic sensing. As for silica PCFs, the works are mainly conceptual to date. In [75], a PMMA PCF with some holes coated by an approximately 70-nm indium tin oxide layer and infiltrated by the analytes is studied. A refractometric sensitivity reaching 2,000 nm/RIU is numerically obtained. More specific possible geometries were studied in [76] with the FEM. The first fiber design is made in PMMA but it contains only one ring of air holes (diameter of 6 μm) near the fiber edge and a single hole (diameter of 14.4 μm) in the core region. Other fiber designs contain more rings of holes between the core and the surrounding one. A silver film of approximately 40 nm was considered as a coating around the fibers. The future will tell us whether experimental configurations can emerge from these conceptual works, provided that they can bring practical assets.

This review of plasmonic polymer optical fiber sensing configurations confirms that the prime interest of their development remains to facilitate the production process and decrease its cost compared to their silica counterpart. This is certainly true with U-bent and D-shaped configurations and explains why most experimental reports to date are based on these geometries. Let us note that PMMA fibers are strongly sensitive to humidity, which is not an advantage in biochemical sensing and therefore implies to account for this effect when such fibers are used in practice [77]. There is a specific type of optical fibers that presents bioresorbable capability. It will be the subject of the next section.

5.4 Bioresorbable Optical Fibers

Among all the optical fiber configurations mentioned in this book, the ones that could be used *in situ* and operate in physiological conditions for a certain time and then disappear are probably the most exciting for *in vivo* implementations. To achieve this, optical bioresorbable materials that fully dissolve after use need to be finely engineered and transformed into implantable systems that sense biofluids in real time.

Most implants and surgical tools are made of inert biocompatible materials that could last for years and even the patient's lifetime (heart valves, pacemakers, articular protheses) or short lifetimes (bone growth stimulators, wound healing). Bioresorbable materials therefore enable by definition their spontaneous dissolution (e.g., through hydrolysis) after their needed use, with no side effects for the body. Both electronical and optical devices follow this trend and could lead to new biomedical and environmental applications [78, 79].

In the recent achievements presented in the literature, phosphate-based optical fibers were tested for *in vivo* implementation in a rat for pH measurements. The fiber was spliced with a standard MMF and the tip sensor was prepared using a fluorescent dye through sol-gel chemistry. No signs of hepatotoxicity or nephrotoxicity were found in the liver and kidneys of the experimental animals and the rate of bioresorbability was estimated to be around 0.3 μm/day, which was slower than measured *in vitro* around 1.4 μm/day using regularly refreshed PBS [80]. Other optical fiber sensors in phosphate-based platforms were also tested. In [81], Theodosiou et al. reported femtosecond laser written FBG filters, chirped gratings, and Fabry-Perot cavities for sensing purposes. One of the key applications for bioresorbable materials remains neuroscience as the high complexity and high sensitivity of brain tissue for external tools requires careful handling. The use of optical fibers could therefore be viable and versatile, to bring new ways of molecular sensing, phototherapy, and drug delivery. Implantable fibers with low invasiveness could therefore provide new insights to future study of brain [82].

Alongside more classic materials such as calcium phosphate glass, many different bio-based components have been explored for such applications [83, 84]. Optical fibers made of proteins, agarose, silk and spider fibers, cells, and cellulose have been described over the recent years. Hydrogel-based optical fibers are also a hot topic thanks to their high-water content, highly comparable to the target surrounding medium. The drawbacks of these new technologies remain their optical properties and their fabrication and handling. This domain of research is still considered as a niche by most of well-established optical platforms, but could clearly become a center of interest to move forward technologies to practical setups. It is therefore expected that transpositions to optical fibers of concept already reported with bulk materials ([85], for instance) will emerge in the years to come.

5.5 Fibers Incorporated with Metal Nanoparticles

In this section, we will briefly present what is currently a niche for plasmonic sensing but can certainly grow in the future: the production of fibers with nanoparticles incorporated within their constitutive materials.

Noble-metal nanoparticles encapsulated in a dielectric matrix have attracted sustained interest because of their unique optical and electrical properties. Metal-dielectric nanocompositions show nonlinear and fast optical response close to the SPR frequency as a result of their enhanced third-order optical susceptibilities. To date, they have been applied in all-optical switching devices, ultra-short pulse generation, and optical parametric amplification [86–91]. Several approaches, such as the sol-gel process, metal dielectric co-sputtering deposition, metal-ion implantation into a dielectric matrix, and pulsed laser deposition, have been used to prepare metal-dielectric nanocompositions [92–95].

A substantial body of work has been published on the design and development of metal nanoparticles incorporated in optical fibers. In [96], the modified chemical vapor deposition (MCVD) process and sol-gel process were used to fabricate a fiber preform doped with gold nanoparticles. The same fiber was used in [97] to characterize the visible and infrared photoluminescence produced by the gold nanoparticles when they are optically pumped. A variant was presented in [98] with the inclusion of PbTe quantum dots in the glass matrix.

For plasmonic sensing of the surrounding environment, it is much more relevant to produce a fiber with the cladding incorporated by nanoparticles rather than its core. Such a configuration was reported in [99] and was again produced thanks to the MCVD process. The average diameter of the gold nanoparticles in the cladding region was found to be 3.6 nm. Absorption peaks were measured at 392 and 790 nm and were attributed to the SPR of the incorporated nanoparticles. They result from the single nanoparticles and on the coupling effect between nanoparticles according to the dipole–dipole interactions. The measured SPR sensitivity reached 407 nm/RIU for a fiber length of 20 cm. It decreases with the fiber length, because of the increase of the propagation loss.

As written in the introduction of this section, such incorporated fibers remain a niche for plasmonic sensing, as they are relatively complex to obtain compared to metal-coated optical fibers and do not bring decisive practical advantages. The future will tell us if they can be more decisive in applications such as SERS.

In conclusion, the review provided in this chapter confirms that plasmonic sensing with specialty optical fibers is a growing field that has obviously emerged much more recently than configurations based on more standard optical fibers. Each fiber type brings its own practical advantage, in theory at least as for PCFs, the vast majority of the reports remains very hypothetical. In particular, polarization maintaining fiber allows to preserve the state of polarization, as indicated by their name, which can enhance the overall robustness although it is not detrimental in standard optical fibers in practice. PCFs bring numerous design flexibilities but they were mainly conceptually studied so far, certainly due to the additional complexity of their practical implementation. POFs can be more easily handled with less specific equipment to produce plasmonic sensors. There is no doubt that research and development activities on specialty fibers will continue to grow in the years to come. What can be doubted is their real potential to beat the degrees of performance respectively reached by unclad multimode optical fibers and grating assisted single-mode optical fibers. Hence, purposeful research in forthcoming years should deal with improvements (mostly in terms of data extraction and interrogation schemes) to bring them to wide acceptance.

Last but not least, we feel important to mention that specialty fibers encompass so many design opportunities that there are a number of unique configurations that can be hardly attached to one of the categories listed above and were therefore not covered in this chapter. Some will be mentioned in Chapter 10.

References

[1] Agrawal, G., *Fiber-Optic Communication Systems*, New York: John Wiley & Sons, 2002.

[2] Rogers, A., Y. Zhou, and V. Handerek, "Computational Polarization-Optical Time Domain Reflectometry for Measurement of the Spatial Distribution of PMD in Optical Fibers," *Proceedings OFMC 97*, Teddington, U.K., September 1997.

[3] Jones, R., "A New Calculus for the Treatment of Optical Systems: VII Properties of the N-Matrices," *Journal of the Optical Society of America*, Vol. 38, 1948, pp. 671–685.

[4] Rashleigh, S., "Origins and Control of Polarization Effects in Single-Mode Fibers," *Journal of Lightwave Technology*, Vol. 1, 1983, pp. 312–331.

[5] Kaminow, I., "Polarization in Optical Fibers," *IEEE Journal of Quantum Electronics*, Vol. 17, 1981, pp. 15–22.

[6] Noda, J., K. Okamoto, and Y. Sasaki, "Polarization-Maintaining Fibers and Their Applications," *Journal of Lightwave Technology*, Vol. 4, 1986, pp. 1071–1089.

[7] Steel, M. J., et al., "Symmetry and Degeneracy in Microstructured Optical Fibers," *Optics Letters*, Vol. 26, 2001, pp. 488–490.

[8] Yeh, C., "Elliptical Dielectric Waveguides," *Journal of Applied Physics*, Vol. 33, 1962, pp. 3235–3243.

[9] Sasaki, Y., et al., "Polarization-Maintaining and Absorption-Reducing Fibers," *Optical Fiber Communication*, paper ThCC6, Optical Society of America, 1982.

[10] Varnham, M. P., et al., "Single-Polarisation Operation of Highly Birefringent Bow-Tie Optical Fibres," *Electronics Letters*, Vol. 19, 1983, pp. 246–247.

[11] Slavik, R., J. Homola, and J. Ctyroky, "Novel Spectral Fiber-Optic Sensor Based on Surface Plasmon Resonance," *Sensors and Actuators B*, Vol. 74, 2001, pp. 106–111.

[12] Piliarik, M., et al., "Surface Plasmon Resonance Sensor Based on Single-Mode Polarization-Maintaining Optical Fiber," *Sensors and Actuators B: Chemical*, Vol. 90, 2003, pp. 236–242.

[13] Carrara, S. L. A., B. Y. Kim, and H. J. Shaw, "Elasto-Optic Alignment of Birefringent Axes in Polarization-Holding Optical Fiber," *Optics Letters*, Vol. 11, 1986, pp. 470–472.

[14] Digonnet, M. J. F., et al., "Measurement of the Core Proximity of Polished Fiber Substrates and Couplers," *Optics Letters*, Vol. 10, 1985, pp. 463–465.

[15] Tomyshev, K. A., et al., "Polarization Stable Plasmonic Sensor Based on Tilted Fiber Bragg Grating," *Proceedings of the 25th International Conference on Optical Fiber Sensors*, Vol. 10323, 2017.

[16] Hosoki, A., et al., "Surface Plasmon Resonance Sensor Using a Polarization-Maintaining Fiber on a Hetero-Core Optical Fiber Structure with Gold Thin Film," *Optics Express*, Vol. 30, 2022, pp. 35348–35360.

[17] Knight, J. C., et al., "All-Silica Single-Mode Optical Fiber with Photonic Crystal Cladding," *Optics Letters*, Vol. 21, 1996, pp. 1547–1549.

[18] Russell, P., "Photonic Crystal Fibers," *Science*, Vol. 299, 2003, pp. 358–362.

[19] Large, M. C. J., et al., *Microstructured Polymer Optical Fibres*, New York: Springer, 2007.

[20] Berghmans, F., et al., "Challenges in the Fabrication of Fiber Bragg Gratings in Silica and Polymer Microstructured Optical Fibres," *Laser and Photonics Reviews*, Vol. 1, 2014, pp. 27–52.

[21] Monfared, Y. E., "Overview of Recent Advances in the Design of Plasmonic Fiber-Optic Biosensors," *MDPI Biosensors*, Vol. 10, 2020, p. 77.

[22] Kuhlmey, B. T., K. Pathmanandavel, and R. C. McPhedran, "Multipole Analysis of Photonic Crystal Fibers with Coated Inclusions," *Optics Express*, Vol. 14, 2006, pp. 10851–10864.

[23] Hautakorpi, M., M. Mattinen, and H. Ludvigsen, "Surface-Plasmon-Resonance Sensor Based on Three-Hole Microstructured Optical Fiber," *Optics Express*, Vol. 16, 2008, pp. 8427–8432.

[24] Akowuah, E. K., et al., "Numerical Analysis of a Photonic Crystal Fiber for Biosensing Applications," *IEEE Journal of Quantum Electronics*, Vol. 48, 2012, pp. 1403–1410.

[25] Zhao, Y., Z. Q. Deng, and J. Li, "Photonic Crystal Fiber Based Surface Plasmon Resonance Chemical Sensors," *Sensors and Actuators B: Chemical*, Vol. 202, 2014, pp. 557–567.

[26] Hassani, A., and M. Skorobogatiy, "Design of the Microstructured Optical Fiber-Based Surface Plasmon Resonance Sensors with Enhanced Microfluidics," *Optics Express*, Vol. 14, 2006, pp. 11616–11621.

[27] Hassani, A., and M. Skorobogatiy, "Design Criteria for Microstructured Optical Fiber Based Surface Plasmon Resonance Sensors," *Journal of the Optical Society of America B*, Vol. 24, 2007, pp. 1423–1429.

[28] Hassani, A., and M. Skorobogatiy, "Photonic Crystal Fiber-Based Plasmonic Sensors for the Detection of Bio-Layer Thickness," *Journal of the Optical Society of America B*, Vol. 26, 2009, pp. 1550–1557.

[29] Yu, X., et al., "A Selectively Coated Photonic Crystal Fiber Based Surface Plasmon Resonance Sensors," *Journal of Optics*, Vol. 12, 2010, pp. 015005–015011.

[30] Rifat, A. A., et al., "Highly Sensitive Selectively Coated Photonic Crystal Fiber-Based Plasmonic Sensor," *Optics Letters*, Vol. 43, 2018, pp. 891–894.

[31] Shuai, B., et al., "A Multi-Core Holey Fiber Based Plasmonic Sensor with Large Detection Range and High Linearity," *Optics Express*, Vol. 20, 2012, pp. 5974–5986.

[32] Wang, F., et al., "A Highly Sensitive Dual-Core Photonic Crystal Fiber Based on a Surface Plasmon Resonance Biosensor with Silver-Graphene Layer," *Plasmonics*, Vol. 12, 2016, pp. 1–7.

[33] Bing, P., Z. Li, and J. Yao, "Effects of Heterogeneity on the Surface Plasmon Resonance Biosensor Based on Three-Hole Photonic Crystal Fiber," *Optical Engineering*, Vol. 52, 2013, pp. 532–543.

[34] Bing, P., et al., "Surface Plasmon Resonance Biosensor Based on Large Size Square-Lattice Photonic Crystal Fiber," *Journal of Modern Optics*, Vol. 63, 2013, pp. 793–797.

[35] Dash, J. N., and R. Jha, "Graphene-Based Birefringent Photonic Crystal Fiber Sensor Using Surface Plasmon Resonance," *IEEE Photonics Technology Letters*, Vol. 26, 2014, pp. 1092–1095.

[36] Dash, J. N., and R. Jha, "On the Performance of Graphene-Based D-Shaped Photonic Crystal Fibre Biosensor Using Surface Plasmon Resonance," *Plasmonics*, Vol. 10, 2015, pp. 1123–1131.

[37] Luan, N., et al., "Surface Plasmon Resonance Sensor Based on D-Shaped Microstructured Optical Fiber with Hollow Core," *Optics Express*, Vol. 23, 2015, pp. 8576–8582.

[38] Gangwar, R. K., and V. K. Singh, "Highly Sensitive Surface Plasmon Resonance Based D-Shaped Photonic Crystal Fiber Refractive Index Sensor," *Plasmonics*, Vol. 12, 2016, pp. 1367–1372.

[39] Lu, J., et al., "D-Shaped Photonic Crystal Fiber Plasmonic Refractive Index Sensor Based on Gold Grating," *Applied Optics*, Vol. 57, 2018, pp. 5268–5272.

[40] Monfared, Y. E., et al., "Quasi-D-Shaped Fiber Optic Plasmonic Biosensor for High-Index Analyte Detection," *IEEE Sensors Journal*, Vol. 21, 2019, pp. 17–23.

[41] Liu, C., et al., "Symmetrical Dual D-Shape Photonic Crystal Fibers for Surface Plasmon Resonance Sensing," *Optics Express*, Vol. 26, 2018, pp. 9039–9049.

[42] An, G., et al., "D-Shaped Photonic Crystal Fiber Refractive Index Sensor Based on Surface Plasmon Resonance," *Applied Optics*, Vol. 56, 2017, pp. 6988–6992.

[43] Chen, Y., et al., "Experimental Realization of D-Shaped Photonic Crystal Fiber SPR Sensor," *Journal of Physics D*, Vol. 50, 2016, p. 025101.

[44] Erdmanis, M., et al., "Comprehensive Numerical Analysis of a Surface-Plasmon-Resonance Sensor Based on an H-Shaped Optical Fiber," *Optics Express*, Vol. 19, 2010, pp. 13980–13988.

[45] Han, H., et al., "A Large Detection-Range Plasmonic Sensor Based on an H-Shaped Photonic Crystal Fiber," *Sensors*, Vol. 20, 2020, p. 1009.

[46] Gomez-Cardona, N., E. Reyes-Vera, and P. Torres, "High Sensitivity Refractive Index Sensor Based on the Excitation of Long-Range Surface Plasmon Polaritons in H-Shaped Optical Fiber," *Sensors*, Vol. 20, 2020, p. 2111.

[47] Wong, W. C., et al., "Photonic Crystal Fiber Surface Plasmon Resonance Biosensor Based on Protein G Immobilization," *IEEE Journal of Selected Topics in Quantum Electronics*, Vol. 19, 2013, p. 4602107.

[48] Kostecki, R., et al., "Silica Exposed-Core Microstructured Optical Fibers," *Optical Materials Express*, Vol. 2, 2012, pp. 1538–1547.

[49] Klantsataya, E., et al., "Surface Plasmon Scattering in Exposed Core Optical Fiber for Enhanced Resolution Refractive Index Sensing," *MDPI Sensors*, Vol. 15, 2015, pp. 25090–25102.

[50] Eggleton, B. J., et al., "Cladding-Mode-Resonances in Air-Silica Microstructure Optical Fibers," *Journal of Lightwave Technology*, Vol. 18, 2000, pp. 1084–1100.

[51] Chen, C., et al., "Sensitivity of Photonic Crystal Fiber Modes to Temperature, Strain and External Refractive Index," *Optics Express*, Vol. 16, 2008, pp. 9645–9653.

[52] Rusyakina, O., et al., "Plasmon-Enhanced Refractometry Through Cladding Mode Excitation by a Fiber Bragg Grating in Photonic Crystal Fiber," *Journal of Lightwave Technology*, Vol. 40, 2022, pp. 1121–1129.

[53] Hu, D. J. J., and H. P. Ho, "Recent Advances in Plasmonic Photonic Crystal Fibers: Design, Fabrication and Applications," *Advances in Optics and Photonics*, Vol. 9, 2017, pp. 257–314.

[54] Danlard, I., and E. K. Akowuah, "Assaying with PCF-Based SPR Refractive Index Biosensors: From Recent Configurations to Outstanding Detection Limits," *Optical Fiber Technology*, Vol. 54, 2020, p. 102083.

[55] Zubia, J., and J. Arrue, "Plastic Optical Fibers: An Introduction to Their Technological Processes and Applications," *Optical Fiber Technology*, Vol. 7, 2001, pp. 101–140.

[56] Ziemann, O., et al., "Optical Fibers," Chapter 2 in *POF Handbook*, New York: Springer, 2008.

[57] Christopher, C., A. Subrahmanyam, and V. V. R. Sai, "Gold Sputtered U-Bent Plastic Optical Fiber Probes as SPR- and LSPR-Based Compact Plasmonic Sensors," *Plasmonics*, Vol. 13, 2018, pp. 493–502.

[58] da S. Arcas, A., et al., "Surface Plasmon Resonance and Bending Loss-Based U-Shaped Plastic Optical Fiber Biosensors," *MDPI Sensors*, Vol. 18, 2018, p. 648.

[59] Jiang, S., et al., "A Novel U-Bent Plastic Optical Fibre Local Surface Plasmon Resonance Sensor Based on Graphene and Silver Nanoparticles Hybrid Structure," *Journal of Physics D: Applied Physics*, Vol. 50, 2017, p. 165105.

[60] Gasior, K., T. Martynkien, and W. Urbanczyk, "Effect of Constructional Parameters on the Performance of a Surface Plasmon Resonance Sensor Based on a Multimode Polymer Optical Fiber," *Applied Optics*, Vol. 10, 2014, pp. 8167–8174.

[61] Muñoz-Berti, V. M., et al., "Low Cost Plastic Optical Fiber Sensor Based on Surface Plasmon Resonance," *Proceedings of SPIE*, Vol. 7653, 2010, p. 765327.

[62] Cennamo, N., et al., "Low Cost Sensors Based on SPR in a Plastic Optical Fiber for Biosensor Implementation," *MDPI Sensors*, Vol. 11, 2011, pp. 11752–11760.

[63] Shouzhen, J., et al., "A novel U-bent plastic optical fibre local surface plasmon resonance sensor based on a graphene and silver nanoparticle hybrid structure," *J. Phys. D: Appl. Phys.* Vol. 50, 2017, 165105.

[64] Zeni, L., et al., "A Portable Optical-Fibre-Based Surface Plasmon Resonance Biosensor for the Detection of Therapeutic Antibodies in Human Serum," *Scientific Reports*, Vol. 10, 2020, p. 11154.

[65] Liu, L., et al., "An Enhanced Plastic Optical Fiber-Based Surface Plasmon Resonance Sensor with a Double-Sided Polished Structure," *MDPI Sensors*, Vol. 21, 2021, p. 1516.

[66] Cao, S., et al., "Highly Sensitive Surface Plasmon Resonance Biosensor Based on a Low-Index Polymer Optical Fiber," *Optics Express*, Vol. 26, 2018, pp. 3988–3994.

[67] Gasior, K., et al., "A Surface Plasmon Resonance Sensor Based on a Single Mode D-Shape Polymer Optical Fiber," *Journal of Optics*, Vol. 19, 2017, p. 025001.

[68] Xiong, Z., et al., "Highly Tunable Bragg Gratings in Single-Mode Polymer Optical Fibers," *IEEE Photonics Technology Letters*, Vol. 11, 1999, pp. 352–354.

[69] Broadway, C., et al., "Toward Commercial Polymer Fiber Bragg Grating Sensors: Review and Applications," *Journal of Lightwave Technology*, Vol. 37, 2019, pp. 2605–2615.

[70] Min, R., "Fabrication and Application of Polymer Optical Fiber Grating Devices," in S. W. Harun, *Application of Optical Fiber in Engineering*. London, United Kingdom: IntechOpen, 2021, https://www.intechopen.com/chapters/73847.

[71] Jiang, J., et al., "Recent Achievements on Grating Fabrications in Polymer Optical Fibers with Photosensitive Dopants: A Review," *Polymers*, Vol. 14, 2022, p. 273.

[72] Hu, X., P. Mégret, and C. Caucheteur, "Surface Plasmon Excitation at Near-Infrared Wavelengths in Polymer Optical Fibers," *Optics Letters*, Vol. 40, 2015, pp. 3998–4001.

[73] van Eikelenborg, M. A., et al., "Microstructured Polymer Optical Fibre," *Optics Express*, Vol. 9, 2001, pp. 319–327.

[74] Markos, C., et al., "High Tg TOPAS Microstructured Polymer Optical Fiber for Fiber Bragg Grating Strain Sensing at 110 Degrees," *Optics Express*, Vol. 21, 2013, pp. 4758–4765.

[75] Dash, J. N., and R. Jha, "SPR Biosensor Based on Polymer PCF Coated with Conducting Metal Oxide," *IEEE Photonics Technology Letters*, Vol. 26, 2014, pp. 595–598.

[76] Lu, Y., et al., "SPR Sensor Based on Polymer Photonic Crystal Fibers with Metal Nanolayers," *Sensors*, Vol. 13, 2013, pp. 956–965.

[77] Webb, D. J., "Fibre Bragg Grating Sensors in Polymer Optical Fibres," *Measurement Science and Technology*, Vol. 26, 2015, p. 092004.

[78] La Mattina, A. A., S. Mariani, and G. Barillaro, "Bioresorbable Materials on the Rise: From Electronic Components and Physical Sensors to In Vivo Monitoring Systems," *Advances in Science*, Vol. 7, 2020.

[79] Pugliese, D., et al., "Bioresorbable Optical Fiber Bragg Gratings," *Optics Letters*, Vol. 43, 2018, pp. 671–674.

[80] Podrazký, O., et al., "In Vivo Testing of a Bioresorbable Phosphate-Based Optical Fiber," *Journal of Biophotonics*, Vol. 12, 2019.

[81] Theodosiou, A., et al., "Femtosecond Laser Written Plane-by-Plane Bragg Grating Sensors in Bioresorbable Phosphate Optical Fibres," *Journal of Lightwave Technology*, Vol. 37, 2019, pp. 2363–2369.

[82] Nazempour, R., et al., "Emerging Applications of Optical Fiber-Based Devices for Brain Research," *Adv. Fiber Mater.*, Vol. 4, No. 1, 2022, pp. 24–42.

[83] Shan, D., et al., "Flexible Biodegradable Citrate-Based Polymeric Step-Index Optical Fiber," *Biomaterials*, Vol. 143, 2017, pp. 142–148.

[84] Wang, Y., et al., "Biocompatible and Biodegradable Polymer Optical Fiber for Biomedical Application: A Review," *Biosensors*, Vol. 11, 2021, pp. 1–30.

[85] Cennamo, N., et al., "An Exo-Friendly Disposable Plasmonic Sensor Based on Bacterial Cellulose and Gold," *MDPI Sensors*, Vol. 19, 2019, p. 4894.

[86] Hache, F., D. Ricard, and C. Flytzanis, "Optical Nonlinearities of Small Metal Particles: Surface-Mediated Resonance and Quantum Size Effects," *Journal of the Optical Society of America B*, Vol. 3, 1988, pp. 1647–1655.

[87] Haglund, R. F., et al., "Picosecond Nonlinear Optical Response of a Cu:Silica Nanocluster Composite," *Optics Letters*, Vol. 18, 1993, pp. 373–375.

[88] Lin, A., et al., "Fabrication and Third-Order Optical Nonlinearity of Germano-Silicate Glass Optical Fiber Incorporated with Au Nanoparticles," *Proceedings of SPIE*, Vol. 6481, 64810M, 2007.

[89] Papadogiannis, N. A., et al., "Temporal Characterization of Ultra Short Laser Pulses Based on Multiple Harmonic Generation on a Gold Surface," *Applied Physics B*, Vol. 65, 1997, pp. 339–345.

[90] Torounidis, T., M. Karlsson, and P. A. Andrekson, "Fiber Optical Parametric Amplifier Pulse Source: Theory and Experiment," *Journal of Lightwave Technology*, Vol. 23, 2005, pp. 4067–4073.

[91] Radic, S., and C. J. Mckinstrie, "Optical Amplification and Signal Processing in Highly Nonlinear Optical Fiber," *IEICE Transactions on Electronics*, Vol. E88-C, 2005, pp. 859–869.

[92] Wang, W. T., et al., "Resonant Absorption Quenching and Enhancement of Optical Nonlinearity in Au:BaTiO3 Composite Films by Adding Fe Nanoclusters," *Applied Physics Letters*, Vol. 83, 2003, pp. 1983–1985.

[93] Dalacu, D., and L. Martinu, "Temperature Dependence of the Surface Plasmon Resonance of Au/SiO2 Nanocomposite Films," *Applied Physics Letters*, Vol. 77, 2000, pp. 4283–4285.

[94] Pardo-Yissar, V., et al., "Gold Nanoparticle/Hydrogel Composites with Solvent-Switchable Electronic Properties," *Advanced Materials*, Vol. 13, 2001, p. 13201323.

[95] Dhara, S., et al., "Quasiquenching Size Effects in Gold Nanoclusters Embedded in Silica Matrix," *Chemical Physics Letters*, Vol. 370, 2003, pp. 254–260.

[96] Ju, S., et al., "Fabrication and Optical Characteristics of a Novel Optical Fiber Doped with the Au Nanoparticles," *Journal of Nanoscience and Nanotechnology*, Vol. 6, 2006, pp. 3555–3558.

[97] Lin, A., et al., "Visible to Infrared Photoluminescence From Gold Nanoparticles Embedded in Germane-Silicate Glass Fiber," *Optics Express*, Vol. 15, 2007, pp. 6374–6379.

[98] Ju, S., P. R. Watekar, and W.T. Han, "Fabrication of Highly Nonlinear Germane-Silicate Glass Optical Fiber Incorporated with PbTe Semiconductor Quantum Dots Using Atomization Doping Process and Its Optical Nonlinearity," *Optics Express*, Vol. 19, 2011, pp. 2599–2607.

[99] Ju, S., et al., "Surface Plasmon Resonance Characteristics of Optical Fiber Incorporated with Au Nano-Particles in Cladding Region," *Journal of Nanoscience and Nanotechnology*, Vol. 16, 2016, pp. 6308–6312.

CHAPTER 6

Immunosensors

A wide variety of biosensors operate with antibodies or derivatives as bioreceptors and are part of the family of immunosensors. From paper-based rapid tests to enzyme-linked immunosorbent assays (ELISAs), molecular receptors play a tremendous role for both sensitivity and selectivity of bioassays. In this chapter, immunosensors are reviewed and the selection of their associate bioreceptors is explained. Monoclonal or polyclonal antibodies, fragments of antibodies, nanobodies, and affimers are some examples of bioreceptors studied in this chapter. Immobilization techniques on different substrates are also discussed, leaving room for a multitude of sensor combinations. Finally, novel applications of fiber optic immunosensors are proposed, exposing the manifold possibilities of detection that can be built with these latest generation tools.

6.1 Introduction to Biochemical Sensors

The origin of sensors go back to 1906 with the demonstration by Max Cremer of the glass electrode concept. He reported that a concentration of an acid in a liquid solution is proportional to the electric potential that arises between parts of the fluid located on opposite sides of a glass membrane [1]. Three years later, in 1909, Søren Peder Lauritz Sørensen presented the concept of pH sensor based on the measurement of the hydrogen ion concentration. In 1922, the first pH electrode was born thanks to Walter S. Hughes exploiting the potential difference between electrolytes in contact with the two sides of a thin glass wall [2]. Decades later, in 1956, Clark and Lyons revealed the development of an oxygen probe and presented in 1962 at a New York Academy of Sciences Symposium the building process to obtain "more intelligent" electrochemical sensors (pH, potentiometric, or conductometric) [3]. For example, the combination of glucose oxidase enzymes (GOx) with a Clark oxygen-electrode allowed the detection of glucose by measuring the drop in oxygen while the glucose was converted to gluconic acid and hydrogen peroxide. His concept, built on the transformation of a nonresponsive substance into a responsive one using an enzyme, made him the "father of biosensors." His groundbreaking works opened the doors to a series of innovations in the field of biosensors [4].

Over the years, the concept of biosensing has evolved in different ways. About 50 years ago, biosensors were only considered "self-contained analytical devices that respond to the concentration of chemical species in biological samples" [5]. This straitjacket description was clearly wrong, because any physical or chemical sensor operating in biological samples could then be considered as a biosensor. The definition therefore evolved over decades. According to the International Union of Pure and Applied Chemistry (IUPAC) in 1999, the term "biosensor" had been, until then, defined as "an independently integrated receptor transducer device, which is capable of providing selective or semi-quantitative analytical information using biological recognition element" [5]. This modified definition followed one from 1992 when it was described as [6]: "a device that uses specific biochemical reactions mediated by isolated enzymes, immunosystems, tissues, organelles, or whole cells to detect chemical compounds usually by electrical, thermal or optical signals."

Biosensors are therefore based on a two-component system, which involves (Figure 6.1):

1. A ligand, as recognition element against the selected target;
2. A transducing system, which converts the signal into legible data.

Different types of receptors and transducers are listed in Table 6.1. They can be combined in different structures and shapes to monitor many types of analytes. Biochemical sensors can therefore be classified by their receptors (DNA, RNA, antibodies, proteins, cells, enzymes, polymers) or transducing systems. The latter can be electrochemical, such as for the amperometric or potentiometric sensors where the receptors are immobilized on an electrode to measure current or voltage changes. It can be optical, for which absorbance, luminescence, and/or light scattering are parameters of interest. SPR-based biosensors described in previous chapters belong to this category. Thermometric or calorimetric transducers measure a temperature

Table 6.1 Receptors and Transducers for Biosensors

Receptors	*Transducers*
Organisms, tissues, or cells	Potentiometric
Organelles	Mechanical
Membranes	Acoustic
Enzymes	Amperometric
Antibodies	Conductimetric
Proteins	Impedimetric
Receptors	Calorimetric
Nucleic acids and aptamers	Optical
Organic molecules	
Molecularly imprinted polymers (MIPs)	

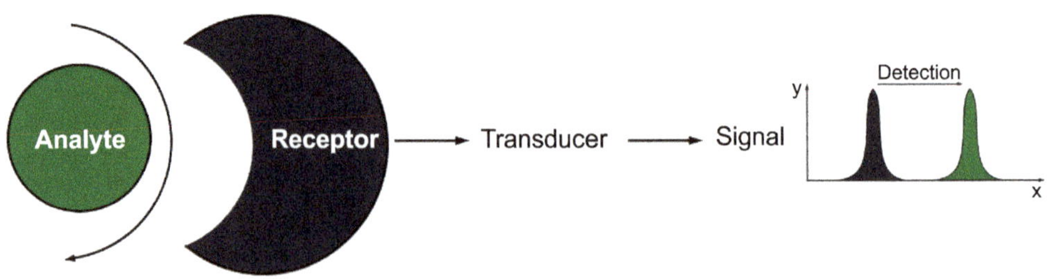

Figure 6.1 General outline of a biosensor describing its operating principle.

change due to an affinity reaction between the surface receptors and the analytes. Piezoelectric sensors are based on specific materials such as quartz that enable a measurement of small mass changes due to modifications in their resonant frequency [7, 8]. Finally, magnetic biosensors make use of magnetic particles or crystals to detect biological interactions by changes of their magnetic properties. This chapter focuses on immunosensors, based on the use of immunoglobulins (antibodies).

The elaboration of biosensing tools need to target relevant molecules or cells, often called biomarkers. A biomarker is a biological indicator of some state or condition that is useful to examine biological processes, pathogenic cases, and pharmacological responses or to help for therapeutic and surgical interventions. Some biomarkers can lead to a direct diagnosis due to their presence or absence, while others are dependent on their level of expression. Most diagnostic tests also rely on a screening of multiple biomarkers to increase the specificity and the level of details of the analysis. Many of them are related to a cancer diagnosis and cancer care or viral, bacterial, or parasitic diseases [9].

Cancers are the second most common cause of mortality over the world, and their early detection is strongly correlated with positive and faster patient outcomes. Nowadays, the major cancer indicators are morphological or histological. However, myriads of molecular biomarkers are used routinely to refer the diagnosis (e.g., prostate-specific antigen (PSA), carcinoembryonic antigen (CEA), epidermal growth factor receptor 2 (HER2 or ErbB2), cytokeratins (CKs), thyroid transcription factor-1 (TTF1), among others). The determination of a positive or negative expression status for certain biomarkers directly influences the diagnosis, the decision for a suited treatment, or the prognosis [10, 11].

New biomarker discoveries are therefore very important, but, despite intensified interest and investment, few novel biomarkers are used in clinical practice and their rate of introduction is decreasing. We can estimate a rate of effective biomarkers (approved by the U.S. Food and Drug Administration (FDA)) to an average of only 1 per year since 1998. It clearly reflects the long and difficult path from the identification of a primary candidate molecule to its validation for clinical assays [12]. In order to identify new biomarkers, the first step consists of sampling. The nature of the collected samples plays a key role for all the following steps of the process. The samples can be body fluids, cell lines, plasma or blood, or even organs or pieces of organs. The primary comparison between cancerous cases and healthy samples needs to be done at a first glance. This is often performed due to referenced biobanks or collected biopsies preserved in liquid nitrogen by hospitals and research centers. Molecular extraction (proteins, lipids, DNA) and histological staining are then performed to ensure high-performance analyses. For instance, protein extracts may follow liquid chromatography and mass spectrometry to determine their composition, while histological data need to be compared with many other samples or databases.

The second step is the qualification and confirmation of the biomarker candidates. As the selection is bottleneck for the analysis, this step is performed on a higher number of patient samples. Complementary techniques are also needed to confirm the first results obtained. Then focused experiments begin on this selection to obtain more information about the quantification and the ratio of patient samples that follows the trend for the (over)expression of the biomarker candidates in tumoral cases. Some of them will not fill the requirements for further studies and will be stopped. Others may confirm their trend and become selected candidates. However, a third step is still needed to ensure the selection of highly represented biomarkers. Indeed, the heterogeneity of the cases and analyzed population must be considered. The final step therefore requires thousands of patients to obtain the agreements of the FDA and to be sufficiently robust for consideration in the biomedical field (Figure 6.2).

For all these reasons, the development of clinical assays is expensive and prickly on ethics. It is the main reason why only one of the biomarker candidates is usually selected to pass the last step, even if more of them seem relevant.

High throughput analyses need to be performed to ensure a sufficient screening of the population with quick response time, and this often relies on laboratory-based immunoassays. The whole process can lead to the selection of an efficient biomarker, after many years of research and hundreds of thousands of dollars spent in development. It may also fail its qualification at the end of the process, if the selection does not meet the criteria required by the FDA or other international health agencies. It is partly because of the length and cost of the process that few studies are completed and lead to a productive end. As hope for rapid improvement in selection processes, deep learning and artificial intelligence will undoubtedly become game-changers in the years to come to fully exploit the information hidden in histological databases, but also to simplify and enrich clinical decision-making [13, 14]. Finally, once the analyte is defined as a target, whether it is for a cancer diagnosis, pathogenic disease detection, or the determination of a water pollutant, it requires means of detection. This is the reason why biochemical sensors make use of bioreceptors, such as antibodies.

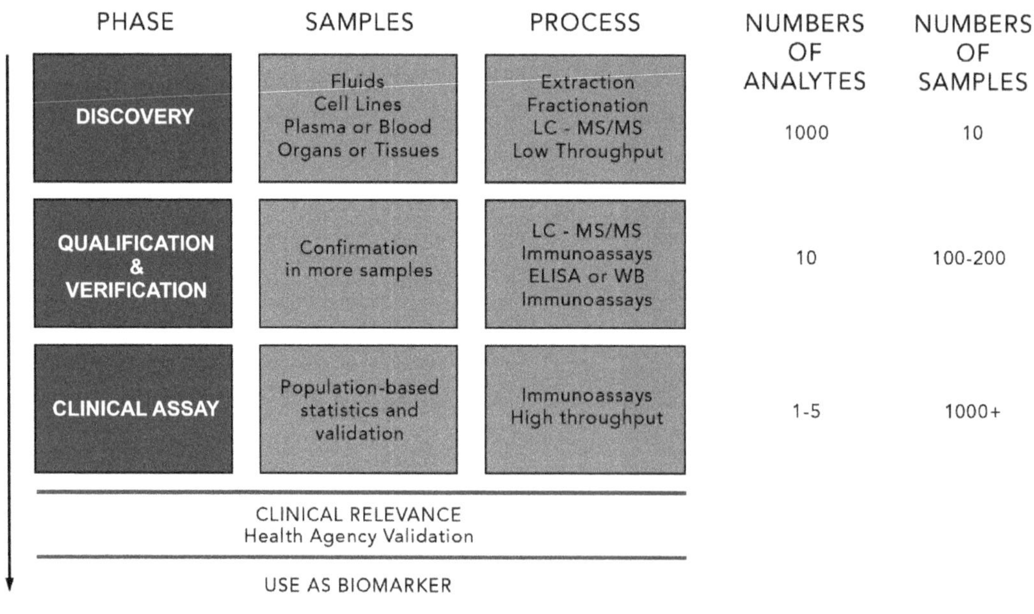

Figure 6.2 Schematic process of biomarkers discovery.

6.2 Antibodies

We are all prone to attacks by other organisms, including micro-organisms such as bacteria and viruses. In higher animals, pathogens may pass through the skin and mucous membranes (first-line defense) and then be detected and destroyed by the immune system. It is important to distinguish two types of immunity:

1. The cellular immunity, which is mediated by T lymphocytes developed inside the thymus (a gland located in the neck or upper thorax region of vertebrates, which produces T-cells and which disappears or becomes vestigial after puberty). This immunity guards against virally infected cells and parasites, fungi, and foreign tissues (e.g., graft rejection).
2. The humoral immunity mediated by antibodies (or immunoglobulins) produced by B lymphocytes, matured within the bone marrow. This immunity is a barrier against bacterial infections and extracellular phases of viral infections, among others.

The mechanisms that trigger the immune system are complex but can be simplified as follows. While foreign macromolecules (proteins, DNA or RNA, lipids, polysaccharides) considered as antigens enter inside the body, they can be recognized by some B lymphocytes, known until binding as naïve cells. These B-cells originate from the hematopoietic stem cells in the bone marrow and migrate into lymphoid organs such as the spleen and lymph nodes. Mature B-cells randomly display immunoglobulins (receptors) anchored in their plasma membrane so that some specific antigens can bind. While this happens, the antigen-receptor complex enters inside the B-cell and is degraded by several enzymes inside endosomes or lysosomes. This B-cell is consequently activated by this interaction, proliferates, and differentiates into an antibody-secreting effector cell, more commonly known as plasma cell. Activated B lymphocytes therefore generate other circulating cells that secrete large amounts of antibodies directed against the antigen. Antibodies essentially act in three different ways: (1) by neutralizing the antigen such as toxins to avoid their binding to host cells, (2) by labeling the antigens for destruction by other actors of the immune system (for example, the Fc regions of antibodies interact with Fc receptors present on the surface of phagocytes), and (3) the activation of the C3b complement, which is potent in opsonization (meaning tagging pathogens, antigen-antibody complexes, and apoptotic cells) and helps the absorption of pathogens, due to receptors present at the surface of phagocytes. Several antigen fragments provided by its degradation are also displayed on the surface of B-cells, which become antigen-presenting cells (APCs) and secrete cytokines, known as cell signaling molecules. The help of some T-cells known as T-helper is sometimes needed for the T-cell-dependent activation system, in correlation with the expression of MHC class II molecules, which will not be further described here. B-cells usually present short lifetime (roughly 2 or 3 days if they are not more stimulated by their specific antigen), but few of them called "memory cells" are able to target the same antigen weeks or years later to enable a higher and faster immune response. The underlying mechanisms of this secondary response are highly studied in immunology, especially to develop new generations of vaccines [15]. Some interesting reference books to deepen this topic are available, such as *Cellular and Molecular Immunology* by Abbas et al. and *Janeway's Immunobiology* by Janeway.

As mentioned earlier, antibodies are natural proteins produced in animals in response to antigenic stimuli. These immunoglobulins (Ig) are composed of two heavy chains and two light chains, connected by means of disulfide bonds, finally leading to Y-shape structures. Both chains are characterized by constant and variable parts [16]. The variable parts (paratopes) are located at the end of the two branches of the antibody and have a strong affinity for a sequence and conformation of a part of a target molecule called the epitope, and usually ranging from 5 to 15–20 amino acids when the antigen is a protein (see Table 6.2). One given antibody is therefore highly specific and selective to its corresponding epitope. The Fc portion of the antibody (F means fragment and c means that it easily crystallizes) consists of the C-terminal part of the two heavy chains. Both arms of the antibodies are also connected to the structure by a flexible hinge region allowing the detection of targets within a certain range of angles (3-D detection). These arms also present three short segments that provide the antibody with its ability to recognize antigens. These three hypervariable sequences concentrate the amino-acid variations between one antibody to another and constitute its binding site. Finally, different types of antibodies exist (Figure 6.3). IgGs are the most common antibodies and are equally distributed between the extravascular fluid and blood. They are secreted as a response to an antigen. This is also the reason why they are among the most used receptors in bioassays. IgAs are essentially present in the intestinal tract and target pathogens, targeting antigenic sites to avoid their anchoring to the epithelial

Table 6.2 Useful Keywords and Definitions Related to Antibodies and Biofunctionalization

Adsorption	The passive bonding of reactants (receptors, analytes) onto a surface or a solid phase during incubation.
Washing	The flooding of the sensing surface with a buffer (often called washing buffer) to separate nonspecific (other molecules) and unbound targets from the surface of the sensor.
Buffer	Biological buffers are solutions keeping a constant pH in a certain range, by taking up protons released during reactions or releasing protons when consumed by reactions. Different buffers exist and are composed of bicarbonates, or proteins in biological fluids. Buffers such as phosphate buffer saline (PBS) are often used because they play the role of pH-stabilizers (usually between 7.0 and 7.5) in addition to keeping isotonicity, which is important to keep biological integrity.
Antibody (Ab)/antigen (Ag)	Mainly proteins by nature, antibodies are antigenic and often commercially available, produced when target proteins (antigens) are injected into animals. They are produced *in vitro* using the hybridoma methods using fused myeloma cells and selected B-lymphocytes.
Paratope (antibody)/ epitope (antigen)	Binding regions of an antibody and an antigen, respectively.
Blocking	Surface passivation using blocking agents to improve the specificity and avoid the detection of other analytes present in complex media.
Apoptotic cell	Biochemical events lead to the modification of the cells to achieve apoptosis, a programmed cell death that happens in multicellular organisms. It is the opposite phenomenon of the necrosis, where the cell death results from a cellular affection or injury.

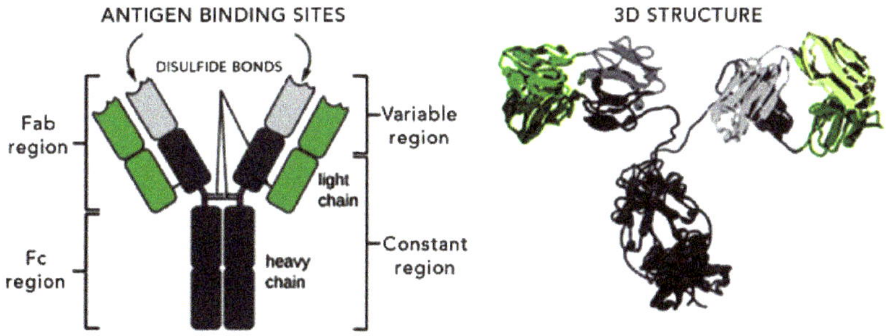

TYPES AND CHARACTERISTICS OF ANTIBODIES

TYPE	ABUNDANCE	MAJOR ROLE	SHAPE
IgG	> 75% 10 to 12 g/L	HUMORAL IMMUNITY BINDING TO PATHOGENS	Y
IgM	10%	FIRSTLY PRODUCED UPON ANTIGEN INVASION PRIMARY RESPONSE	
IgA	< 15%	IMMUNITY IN MUCOSAL TISSUES FORMS DIMERS AFTER SECRETION	
IgD	< 1%	SIGNALS B CELLS TO BE ACTIVATED IMMUNE RESPONSE STIMULATION	Y Variable
IgE	Variable	ALLERGIC REACTION	Y

Figure 6.3 Type, abundance, role, and shape of antibodies.

surface of the digestive tract. IgEs are only present in trace in blood and are also involved in allergic reactions.

The immune system of animals and humans therefore naturally supplies an antibody production against markers of interest. Once the target is identified and available (e.g., a biomarker), it can be injected into animals such as mice or rabbits to exploit their antibodies against this target. Human antibodies can also be produced in mice, using specific mice strains generated by introducing segments of human immunoglobulin loci into the germlines deficient in mouse antibody production. This results from a gene-targeting modification of mice strains. Genetic and biological manipulations in a broader extent have made huge progress since the 1970s, and it is now possible to investigate and use these genetic technologies to produce efficient antibodies, for biosensing purposes but essentially for therapeutic scopes (drug targeting and biotherapeutics). Considering this, it is important to mention that antibodies can be polyclonal or monoclonal depending on their production method, as discovered in 1975. Polyclonal antibodies are obtained from several B-lymphocytes lines, initially coming from suitable selected types of animals such as goats, mice, and rabbits. They specifically bind to their antigens, but at different epitopes. Monoclonal antibodies are produced from only one single line of isolated

B-lymphocytes and are selected to only bind to one epitope of the target. The use of monoclonal or polyclonal antibodies is consequently a supplementary parameter to consider, avoiding possible specificity troubles [15, 17].

How do we produce antigens for injection and how do we purify their related antibodies? Let us emphasize that the upstream production of antibodies is now industrially performed in huge volumes of mammalian cell cultures, followed by numerous downstream processing techniques. At the emergence of antibodies' production and commercialization at the end of the 1970s and beginning of the 1980s, yields were extremely poor (approximately 100 mg/L) and 1g of antibodies was worth far more than its weight in gold. Today, the large-scale production of immunological tests and immunotherapies require enormous quantities in terms of production every year, with yields often above 5 g/L and manufacturing plants of 10,000 liters or even larger. One single cell is now exploited to produce more than 20 pg of antibodies per day, and major companies have reported a cell concentration of more than 20 million cells/mL in fed-batch processes [18]. The origins of the *in vitro* production of antibodies come from the use of the hybridoma method. It consists of the fusion of myeloma cells (cancer cells) and isolated B-lymphocytes. This cellular fusion yields an hybridoma cell line, which can be cultured *in vitro* and benefits both from the proliferation and lifetime of the tumor cells, and the production of antibodies coming from the B-cell (Figure 6.4). After the production phase, cells are centrifuged and separated from the culture medium containing released antibodies. This rich and complex supernatant is then sent to purification, downprocessing. Purification techniques are essential to separate antibodies (and proteins or antigens in a broader extent) from their matrix and are essentially based on chromatography columns (size exclusion, ion exchange, metal binding, affinity, high pressure), precipitation in specific ionic conditions, electrophoresis, and ultracentrifugation, among others. Due to all these techniques and other emerging

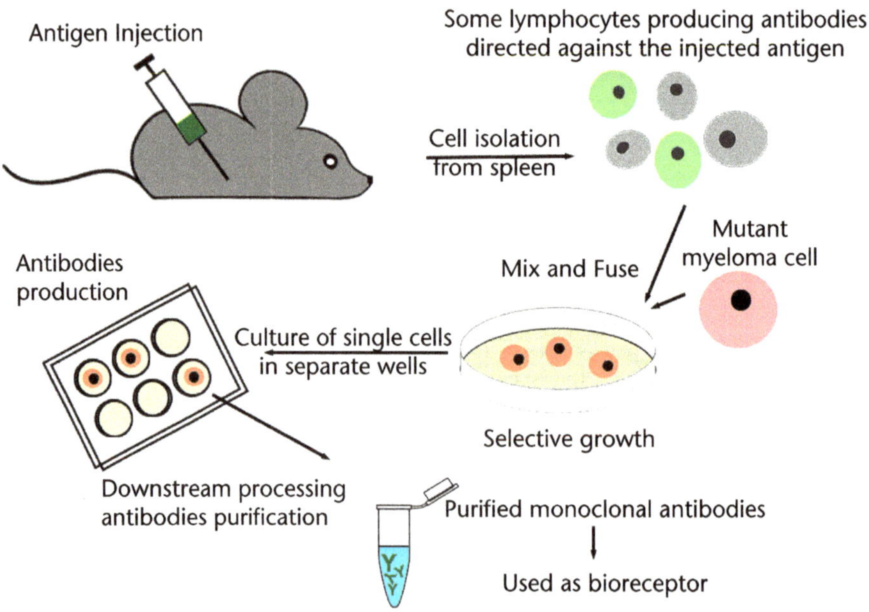

Figure 6.4 General outline for the production of monoclonal antibodies.

purification processes, it is possible to manufacture a customized product for a specific application and order a purified and commercially available antibody/target couple to develop a detection method or a targeted therapy [18, 19].

Besides their natural protective role, antibodies are engineered and selected to target analytes of interest with a high affinity. This affinity is considered the strength of binding between two components and is quantified by the K_D value, which represents the equilibrium dissociation constant. The lower the K_D, the higher the affinity of the antibody for its target (Figure 6.5). Most antibodies present a strong affinity for their targets and oscillate between 10^{-7} and 10^{-9} M. As a reference, one of the highest interaction affinities is shown by the well-known couple (strept) avidin/biotin, which is of the order of 10^{-14} M. This affinity is highly dependent on the experimental conditions, so bioreceptors such as antibodies are sensitive to pH and temperature changes. These changes may provoke alteration of their 3-D conformation and, consequently, affect the integrity of the binding sites. Physiological conditions are therefore needed to ensure sufficient pH-stabilization around 6.5 to 7.5 and to exploit their highest affinity against the target. It is also known that the production and purification of antibodies is a very sensitive process with many parameters that come into consideration. The dissociation constant is experimentally determined by techniques such as the ELISA, the most popular [20], kinetic exclusion assays, or SPR using a benchmarking platform such as the Biacore. Other more recent techniques such as scanning ellipsometry for the real-time analysis of microarrays based on polarization-modulated oblique-incidence reflectivity difference (OI-RD) also exist and are used in practice [21]. The essence of these assays is to target antigens immobilized on a substrate with antibodies incubated at a series of concentrations to obtain sufficient data to calculate the affinity (Figure 6.5). The K_D value is given in the product datasheet of most commercial antibodies, usually without any other notice about the expected binding results. It is therefore crucial to consider the quality of the antibodies and verify their effectiveness against the target to ensure high (or at least sufficient) binding efficiency [22–24].

The use of antibodies to highlight the presence of a target molecule is called an immunoassay (IA). Many techniques rely on this antibody/antigen interaction. In fact, their use began with radioimmunoassay (RIA) and with the ELISA. The first RIA was developed by Rosalyn Sussman Yalow and colleagues, who won the Nobel

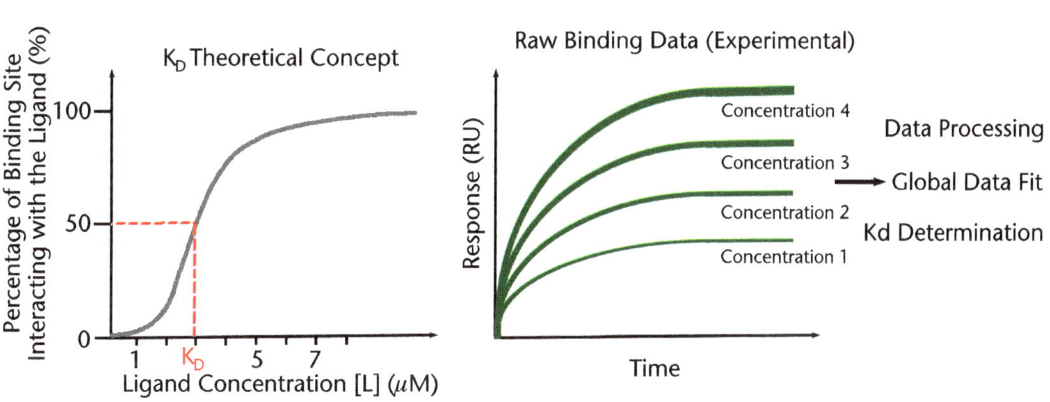

Figure 6.5 Understanding the K_D determination.

Prize for Medicine in 1977. Radioactive isotopes of iodine such as those present in tyrosine residues were used to label antigens and were detected by adding antibodies and by measuring the resulting gamma rays emitted by their interaction [25]. ELISA became a high-standard routine technique to highlight the presence of targets in biological samples. In a simple or direct ELISA, antigens are immobilized in wells (often 96-well microtiter plates). Then antibodies are added so they can bind to their targets. These antibodies are coupled with an enzyme so when the substrate is added into the wells, the assay can reveal their presence by producing a color change or an emission of light, detected within a calibrated spectrometer (see more in Chapter 8). Different ELISA configurations can be implemented, such as sandwich assays where antibodies are first attached to the wells or competitive assays when antigen inhibitors are added, among others. In the case of a sandwich assay, samples possibly containing the target antigen are added in the wells covered by a layer of antibodies. After rinsing, secondary antibodies are supplemented to bind to the already trapped antigens. Specific signal amplifiers are also used to improve the level of detection in specific cases. Over the years, the ELISA rapidly became a laboratory gold standard for biochemical detections and took the front of the stage in comparison with RIAs, which deal with radioactive components and are therefore much more complex. Nowadays, RIAs are strictly limited to some applications. Colorimetric IAs, chemiluminescent IAs, fluorescent IAs, electrochemical IAs, SPR-based IAs, and paper-based IAs can also be cited as some examples in this antibody-based category. Indeed, although the ELISA is the most popular immunoassay, other widely used techniques exist [26, 27].

Among these other techniques, paper-based diagnostic tests are particularly essential for the fast detection of tropical diseases such as malaria, Zika, or dengue in developing countries with poor access to laboratories. These tests, also called rapid diagnostic tests (RDTs), strip-tests, or even auto-tests, have become a key asset in the fight against pandemics such as the Covid-19 crisis, as the result is obtained within 10 to 30 minutes [28, 29]. Well-known pregnancy tests also take the advantage of antibodies to reveal the presence of a hormone (hCG) in urine by lateral flow assay (LFA) on a paper band [30]. The key points of this technology are high versatility and strong user-friendliness level. They can be read out by the naked eye due to the migration of antibody conjugates (often with complexes involving gold nanoparticles, AuNPs) on a nitrocellulose membrane that causes a color change during the migration on a test line (Figure 6.6). Several types of paper-based LFAs were designed (e.g., the printing of antibodies on the paper substrate, the use of glass fiber as substrate, different functionalization techniques to form nanoparticular complexes). These simple components are perfectly suited for mass production at low cost and are easily available to a large public. These sensing facilities are considered as point-of-care (POC). Some of the few weaknesses of paper-based LFAs are probably the lack of sensitivity for certain types of samples and variability against environmental factors such as humidity or temperature changes. In most cases, the samples (e.g., drop of blood or saliva) need to be pre-treated/diluted using a solution given by the manufacturer inside a testing kit before migration on the paper. While classical LFAs are considered a 1-D assay, newer multistep assays present better fluid control in paper-based devices. Different paper geometries were developed with 2-D or even 3-D networks (origami assays), following procedures to fold the strip test

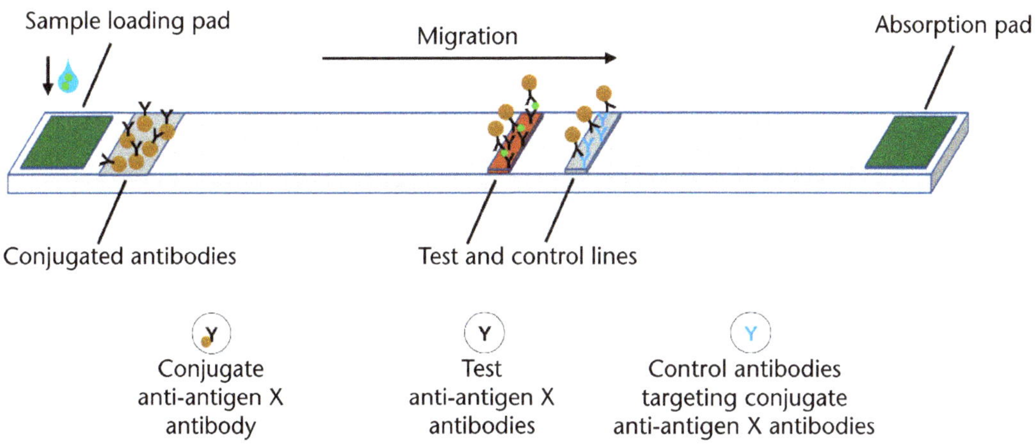

Figure 6.6 General outline of a paper-based lateral flow immunoassay.

in such a way that fluidic channels, pores of different sizes, or other electrochemical or magnetic components come into contact when performing the assay [31, 32].

In addition to these colorimetric LFAs, more advanced structures such as paper-based electrochemical biosensors have shown interesting sensitivities and appear to become competitive alternatives in certain fields. They consist of designing small electrodes printed on paper and the biodetection response is collected via an electronic chip or adaptor for direct transmission to a computer or mobile phone. The readout is qualitative (or semi-quantitative) but not quantitative. It is therefore necessary to identify the suited detection technique by balancing the advantages and disadvantages to suit the final application requirements.

Among all these possible ways to detect molecules of interest, the nuance between biochemical assays and biosensors is not always well defined in the literature and the two concepts are often confused or interconnected. Based on their distinct definition, biosensors are self-contained integrated devices, so the biological element is in direct contact with a transducer that leads to the sensing signal, while biochemical assays rely on bioreceptors or biological activities but do not necessarily embed the detection method. For example, the ELISA is an assay that requires microtiter plates where the receptors and the biosamples interact, but it also needs an external quantification method such as a spectrometer.

6.3 Nanobodies

Along with the use of antibodies, the development of other binding proteins such as DARPins [33], monobodies, and affibodies have proved to be useful in many antibody-like applications [34, 35]. Among these molecular alternatives, the use of camelid heavy-chain antibody fragments commonly named nanobodies (Nbs) quickly became popular. Animals such as alpacas, dromedaries, camels, and llamas produce immunoglobulins with only one heavy chain (HCAbs). It is structured in two constant regions, one hinge region, and antigen binding site (noted VHH). Some of these single-domain antibodies can also be produced from some species of cartilaginous fishes or sharks.

Nanobodies are among the shortest available antibody-like bioreceptors (<3 nm and around 13 to 15 kDa, in comparison with classical antibodies around 15 nm and 150 kDa). They consist of an antibody fragment of HCAbs made of a monomeric binding domain. They possess higher resistance against heat and provide higher stability against detergents. They also seem more soluble in water than conventional antibodies. The major asset of Nbs is their ability to bind to smaller or hidden antigens not accessible to whole antibodies. They are therefore studied to be used as bioreceptors to target small analytes or to improve site-directed immobilization techniques but are mainly investigated as therapeutics. Nbs are first choice molecules to be tested *in vivo* as they are expected to cross the blood-brain barrier and penetrate solid tumors more easily than conventional antibodies, which would help for the development of new targeted therapies and diagnostic strategies [36, 37].

Nbs are essentially produced by recombinant technology raising a VHH region. This technology allows the direct fusion of tags in their tertiary structures, such as histidine-tags (His-tags) or fluorescent markers such as the green fluorescent protein (GFP). The critical step for their production is certainly the acquisition of libraries containing all the required genetic information to produce high affinity Nbs. The generation of libraries is often performed through the immunization of Camelidae, from which lymphocytes are isolated (Figure 6.7). The mRNA extraction and amplification of the sequences encoding for mature Nbs repertoire by PCR are then required to be further inserted in a cloning vector. The selection of target Nbs is then commonly performed by the phage display technology. Restriction enzyme sites are also added so that they can be inserted into a phagemid cloning vector. The VHH library is then produced by infecting the phage expression system. Finally, to select the wanted Nbs, a phage selection is performed. This is often achieved by ELISA using the specific antigen as a target. This technique is called biopanning or panning and several rounds of selection are often needed to ensure strong Nbs selection. These steps are routine biological techniques and are relatively fast and

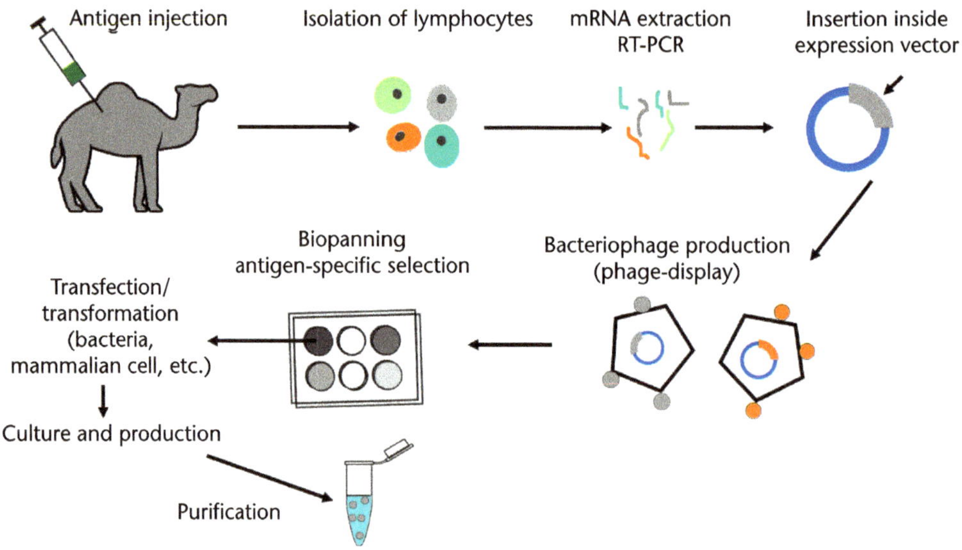

Figure 6.7 General outline for the production of nanobodies (NBs).

low-cost in comparison to the production of conventional antibodies (both monoclonal and polyclonal) [38].

6.4 Affimers

A third and final major category of antibody-like receptors is concerning affimer molecules highlighted around 2015. They are small proteins that bind to target proteins with affinity in the 10^{-9} M range. Affimers tend to be assimilated by use as a common name, but it originates from a registered trademark from Avacta Life Sciences Ltd. The proteins designed and developed by Avacta mimic the characteristics of monoclonal antibodies with optimized stability and lower size in comparison with conventional Abs. They are derived from the cysteine protease inhibitor family of cystatins and display two peptide loops that can be optimized to bind to targets of interest [35].

They are produced faster than antibodies using the technology of phage display libraries and their main applications remain cell therapies and therapeutics. Recent works investigated the use of affimers as bioreceptors for biosensing purposes but relatively few studies exist to establish a fair comparison concerning their advantages or disadvantages compared to more conventional receptors at the moment [39, 40].

6.5 Application of Immunosensors

Biosensors have known cutting-edge improvements in the recent years in terms of technology, miniaturization, versatility, response time, and sensitivity. Optical fibers biosensors are part of the game and have undergone multidisciplinary progress in both biochemical sciences and photonics engineering. As grating-assisted optical fibers are highly sensitive refractometers, they possess intrinsic properties that make them suited to detect molecular events happening on their surface. In that philosophy, bare modal sensors such as TFBGs or LPFGs and their etched versions, FBGs and etched FBGs, ex-TFGs, made in silica or polymer fibers, can be functionalized with bioreceptors through different ways. Indeed, the immobilization of receptors on glass and polymeric materials has been the subject of an abundant literature, especially for the functionalization of chips and site-directed anchoring. Reliable biochemical immobilization processes therefore exist through covalent or noncovalent bindings. While covalent bindings present higher stability, passive adsorption on surfaces remains highly exploited in biochemical assays such as for ELISA microtiter plates and leads to sufficient binding forces to operate in successive solutions and samples. Surface treatment such as plasma, chemical modifications, or UV-O_3 exposure coupled with deep surface drying are often needed to ensure sufficient molecular adhesion for physical adsorption.

6.5.1 Immobilization Strategies

Among the many physical or biochemical immobilization techniques available and extensively studied in the literature over the last decade, four main categories of immobilizations can be highlighted: covalent or noncovalent binding; and random or site-directed immobilization (Figure 6.8) [41, 42]. As aforementioned, the use of covalent bonds is highly stable, but adsorption processes may be sufficient for certain applications. The 3-D structure of antibodies also shows that the orientation of the antigen binding sites is important. Despite large amounts of antibodies immobilized, their bioactivity can be low due to poorly controlled immobilization. Although solutions exist, especially using chemical or protein intermediates or targeting certain amino acids more frequently present in the Fc region, their characterization remains challenging. As the orientation is influenced by the surface chemistry, it must be considered that an average of antibodies is well (or not well) oriented for detection purposes. The mass effect conducted via the total number of available receptors makes it possible to obtain an effective antibody/antigen detection. Surface functionalization is therefore a key parameter to consider, which may explain possible variability in the experimental biosensing results [43, 44].

The simplest surface functionalization remains the random adsorption of antibodies. It does not require multiple steps or chemical reactions, but results in a low stability and random orientation and coverage of the bioreceptors along the sensing surface. This is particularly true on nonplanar substrates such as cylindrical optical fibers. To overcome this issue, one of the most popular immobilization methods remains the use of the amine coupling, by modifying the sensor surface with a self-assembled monolayer (SAM) or a polymeric matrix (e.g., dextran). A SAM is a monolayer formed of alkyl chains with functional groups enabling the coupling of molecules such as carboxyl groups. The latter enables the formation of a peptide bond (amide) with amino groups randomly present on the antibody surface. This technique allows the formation of a strong bond between receptors and the surface (often gold coating or noble metal/particles in the case of plasmonic sensors). The binding chemistry often relies on the carbodiimide crosslink through NHS/EDC (N-hydroxysuccinimide and carbodiimide) in specific buffered conditions (Figure 6.9). The control of the pH is mandatory to target specific amino acids of antibodies to bind on the surface. The deprotonation of many reactive sites is indeed affected by the pH. Selecting the adapted buffer is therefore needed, depending on the target molecular groups. Most proteins (including antibodies) contain exposed lysines and carboxyl groups. EDC reacts with available carboxyl groups (e.g., exposed by the SAM on the optical fiber surface) to form an o-acylisourea intermediate. This molecule then reacts with a primary amine (e.g., exposed by lysine residues of antibodies) to form a stable amide bond and a soluble by-product. As the primary amine residues are positively charged at physiological conditions, they are usually on the outer surface of proteins and accessible for conjugation without need for denaturation. N-hydroxysuccinimide esters are reactive groups formed

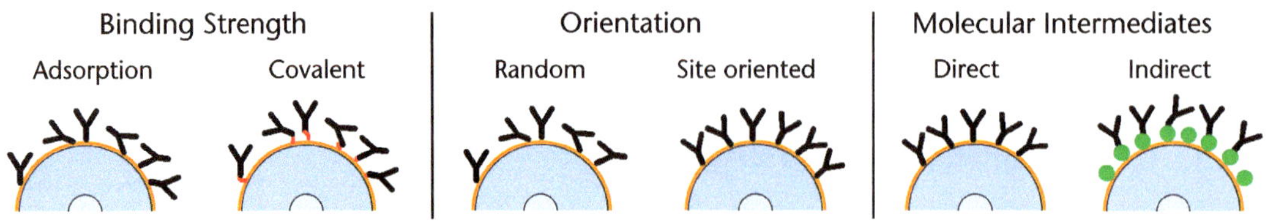

Figure 6.8 Scheme of different types of immobilization strategies.

by EDC activation and considered crosslinkers, as they react with primary amines in slightly alkaline conditions, typically at pH 7.2 to 8.5. However, their relatively short lifetime due to the hydrolysis of NHS needs to be considered in the immobilization protocol. We therefore end up with a two-step reaction where EDC is most efficient in acidic conditions pH approximately 4.5 up to pH approximately 7.5 at their lowest efficiency, while NHS is added in most EDC protocols to improve the coupling efficiency. EDC couples to NHS forming an NHS ester that is much more stable than the o-acylisourea intermediate, allowing an efficient conjugation to primary amines at physiologic pH (7.0–7.5). Their combination is therefore a trade-off between the required pH and the reaction efficiency. Another possible chemistry is the use of glutaraldehyde promoting imine bonds between the antibody and amine groups present in the substrate [45–47].

These options do not regulate the orientation of the antibodies, which can further be investigated. Many strategies have indeed been studied such as the incubation of antibodies with antigens before their immobilization onto mixed SAMs, so that the antigen-binding sites cannot bind on the surface [48].

Once the immobilization is performed, pH changes can be exploited to release the antigen from the binding site and regenerate the receptors while covalent bonds with the sensor surface remain intact. Some investigations also show that it is possible to break the disulfide bridges between the two branches of the antibody, so that these thiol groups can directly bind to the metal surface and present their antigen binding

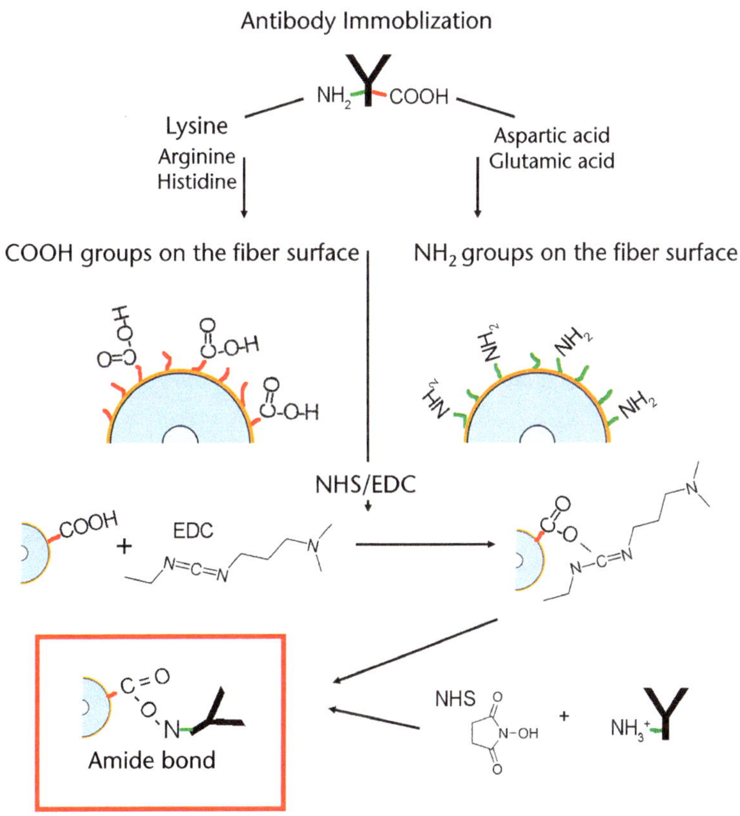

Figure 6.9 NHS/EDC chemistry for covalent immobilization of antibodies.

site to the surrounding medium. Finally, the use of calixarene derivates or proteins such as proteins A or G target the Fc region of the antibodies and therefore help for their orientation. The intermediates, protein A, G, or recombinant fusion proteins A/G, possess specific IgG binding domains. They were initially used to purify and detect IgGs in chromatography, but are now considered as molecular building blocks to orientate the antibody by capturing its constant region and improve the ratio of well-oriented antibodies on the sensor surface [44, 49–51].

The advantage of plasmonic devices is that the binding of molecules on the sensor surface can be monitored in real time and assess the control of the functionalization, based on the SPR shift. As an example, Figure 6.10 shows the immobilization process using the SAM formation with mercaptoundecanoic acid (MUA) overnight in ethanol and the NHS/EDC protocol for IgGs binding. It is followed by a bovine serum albumin (BSA) blocking with rinsing steps performed in PBS at pH 7.2. This measurement was performed on a gold-coated unclad multimode fiber (core diameter of 400 μm) [52].

Alternatives such as Fourier transform infrared spectroscopy (FTIR), atomic force microscopy (AFM), or fluorescent microscopy using tagged antibodies or intermediates can be used to validate the functionalization of the sensor surfaces. However, these techniques often lead to poor reproducibility on cylindrical surfaces such as the ones of optical fibers. Planar substrates metallized in the same way as the optical fiber samples are often used in practice to perform these tests, and the results are transposed from one type of surface to another, without considering the curvature factor. The use of two validation techniques is therefore recommended to reinforce the relevance of the experimental data and to ensure the immobilization protocol.

Once the analytes enter in contact with the immobilized receptors on a plasmonic surface, a small refractive index change occurs and allows a label-free detection, which means that no enzymatic or fluorescent labeling is needed. To avoid

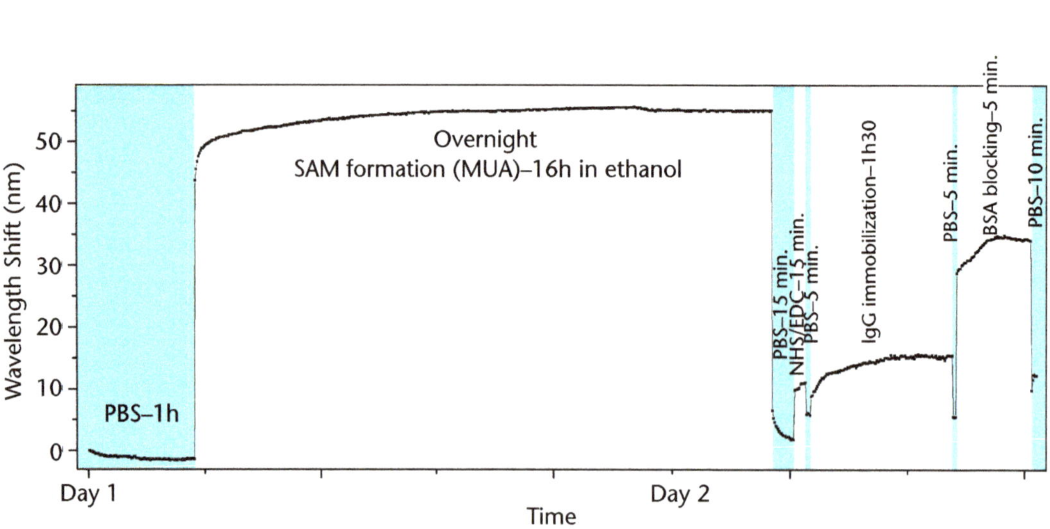

Figure 6.10 Sensorgram obtained from a gold-coated unclad multimode fiber reporting the SPR wavelength shift as function of time. The immobilization protocol was processed overnight in mercaptoundecanoic acid (MUA) in ethanol, followed by the NHS/EDC chemistry for the covalent binding of IgGs. The blocking was performed using BSA.

the detection of other analytes (nonspecific) that may be present within the sample, surface blocking is mandatory. Blocking agents such as BSA, milk casein, fish gelatin, or others need to be handled and well balanced to ensure high surface sensitivity while maintaining sufficient antigen-binding sites free [53]. The ratio between the well-oriented antibodies and a sufficient blocking efficiency drives the performance of the whole system, while establishing the rate of false-positive or false-negative detections due to other elements that may interact with the sensing surface or be released from the surface in the surrounding medium.

6.5.2 Immunosensors

The number of optical fiber-based immunosensors is increasing and presents a high interest for environmental sensing and biomedical tools. In the recent years, high performance detection of biomarkers has been reported and many sensing strategies have been highlighted as alternatives to more traditional biosensors. Some of these immunosensors are listed in Table 6.3.

6.5.2.1 Protein Detection

The first-choice targets when using antibodies are certainly proteins, although they may have affinity for molecules of different nature, as described in Section 6.2. The detection of proteins depends on the binding efficiency (target/receptors) but also on the specificity. It is therefore important to verify the detection against cross-interactions with some elements of the buffer (serum, blood, tap water) or other proteins (almost identical to the target ones or presenting the same epitope). In most cases, adapted blocking solutions exist (e.g., BSA or casein), and commercial monoclonal or polyclonal antibodies are validated for these specific immunosensing applications.

The main advantage of optical fiber-based technologies is essentially its live acquisition in ultrasmall sampling volumes or even directly *in situ*. The binding dynamics can therefore be assessed through the use of sensorgrams (Figure 6.11(a)). Control experiments are mandatory because many research studies are carried out on several optical fiber platforms and for various applications (ELISA, commercial SPR devices, fluorescent assays). In Figure 6.11(b), the response of a paper-based lateral flow assay for the detection of *Plasmodium falciparum* histidine-rich protein 2 (*Pf*HRP2) is shown. It was performed on the same cell cultures studied with a plasmonic unclad optical fiber immunosensor [52].

To fully exploit these fiber sensors, it is also important to carry out calibrations and compare their performance with other available techniques. This part is often bottleneck because the reproducibility of the sensors developed "artisanally" for research purposes and in small numbers is difficult to compare with mass-produced commercial sensors. This being said, it is essential to get a high-performance/quality ratio and develop robust laboratory procedures to obtain comparable detection batches with reference techniques and sufficient statistical analysis to significantly validate the experimental results. For example, it is necessary to ensure that the batches of proteins and antibodies used for validation are the same for all the experiments carried out on the sensors. Usual precautions at all stages of sensor manufacturing help to limit the variations related to this small-scale production.

Table 6.3 Biosensing Applications Using Optical Fiber-Based Immunosensors

Optical Fiber Type	*Receptor/Target*	*Bulk Refractive Index Sensitivity (nm/RIU)*	*Limit of Detection (LOD)*	*Reference*
Unclad fiber	Antibody/TNF-α and MMP-3	NA	34 pg/mL or 1.6 pM	[60]
Unclad fiber	Antibody/Cytokine IL-6	NA	1 pg/mL	[61]
Unclad fiber	Antibody/Ara h1 peanut allergen/NPs amplification	5×10^{-5} RIU (resolution)	Label-free: 9 μg/mL, amplification: 0.21 μg/mL	[62]
Unclad fiber	Antibody/C-reactive protein (CRP)	2,659.64 nm/RIU	NA	[63]
Unclad fiber	Antibody/Infliximab (IFX)/NPs amplification	NA	0.3 ng/mL in buffer, 2.2 ng/mL in 100-fold diluted serum	[64]
Unclad fiber	Antibody/PfHRP2/amplification antibodies	NA	1 μg/mL and 0.1 μg/mL	[52]
Unclad fiber	Antibody/*Cymbidium mosaic virus* (CymMV) and *Odontoglossum ringspot virus* (ORSV)	NA	42–48 pg/mL	[65]
Unclad fiber	Antibody/cortisol	NA	1 pg/mL	[66]
Unclad fiber with porous layers	Antibody/Hepatitis B antigens (HBsAg) and Hepatitis B e antigen (HBeAg)	NA	Limit of quantification (LOQ) of 0.3 fg/mL and 0.01 fg/mL	[67]
U-bent fiber	Antibody/Ciproflaxin antibiotic (CIP)	NA	3.3 pg/L	[68]
Etched fiber	Antibody/Salmonella species	275.86 nm/RIU	247 CFU/mL	[57]
Etched fiber	Antibody/*Escherichia coli*	NA	94 CFU/mL	[59]
Etched fiber	Antibody/IL6 – PCT and CRP biomarkers	NA	1.05 pg/mL (IL-6), 10.6 pg/mL (PCT), 29.4 ng/mL (CRP)	[69]
Etched fiber	Antibody/phthalate esters (PAE)	NA	19–51 ng/L	[70]
Tapered fiber	Antibody/DENV E protein (Dengue virus)	NA	1 pM (5.02 nm/nM)	[71]
Fiber tip	Antibody/NS1 antigen (Dengue)	NA	1.54 nM (0.074 μg/mL)	[72]
Side-polished fiber	Antibody/Staphylococcal enterotoxin B	3,150 nm/RIU	10 ng/mL	[73]
Side-polished fiber	Antibody/*Legionella pneumophila*	NA	10 CFU/mL	[58]
D-shape fiber	Antibody/Anti-IgG antibody	NA	0.15 ng/mL or 1 fM	[74]
D-shape fiber	Antibody/Naphtalene (NAPHTA)	NA	0.76 ng/mL	[75]

Table 6.3 *(Continued)*

Optical Fiber Type	*Receptor/Target*	*Bulk Refractive Index Sensitivity (nm/RIU)*	*Limit of Detection (LOD)*	*Reference*
LPFG	Antibody/Anti-IgG	>2,000 nm/RIU	~10 pM	[76]
LPFG	Antibody/Anti-IgG	~1,309 nm/RIU	70 μg/L (460 pM)	[77]
LPFG	Antibody (IgY)/*Streptococcus aureus*	NA	33 CFU/mL	[78]
LPFG	Biotin/Streptavidin anti-IgM/ Immunoglobulin M	NA	0.02 mg/mL	[79]
LPFG	Antibody/AmpC β-lactamase	1,500 nm/RIU	1 ng/mL or 1.5 pM	[80]
LPFG	Antibody/H5N1 virus	~832 nm/RIU (reported dual peak separation)	1.05 ng/mL	[81]
LPFG	Antibody/Norovirus like particles (VLPs)	~2,000 nm/RIU	~1 ng/mL	[82]
Ex-TFG	Antibody/Newcastle disease virus (NDV)	180 nm/RIU	~25 pg/mL	[83]
Ex-TFG	Antibody/PDL-1	160 nm/RIU	1 pg/mL	[84]
Ex-TFG	Antibody (with Prot. A)/NT proBNP biomarker	150–160 nm/RIU	0.5 ng/mL	[85]
TFBG	Antibody/Cytokeratin 17	NA	14 pM	[86]
TFBG	Antibody/Cytokeratin 7 peptide	NA	0.4 nM	[87]
TFBG	Antibody/cortisol	~566 nm/RIU	~0.1 ng/mL	[88]
TFBG	Protein A/antibody (+ using GO layers)	NA	0.8 nM	[89]
TFBG	Antibody/CK17	9.88 nm/RIU	1 ng/mL	[90]
Taper interferometer cascaded with an FBG	Antibody/HER2 protein	2,333 nm/RIU	2 ng/mL	[91]
Mach-Zehnder interferometer (MZI)	Antibody (Protein A)/Human IgG	13,936 nm/RIU	47 ng/mL	[92]
Photonic crystal fiber (PCF)	Antibody/Alpha fetoprotein (AFP)	NA	0.1 ng/mL	[93]
PCF	Antibody (Prot. A)/Human IgG	4,649.8 nm/RIU	10 ng/mL	[94]
Phase-shifted microfiber Bragg grating	Antibody/Cardiac troponin I (cTn-I)	~32.57 nm/RIU	0.03 ng/mL	[95]
Fiber light-coupled optofluidic waveguide (FLOW)	Antibody/Protein p53	NA	10 fg/mL	[96]
Spherical fiber tip	Antibody/CD44	95.76 dB/RIU	17 pM	[97]

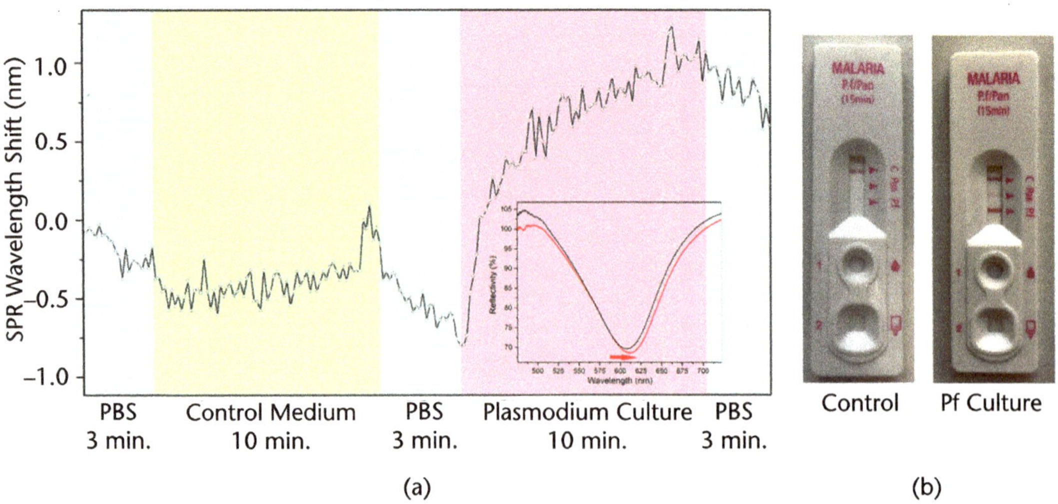

Figure 6.11 (a) Sensorgram obtained from a plasmonic unclad optical fiber (400-μm core-diameter gold-coated fiber) functionalized with anti-HRP2 antibodies. The sensor was first immersed into a control medium and then immersed in the *Plasmodium falciparum* culture, containing HRP2 proteins. (b) The same samples were tested on commercial paper-based lateral flow assays. (Reproduced from [52], Creative Commons Public Domain.)

Following on from the *Pf*HRP2 detection example, different biosensors were used for the detection of spiked proteins at different concentrations in PBS buffer (Figure 6.12(a)). These same biosensors were then immersed in a solution containing secondary antibodies to perform a sandwich assay and consequently improve the detection threshold by amplifying the modification of the surface refractive index (Figure 6.12(b)). This calibration curve was first carried out with wide-step concentration ranges and then refined with intermediate concentrations to determine the dynamic range of detection more precisely.

6.5.2.2 Virus and Cell Detection

The detection of viruses is a hot topic at a time where the world faces the Covid-19 pandemic. The major effect of the worldwide deployment of viral detection tests for a Covid diagnosis, in particular, paper-based assays, has brought biosensors to light to a large public. Most viruses are targeted through their membrane, envelope, or nucleocapsid proteins and can be directly detected using immunosensors. However, other technologies based on the detection of viral DNA or RNA traces after amplification using (RT)-PCR-like methods also exist and are more suited to detect small concentrations of viral presence. A boost has been given to develop virus biosensors given the urgency and relevance of these studies. Recent adaptation of optical fiber based platforms are therefore topical and many research papers will be dedicated to them in the coming months and years [54–56].

However, the detection of cellular pathogens in water, food, or biomedical samples is a growing field of research. The use of antibodies to target bacteria has indeed been recently reported in several works. Achievements for the detection of *Salmonella typhimurium* using an etched optical fiber-based SPR platform [57]

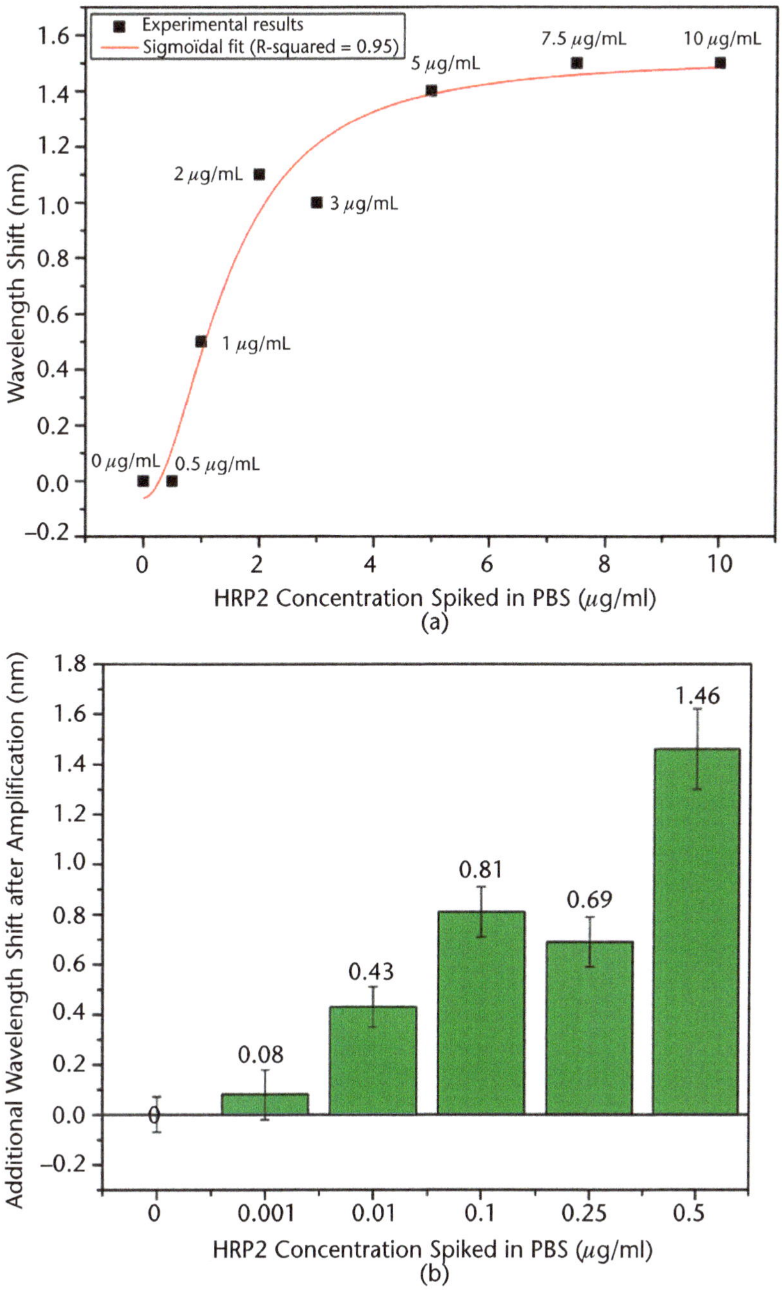

Figure 6.12 (a) Label-free detection of *Pf*HRP2 spiked in PBS at different concentrations, each experimental data point represents the data obtained by a different optical fiber probe. (b) Shift observed after immersion in secondary antibodies to perform a sandwich assay. Mean ± standard deviation (n = 3 different fibers for each concentration). (Reproduced from [52], Creative Commons Public Domain.)

and side-polished fibers for the detection of *Legionella pneumophila* [58] and *E. coli* [59] can be cited.

6.5.2.3 Critical Review

Table 6.3 summarizes the main performance indicators of miscellaneous configurations of optical fiber immunosensors. Given the important number of publications in this field, the table is certainly not exhaustive but shows relevant configurations for which the limit of detection is given.

From Table 6.3, we can conclude that the biosensing results along with the sensors' sensitivities vary enormously depending on the type of approach and the chosen detection strategy. Many parameters come into consideration for these SPR sensors: quality of the substrate, surface cleanliness, homogeneity and adhesion of the metallic layer, strength of the bioreceptors anchoring, antibody/antigen binding efficiency, the presence of nonspecific molecules in the medium tested, the efficiency of the surface blocking, the sample volume and concentration range, and the use of a microfluidic or static system, among many others. Most biosensing studies need more samples and control tests to assess the reproducibility of the presented results.

It is therefore very difficult to extract relevant information from a constantly growing collection of research articles, where the focus is more often oriented on the performances and limits of detection rather than on the robustness of the technique on field. Aside from these works oriented towards publication, other teams work more closely in the lab towards enabling constant sensing performances for mass-produced sensors. We believe that the major advances made over the past few years will clearly cause an acceleration in the development of these sensors and arouse growing interest, especially in areas where traditional sensors show their limits. For instance, the detection of biomarkers and pollutants *in situ* at high throughput and their characterization using optical fiber immunosensors are among the most promising technologies. It is now time to test these sensors under standardized experimental conditions approaching their target use. This stepping stone will be mandatory to move optical fiber-based biosensors forward, from scientific prospects towards factual uses of the technology and its industrialization.

6.5.3 SPR Signal Analysis

After the aforementioned considerations about surface functionalization and biodetection, it is important to discuss about the data analysis and the typical information provided by SPR sensorgrams. As presented in Chapters 2 and 3, SPR devices lead to high refractive index sensitivity, but many biochemical parameters strongly correlated with RI surface changes can be extracted from SPR data, especially:

- *Affinity:* How strongly do the analytes bind to receptors?
- *Kinetics:* How fast do the analytes bind to receptors?
- *Concentration:* How much do the analytes bind to receptors?
- *Specificity:* How specific is the interaction?

The typical molecular binding response is dynamic and can therefore be monitored over time (sensorgram). It provides different characteristics, as depicted in

Figure 6.13. First, it is important to consider that the running buffer matches with the sample buffer to minimum bulk refractive index changes and, therefore, signal jumps. Usually, PBS is used to dilute antibodies or proteins, but its composition can slightly change from one provider to another, especially in terms of salt concentration. The sample dilutions can therefore induce a slight refractive index shift (from the prepared PBS and from the one coming from provided molecules) and lead to misinterpretation of binding events. Also, from one target to another, the levels of bioreceptors immobilized and timing needed for binding can be different. Pretests need to be performed with the simplest configurations as possible to evaluate both binding time and binding efficiency.

While the determination of immobilized materials seems straightforward using the binding shift before binding and after rinsing, both in the same buffer, the determination of kinetics requires more data. Indeed, the determination of kinetics requires a sufficient curvature level of the sensorgram to clearly identify the association/dissociation phases. It also requires reaching a steady state during the sample injection. To enhance the performance of kinetics analyses, slight bulk refractive index changes can be compensated.

The determination of affinity constants requires at least three to five analyte concentrations. The range should be chosen so that the first one is a blank, and the last one at least more than twice the value of the K_D, to reach sufficient curvature of the sensorgram. The injection sequence can lead to single-cycle experiments (using a surface regeneration in-between each concentration) or in separate cycles (each concentration after each other without regeneration), but the last one is more likely to present signal drift. This being said, the use of antibodies usually generates avidity effects (multivalent analytes) so molecules dissociating at a point of the surface can associate again at other locations, leading to longer dissociation times and

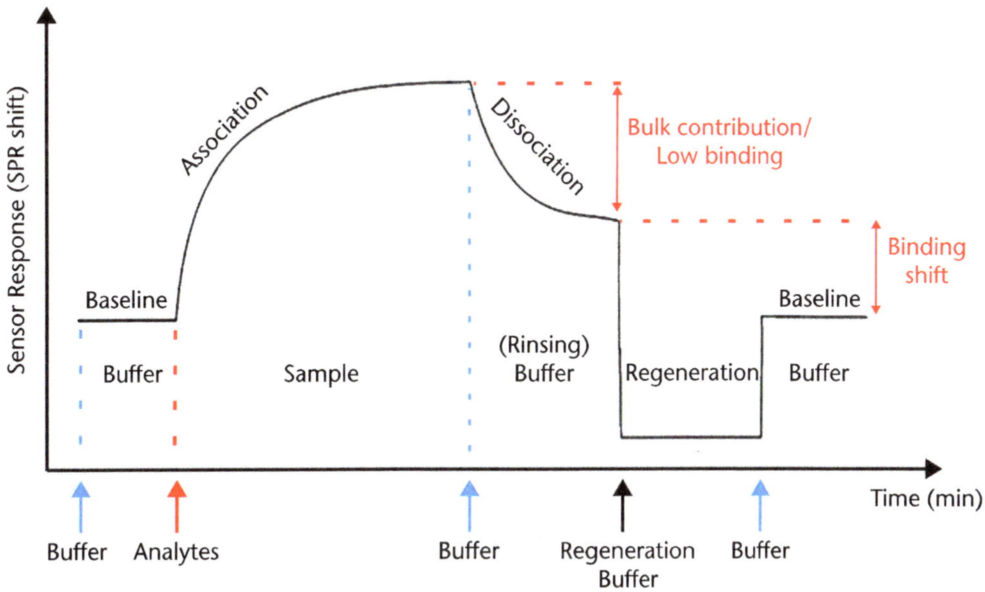

Figure 6.13 Typical SPR signal (sensorgram) for the binding of analytes over time with surface regeneration.

slower dissociation rates. This effect can be reduced using spacers or PEG molecules during the bioreceptors immobilization or by reducing the concentration of receptors on the surface, consequently reducing the probability for the analyte to bind multiple times. The comparison of data from one run to another is possible but the experimental conditions need to be carefully reproduced to avoid any additional drift or parameter change. In most commercial programs, the K_a/K_d determination is automated and is based on fitting curves plotted over the experimental dataset. Although very practical and improved over the last decade, the determination of such parameters requires a high diligence and remains objectively tedious.

The specificity and the concentration of bound analytes can also be determined through SPR due to the shift observed with similar analytes in terms of molecular weight (nonspecific binding evaluation) and by estimation of surface coating (size of receptors/analytes versus target concentration injected). It should also be pinpointed that it is always important to determine the volume in contact with the surface of the biosensor. For example, detecting a low concentration only makes sense by specifying the injected volume and the flow rate. Thus, detecting a low concentration for hours with huge volumes makes no more sense than detecting a high concentration in a low volume. The race for the detection of the lowest concentration is therefore often misleading since a lot of other settings come into play.

One last thing about the reference level is that most SPR devices suffer from baseline drift over time. This drift is usually compensated by a reference channel always soaked into buffer. This response is subtracted from the test channels. In most optical fiber-based plasmonic sensors, this reference measurement cannot be implemented and other spectral (or experimental) tricks need to be performed to get rid of an unstable signal in the buffer. A continuous and regulated flow rate possibly limits this effect by itself, which could be related to the accumulation of salts from the buffer on the metal surface or be assigned to a delayed surface wetting effect. For that same reason, we recommend soaking the plasmonic surface in buffer for a long moment before its use to stabilize the signal and improve reproducibility. Although the reference-subtracted sensorgram provides the actual binding response, it is mandatory to analyze both reference and active spectra individually. For example, the reference surface (control channel) may provide information about nonspecific binding, which cannot be subtracted from the reference channel in buffer because the surface functionalization is often not performed on that channel and therefore may affect the binding properties. In most commercial devices, it is therefore needed to perform a test with a reference surface (not functionalized, with buffer only), a surface functionalized against nontarget molecules (control), and a nonfunctionalized but blocked surface with the presence of nontarget molecules. These results will provide more evidence about the surface adsorption and nonspecific binding in different situations.

Once the experiment cycle ends, the regeneration step helps to remove all bound analytes (or nonspecific molecules) from the surface while keeping covalently immobilized bioreceptors attached. An efficient regeneration process is crucial to keep a high binding activity of the surface and be able to reuse it for further analysis. Most commercial chips can be reused tens (even hundreds) of times due to regeneration. Also, different surfaces such as those relying on intermediate interactions with protein A or secondary binders will need to start again the functionalization after the regeneration. In these particular cases, the bounds are not covalent and may be affected by the process, removing all receptors from the surface. The idea of the regeneration is to us a condition affecting the nature of the binding, such as acidic solutions (glycine-HCl buffer) or basic solutions (NaOH). Other types of regeneration involving detergents such as sodium dodecyl sulfate (SDS) may sometimes be useful to fully recover the initial surface. Optimization can be tested performing different types of regeneration and achieving consistent reference level before and after each cycle. Another trick could be the addition of small concentrations of Tween (usually < 0.01%) in running buffer to prevent from signal drift and to help reducing nonspecific binding.

6.6 Conclusion

Optical fiber immunosensors present many assets. Their use in small volumes and their high resolution/low limit of detection are competitive with most other existing platforms. They are also studied for real-time and remote measurements, especially for environmental sensing and diagnostic purposes. Technological maturation is still necessary to improve reproducibility and feasibility to integrate these biosensors in portable and automated platforms. Their better understanding in standardized ways will certainly allow large-scale use, with robust datasets. The transition between fundamental research and real application in the biosensing field is already beginning and will accelerate in the coming years, in particular due to the democratization of coating and testing platforms and the possibility of developing proper interfaces at lower cost (3-D printers, micro-controllers such as Arduino and Raspberry, open-source coding systems, automated manufacturing processes, miniaturization of microfluidic interfaces). The use of antibodies as bioreceptors is certainly the most widespread technique. To date, the improvement of surface interactions and their handling also take place, while bioengineering plays a key role in renewing integrated immunosensing. Other strategies, such as the use of nucleic acids as bioreceptors, will be described in Chapter 7.

References

[1] Cremer, M., *Uber Die Ursache Der Elektromotorischen Eigenschafter Der Gewebe, Zugleich Ein Beitrag Zur Lehre von Den Polyphasischen Elektrolytketten* (Zeitschrift für Biologie), 1906, p. 47.

[2] Hughes, W. S., "The Potential Difference Between Glass and Electrolytes in Contact with the Glass," *J. Chem. Inf. Model.*, Vol. 102, 1922, pp. 29–45.

[3] Clark, L. C., and C. Lyons, "Electrode Systems for Continuous Monitoring in Cardiovascular Surgery," *Ann. N. Y. Acad. Sci.*, Vol. 102, No. 1, 1962, pp. 29–45.

[4] Yoo, E. H., and S. Y. Lee, "Glucose Biosensors: An Overview of Use in Clinical Practice," *Sensors*, Vol. 10, 2010, pp. 4558–4576.

[5] Thévenot et al., Electrochemical biosensors: recommended definitions and classifications, Biosensors and Bioelectronics 16, 121-131, 2001.

[6] International Union of Pure and Applied Chemistry (IUPAC), *Compendium of Chemical Terminology: IUPAC Recommendations*, 2nd ed., New York: John Wiley & Sons, 1997.

[7] Perumal, V., and U. Hashim, "Advances in Biosensors: Principle, Architecture and Applications," *J. Appl. Biomed.*, Vol. 12, 2014, pp. 1–15.

[8] Turner, A. P. F., I. Karube, and G. S. Wilson, *Biosensors: Fundamentals and Applications*, Oxford, U.K.: Oxford University, 1987.

[9] Mascini, M., and S. Tombelli, "Biosensors for Biomarkers in Medical Diagnostics," *Biomarkers*, Vol. 13, No. 7–8, 2008, pp. 637–657.

[10] Bertozzi, S., et al., "Biomarkers in Breast Cancer," Chapter 1 in *Biomarker: Indicator of Abnormal Physiological Process*, G. Begum, (ed.), Londo, U.K.: InTech Open, 2018.

[11] O'Connor, J. P. B., et al., "Imaging Biomarker Roadmap for Cancer Studies," *Nat. Rev. Clin. Oncol.*, Vol. 14, No. 3, 2017, pp. 169–186.

[12] Rifai, N., M. A. Gillette, and S. A. Carr, "Protein Biomarker Discovery and Validation: The Long and Uncertain Path to Clinical Utility," *Nat. Biotechnol.*, Vol 24, No. 8, 2006, pp. 971–983.

[13] Echle, A., et al., "Deep Learning in Cancer Pathology: A New Generation of Clinical Biomarkers," *Br. J. Cancer*, Vol. 124, No. 4, 2021, pp. 686–696.

[14] Goossens, N., et al., "Cancer Biomarker Discovery and Validation," *Transl. Cancer Res.*, Vol. 4, No. 3, 2015, pp. 256–269.

[15] Roque, A. C. A., C. R. Lowe, and M. Â. Taipa, "Antibodies and Genetically Engineered Related Molecules: Production and Purification," *Biotechnol. Prog.*, Vol. 20, No. 3, 2004, pp. 639–654.

[16] Pohanka, M., "Monoclonal and Polyclonal Antibodies Production—Preparation of Potent Biorecognition Element," *J. Appl. Biomed.*, Vol. 7, No. 3, 2009, pp. 115–121.

[17] Byrne, B., et al., "Antibody-Based Sensors: Principles, Problems and Potential for Detection of Pathogens and Associated Toxins," *Sensors (Switzerland)*, Vol. 9, No. 6, 2009, pp. 4407–4445.

[18] Li, F., et al., "Cell Culture Processes for Monoclonal Antibody Production," *MAbs*, Vol. 2, No. 5, 2010, pp. 466–479.

[19] Jyothilekshmi, I., and N. S. Jayaprakash, "Trends in Monoclonal Antibody Production Using Various Bioreactor Systems," *J. Microbiol. Biotechnol.*, Vol. 31, No. 3, 2021, pp. 349–357.

[20] Beatty, J. D., et al., "Measurement of Monoclonal Antibody Affinity by Non-Competitive Enzyme Immunoassay," *J. Immunol. Methods*, Vol. 100, No. 1–2, 1987, pp. 173–179.

[21] Landry, J. P., et al., "Measuring Affinity Constants of 1,450 Monoclonal Antibodies to Peptide Targets with a Microarray-Based Label-Free Assay Platform," *J. Immunol Methods*, Vol. 417, 2015, pp. 86–96.

[22] Saper, C. B., "A Guide to the Perplexed on the Specificity of Antibodies," *J. Histochem. Cytochem.*, Vol. 57, No. 1, 2009, pp. 1–5.

[23] Couchman, J. R., "Commercial Antibodies: The Good, Bad, and Really Ugly," *J. Histochem. Cytochem.*, Vol. 57, No. 1, 2009, pp. 7–8.

[24] Bordeaux, J., et al., "Antibody Validation," *Biotechniques*, Vol. 48, No. 3, 2010, pp. 197–209.

[25] Walsh, J. H., R. Yalow, and S. A. Berson, "Detection of Australia Antigen and Antibody by Means of Radioimmunoassay Techniques," *J. Infect. Dis.*, Vol. 121, No. 5, 1970, pp. 550–554.

[26] Lequin, R. M., "Enzyme Immunoassay (EIA)/Enzyme-Linked Immunosorbent Assay (ELISA)," *Clin. Chem.*, Vol. 51, No. 12, 2015, pp. 2415–2418.

[27] Crowther, J. R., *The ELISA Guidebook*, 2nd ed., New York: Springer, 2009.

[28] Hristov, D. R., et al., "Designing Paper-Based Immunoassays for Biomedical Applications," *Sensors (Switzerland)*, Vol. 19, No. 3, 2019.

[29] Yakoh, A., et al., "Paper-Based Electrochemical Biosensor for Diagnosing COVID-19: Detection of SARS-CoV-2 Antibodies and Antigen," *Biosens. Bioelectron.*, Vol. 176, 2021, p. 112912.

[30] Kasetsirikul, S., M. J. A. Shiddiky, and N. T. Nguyen, "Challenges and Perspectives in the Development of Paper-Based Lateral Flow Assays," *Microfluid. Nanofluidics*, Vol. 24, No. 2, 2020, pp. 1–18.

[31] Jiao, Y., et al., "3D Vertical-Flow Paper-Based Device for Simultaneous Detection of Multiple Cancer Biomarkers by Fluorescent Immunoassay," *Sensors Actuators B Chem.*, Vol. 306, March 1, 2020.

[32] Colozza, N., et al., "Origami Paper-Based Electrochemical (Bio)Sensors: State of the Art and Perspective," *Biosensor*, Vol. 11, No. 328, 2021, pp. 1–29.

[33] Binz, H. K., et al., "Designing Repeat Proteins: Well-Expressed, Soluble and Stable Proteins from Combinatorial Libraries of Consensus Ankyrin Repeat Proteins," *J. Mol. Biol.*, Vol. 332, No. 2, 2003, pp. 489–503.

[34] Sha, F., et al., "Monobodies and Other Synthetic Binding Proteins for Expanding Protein Science," *Protein Sci.*, Vol. 26, No. 5, 2017, pp. 910–924.

[35] Tiede, C., et al., "Affimer Proteins Are Versatile and Renewable Affinity Reagents," *Elife*, Vol. 6, June 27, 2017.

[36] Goode, J., G. Dillon, and P. A. Millner, "The Development and Optimisation of Nanobody Based Electrochemical Immunosensors for IgG," *Sensors Actuators B Chem.*, Vol. 234, 2016, pp. 478–484.

[37] Alhamoud, Y., et al., "Advances in Biosensors for the Detection of Ochratoxin A: Bio-Receptors, Nanomaterials, and Their Applications," *Biosens. Bioelectron.*, Vol. 141, September 15, 2019.

[38] Salvador, J. P., L. Vilaplana, and M. P. Marco, "Nanobody: Outstanding Features for Diagnostic and Therapeutic Applications," *Anal. Bioanal. Chem.*, Vol. 411, No. 9, 2019, pp. 1703–1713.

[39] Thangsunan, P., et al., "Affimer-Based Impedimetric Biosensors for Fibroblast Growth Factor Receptor 3 (FGFR3): A Novel Tool for Detection and Surveillance of Recurrent Bladder Cancer," *Sensors Actuators B Chem.*, Vol. 326, 2021, p. 128829.

[40] Ferrigno, P. K., "Non-Antibody Protein-Based Biosensors," *Essays Biochem.*, Vol. 60, No. 1, 2016, pp. 19–25.

[41] Makaraviciute, A., and A. Ramanaviciene, "Site-Directed Antibody Immobilization Techniques for Immunosensors," *Biosens. Bioelectron.*, Vol. 50, 2013, pp. 460–471.

[42] Peluso, P., et al., "Optimizing Antibody Immobilization Strategies for the Construction of Protein Microarrays," *Anal. Biochem.*, Vol. 312, No. 2, 2003, pp. 113–124.

[43] Jung, Y., J. Y. Jeong, and B. H. Chung, "Recent Advances in Immobilization Methods of Antibodies on Solid Supports," *Analyst*, Vol. 133, No. 6, 2008, pp. 697–701.

[44] Kausaite-Minkstimiene, A., et al., "Comparative Study of Random and Oriented Antibody Immobilization Techniques on the Binding Capacity of Immunosensor," *Anal. Chem.*, Vol. 82, No. 15, 2010, pp. 6401–6408.

[45] Wang, C., et al., "Different EDC/NHS Activation Mechanisms Between PAA and PMAA Brushes and the Following Amidation Reactions," *Langmuir*, Vol. 27, No. 19, 2011, pp. 12058–12068.

[46] Gao, S., et al., "Oriented Immobilization of Antibodies Through Different Surface Regions Containing Amino Groups: Selective Immobilization Through the Bottom of the Fc Region," *Int. J. Biol. Macromol.*, Vol. 177, 2021, pp. 19–28.

[47] Fung, Y. S., and Y. Y. Wong, "Self-Assembled Monolayers as the Coating in a Quartz Piezoelectric Crystal Immunosensor to Detect Salmonella in Aqueous Solution," *Anal. Chem.*, Vol. 73, No. 21, 2001, pp. 5302–5309.

[48] de Mol, N. J., and M. J. E. Fischer, "Surface Plasmon Resonance: Methods and Protocols," *Life Sci.*, Vol. 255, 2010.

[49] Kanno, S., et al., "Assembling of Engineered IgG-Binding Protein on Gold Surface for Highly Oriented Antibody Immobilization," *J. Biotechnol.*, Vol. 76, No. 2–3, 2000, pp. 207–214.

[50] Jeong, M. L., et al., "Direct Immobilization of Protein G Variants with Various Numbers of Cysteine Residues on a Gold Surface," *Anal. Chem.*, Vol. 79, No. 7, 2007, pp. 2680–2687.

[51] Loyez, M., et al., "Optical Fiber Gratings Immunoassays," *Sensors*, Vol. 19, 2019, p. 2595.

[52] Loyez, M., et al., "PfHRP2 Detection Using Plasmonic Optrodes: Performance Analysis," *Malar. J.*, Vol. 20, No. 1, 2021, pp. 1–9.

[53] Loyez, M., et al., "HER2 Breast Cancer Biomarker Detection Using a Sandwich Optical Fiber Assay," *Talanta*, Vol. 221, January 1, 2021.

[54] Hadi, M. U., and M. Khurshid, "SARS-CoV-2 Detection Using Optical Fiber Based Sensor Method," *Sensors*, Vol. 22, 2022, p. 751.

[55] Nag, P., K. Sadani, and S. Mukherji, "Optical Fiber Sensors for Rapid Screening of COVID-19," *Trans. Indian Natl. Acad. Eng.*, Vol. 5, No. 2, 2020, pp. 233–236.

[56] Qu, J. -H., et al., "Innovative FO-SPR Label-Free Strategy for Detecting Anti-RBD Antibodies in COVID-19 Patient Serum and Whole Blood," *ACS Sensors*, 2022.

[57] Kaushik, S., et al., "A Label-Free Fiber Optic Biosensor for Salmonella Typhimurium Detection," *Opt. Fiber Technol.*, Vol. 46, June 2018, pp. 95–103.

[58] Lin, H. Y., et al., "Development and Application of Side-Polished Fiber Immunosensor Based on Surface Plasmon Resonance for the Detection of Legionella Pneumophila with Halogens Light and 850 nm-LED," *Sensors Actuators A Phys.*, Vol. 138, No. 2, 2007, pp. 299–305.

[59] Kaushik, S., et al., "Rapid Detection of Escherichia Coli Using Fiber Optic Surface Plasmon Resonance Immunosensor Based on Biofunctionalized Molybdenum Disulfide (MoS 2) Nanosheets," *Biosens. Bioelectron.*, Vol. 126, 2019, pp. 501–509.

[60] Huang, Y. C., et al., "Quantification of Tumor Necrosis Factor-α and Matrix Metalloproteinases-3 in Synovial Fluid by a Fiber-Optic Particle Plasmon Resonance Sensor," *Analyst*, Vol. 138, No. 16, 2013, pp. 4599–4606.

[61] Liu, G., et al., "Sensitive Cytokine Assay Based on Optical Fiber Allowing Localized and Spatially Resolved Detection of Interleukin-6," *ACS Sensors*, Vol. 2, No. 2, 2017, pp. 218–226.

[62] Pollet, J., et al., "Fast and Accurate Peanut Allergen Detection with Nanobead Enhanced Optical Fiber SPR Biosensor," *Talanta*, Vol. 83, No. 5, 2011, pp. 1436–1441.

[63] Wang, W., et al., "A Label-Free Fiber Optic SPR Biosensor for Specific Detection Of C-Reactive Protein," *Sci. Rep.*, Vol. 7, No. 1, 2017, pp. 1–8.

[64] Lu, J., et al., "Fiber Optic-SPR Platform for Fast and Sensitive Infliximab Detection in Serum of Inflammatory Bowel Disease Patients," *Biosens. Bioelectron.*, Vol. 79, 2016, pp. 173–179.

[65] Lin, H. Y., et al., "Direct Detection of Orchid Viruses Using Nanorod-Based Fiber Optic Particle Plasmon Resonance Immunosensor," *Biosens. Bioelectron.*, Vol. 51, 2014, pp. 371–378.

[66] Leitão, C., et al., "Cortisol AuPd Plasmonic Unclad POF Biosensor," *Biotechnol. Reports*, Vol. 29, 2021.

[67] Xu, X., et al., "Chemiluminescent Optical Fiber Immunosensor Combining Surface Modification and Signal Amplification for Ultrasensitive Determination of Hepatitis B Antigen," *Sensors*, Vol. 20, 2020, p. 4912.

[68] Lamarca, R. S., et al., "Label-Free Ultrasensitive and Environment-Friendly Immunosensor Based on a Silica Optical Fiber for the Determination of Ciprofloxacin in Wastewater Samples," *Anal. Chem.*, Vol. 92, No. 21, 2020, pp. 14415–14422.

[69] Nie, R., et al., "A Portable Pencil-Like Immunosensor for Point-of-Care Testing of Inflammatory Biomarkers," *Anal. Bioanal. Chem.*, Vol. 412, No. 13, 2020, pp. 3231–3239.

[70] Tang, M., et al., "Development of an Optical Fiber Immunosensor for the Rapid and Sensitive Detection of Phthalate Esters," *Sensors Actuators B Chem.*, Vol. 258, 2018, pp. 304–312.

[71] Mustapha Kamil, Y., et al., "Label-Free Dengue E Protein Detection Using a Functionalized Tapered Optical Fiber Sensor," *Sensors Actuators B Chem.*, Vol. 257, 2018, pp. 820–828.

[72] Camara, A. R., et al., "Dengue Immunoassay with an LSPR Fiber Optic Sensor," *Opt. Express*, Vol. 21, No. 22, 2013, p. 27023.

[73] Slavík, R., J. Homola, and E. Brynda, "A Miniature Fiber Optic Surface Plasmon Resonance Sensor for Fast Detection of Staphylococcal Enterotoxin B," *Biosens. Bioelectron.*, Vol. 17, No. 6–7, 2002, pp. 591–595.

[74] Chiavaioli, F., et al., "Femtomolar Detection by Nanocoated Fibre Label-Free Biosensors," *ACS Sensors*, Vol. 3, No. 5, 2018, pp. 936–943.

[75] Cennamo, N., et al., "Detection of Naphthalene in Sea-Water by a Label-Free Plasmonic Optical Fiber Biosensor," *Talanta*, Vol. 194, 2019, pp. 289–297.

[76] Chiavaioli, F., et al., "Sol-Gel-Based Titania-Silica Thin Film Overlay for Long Period Fiber Grating-Based Biosensors," *Anal. Chem.*, Vol. 87, No. 24, 2015, pp. 12024–12031.

[77] Chiavaioli, F., et al., "Towards Sensitive Label-Free Immunosensing by Means of Turn-Around Point Long Period Fiber Gratings," *Biosens. Bioelectron.*, Vol. 60, 2014, pp. 305–310.

[78] Gan, W., et al., "Rapid and Sensitive Detection of Staphylococcus Aureus by Using a Long-Period Fiber Grating Immunosensor Coated with Egg Yolk Antibody," *Biosens. Bioelectron.*, Vol. 199, March 1, 2022.

[79] Liu, L. L., et al., "Highly Sensitive Label-Free Antibody Detection Using a Long Period Fibre Grating Sensor," *Sensors Actuators B Chem.*, Vol. 271, October 15, 2018, pp. 24–32.

[80] Quero, G., et al., "Long Period Fiber Grating Nano-Optrode for Cancer Biomarker Detection," *Biosens. Bioelectron.*, Vol. 80, 2016, pp. 590–600.

[81] Luo, B., et al., "Dual-Peak Long Period Fiber Grating Coated with Graphene Oxide for Label-Free and Specific Assays of H5N1 Virus," *J. Biophotonics*, Vol. 14, No. 1, 2021.

[82] Janczuk-Richter, M., et al., "Immunosensor Based on Long-Period Fiber Gratings for Detection of Viruses Causing Gastroenteritis," *Sensors (Switzerland)*, Vol. 20, No. 3, 2020, pp. 1–11.

[83] Luo, B., et al., "A Novel Immunosensor Based on Excessively Tilted Fiber Grating Coated with Gold Nanospheres Improves the Detection Limit of Newcastle Disease Virus," *Biosens. Bioelectron.*, Vol. 100, 2018, pp. 169–175.

[84] Luo, B., et al., "Label-Free and Specific Detection of Soluble Programmed Death Ligand-1 Using a Localized Surface Plasmon Resonance Biosensor Based on Excessively Tilted Fiber Gratings," *Biomed. Opt. Express*, Vol. 10, No. 10, 2019, p. 5136.

[85] Luo, B., et al., "Human Heart Failure Biomarker Immunosensor Based on Excessively Tilted Fiber Gratings," *Biomed. Opt. Express*, Vol. 8, No. 1, 2017, p. 57.

[86] Ribaut, C., et al., "Cancer Biomarker Sensing Using Packaged Plasmonic Optical Fiber Gratings: Towards In Vivo Diagnosis," *Biosens. Bioelectron.*, Vol. 92, 2017, pp. 449–456.

[87] Ribaut, C., et al., "Small Biomolecule Immunosensing with Plasmonic Optical Fiber Grating Sensor," *Biosens. Bioelectron.*, Vol. 77, 2016, pp. 315–322.

[88] Leitao, C., et al., "Cortisol In-Fiber Ultrasensitive Plasmonic Immunosensing," *IEEE Sens. J.*, Vol. 21, No. 3, 2020, pp. 3028–3034.

[89] Wang, Q., J. Y. Jing, and B. T. Wang, "Highly Sensitive SPR Biosensor Based on Graphene Oxide and Staphylococcal Protein A Co-Modified TFBG for Human IgG Detection," *IEEE Transactions on Instrumentation and Measurement*, 2018, pp. 1–8.

[90] Loyez, M., et al., "Functionalized Gold Electroless-Plated Optical Fiber Gratings for Reliable Surface Biosensing," *Sensors Actuators B Chem.*, Vol. 280, February 1, 2019, pp. 54–61.

[91] Sun, D., Y. Ran, and G. Wang, "Label-Free Detection of Cancer Biomarkers Using an In-Line Taper Fiber-Optic Interferometer and a Fiber Bragg Grating," *Sensors (Switzerland)*, Vol. 17, No. 11, 2017.

[92] Wang, B. T., and Q. Wang, "An Interferometric Optical Fiber Biosensor with High Sensitivity for IgG/Anti-IgG Immunosensing," *Opt. Commun.*, Vol. 426, November 1, 2018, pp. 388–394.

[93] Liu, X., et al., "Photonic Crystal Fiber-Based Immunosensor for High-Performance Detection of Alpha Fetoprotein," *Biosens. Bioelectron.*, Vol. 91, May 15, 2017, pp. 431–435.

[94] Wang, Q., and B. Wang, "Sensitivity Enhanced SPR Immunosensor Based on Graphene Oxide and SPA Co-Modified Photonic Crystal Fiber," *Opt. Laser Technol.*, Vol. 107, 2018, pp. 210–215.

[95] Liu, T., et al., "A Label-Free Cardiac Biomarker Immunosensor Based on Phase-Shifted Microfiber Bragg Grating," *Biosens. Bioelectron.*, Vol. 100, 2018, pp. 155–160.

[96] Liang, L., et al., "Fiber Light-Coupled Optofluidic Waveguide (FLOW) Immunosensor for Highly Sensitive Detection of p53 Protein," *Anal. Chem.*, Vol. 90, No. 18, 2018, pp. 10851–10857.

[97] Bekmurzayeva, A., et al., "Label-Free Fiber-Optic Spherical Tip Biosensor to Enable Picomolar-Level Detection of CD44 Protein," *Sci. Rep.*, Vol. 11, No. 1, 2021, pp. 1–13.

CHAPTER 7

Nucleic Acid-Based Receptors (DNA and RNA)

Nucleic acids are one of the four major types of macromolecules essential for all known forms of life alongside lipids, proteins, and polysaccharides. Deoxyribonucleic acid (DNA) is a polymer composed of two polynucleotidic chains made of adenine (A), cytosine (C), thymine (T), and guanine (G) that carries genetic information (e.g., inside the nucleus of eukaryote cells as a double helix). However, ribonucleic acid (RNA) is also a polymer made of A, C, and G but with the uracile (U) replacing the thymine that is coding or regulating the expression of genes within cells and is usually present as a single strand folded onto itself rather than a double strand. In recent years, molecular engineering has strongly developed and has enabled the *in vitro* synthesis of such polymers using classical or variant nucleic acids as building blocks, providing a wide range of possible shapes and predictive interaction sites. Their use as receptors is not limited to the binding of complementary DNA or RNA strands but opens the way to multiple sensing approaches, such as for ion, protein, and cell detection using specifically designed nucleic acid receptors. Their use in these configurations initiates the birth of a new era of biosensors, especially for environmental sensing and long-term monitoring due to their intrinsic stability and versatility.

7.1 DNA Receptors

DNA strands have been extensively exploited for analytical assays as receptors due to their main biological activity: their ability to uniquely hybridize to their complementary sequence. These receptors can be synthesized and chemically modified under laboratory settings and are highly stable, so they are suited for regeneration and multiple use on the sensing surface. They can also be combined with polymerase chain reaction (PCR) to provide an amplification of the target DNA or RNA sequence and therefore improve the detection, going as far as detecting the presence of a few initial target strands in solution.

Whereas immunological probes such as antibodies rely on antibody/antigen interactions, nucleic acid probes are based on the hybridization process. Typically, short single strands of DNA (around 20 to 50 base pairs) complementary to a target sequence are synthesized and immobilized on the surface of an assay (microtiter plate) or a biosensor. Once the target DNA is detected by its receptor, different sensing techniques can be implemented. They can be useful to detect specific DNA sequences (e.g., the presence of bacteria, viral DNA) or mutations (e.g., p53 cancer mutations). This hybridization principle offers high accuracy and can be coupled with secondary techniques to validate the results [1, 2]. The integration of fluorescent labels, marked nucleotides, random or determined sequences, intercalating agents, or complementary fluorescent probes are some of the strategies usually achieved in that purpose. DNA engineering and synthesis offer unlimited possibilities for custom-made bioreceptors, and not only to target complementary DNA sequences. Over the last decade, their use has bolstered many sensing applications, especially through biomarker detection for medical diagnosis, food safety, and environmental monitoring. This is typically the case of DNA receptors called aptamers to target ions, proteins, or cells, which will be discussed in more detail in Section 7.3.

To summarize, antibodies, proteins, and peptides are made up of chains of amino acids while nucleic acids are made up of nitrogenous bases such as A, T/U, C, and G. However, nucleic acid analogs also exist such as peptide nucleic acid (PNA, uncharged) or derivatives such as locked nucleic acid (LNA, negatively charged) and phosphorothioate oligonucleotide (PTO, negatively charged). PNA is a well-known DNA analog introduced in the 1990s, which presents unique physicochemical and biochemical properties. These specialty receptors are highly resistant to nuclease and protease activities with high hybridization ability. This gives them a high resistance in media containing these types of enzymes. In practice, classical nucleic acid backbones are replaced with a synthetic peptide chain, commonly formed by repeated structures made of N-(2-amino-ethyl)-glycine and resulting in an achiral molecule called PNA. These PNAs are therefore exempt of pentoses and phosphate groups (commonly found in traditional nucleic acid chains) and are unsensitive to DNAses, which make them perfectly suited for detection in complex media containing enzymatic components. Moreover, the thermal resistance and hybridization stability of DNA/PNA duplex is higher in comparison to DNA/DNA associations due to its uncharged nature.

Despite their bunch of assets, small nucleic acid-based receptors require strong and oriented immobilization on their substrate, especially when it comes to plasmonics, which is extremely sensitive to surface variations. To do so, the immobilization and growth of DNA receptors can be performed following two main strategies. First, the bottom-up method when direct on-surface synthesis of DNA is achieved by standard phosphoramidite chemistry or even further inkjet/photolithography methods. Second, the top-down method when the receptors are first prepared and then deposited on the activated surface. This second method is the most common way to perform the immobilization of receptors because antibodies and nucleic acid-based probes are usually commercially available and ready to use. Then covalent and noncovalent bindings are still possible. The main inconvenience of the noncovalent deposition is the unpredictable orientation. Receptors can adopt random positioning upon binding on the substrate, in addition of possible release by hydrophobic effect, leading to gradual depletion of the adsorbed layer. Covalent chemistry is therefore more stable and enables improvements of the orientation of binding sites. To ensure an effective process, both the nature of the optical fiber surface and its related pretreatment need to be thoroughly studied. For plasmonic sensors, this surface is often made of a noble metal while its intrinsic properties play a major role (plasmonic sensitivity, homogeneity, and roughness of the surface). Depending on the nature and quality of that substrate, different bioconjugation techniques can be investigated and tuned to achieve the highest detection efficiency. Additional layers can also be immobilized in between the fiber surface and the bioreceptors to improve detection properties, especially if the metal surface is supplemented with a thin polymeric film, or other materials such as nanoparticles or hydrogels. All these possibilities and surface chemistries will be discussed in this chapter (Figure 7.1).

7.1.1 Binding DNA Receptors on Glass

Most optical fiber platforms are made of silica (SiO_2), an inert material resistant to important chemical or physical constraints, which keeps its integrity when it enters in contact with water or physiological buffer. Many substrates such as those manufactured at a large scale for DNA chips are therefore made of silica. To enable a strong adhesion of receptors on SiO_2, the surface first needs to be cleaned (by immersion in acidic solution such as piranha, made of H_2SO_4 and H_2O_2, 3:1), and this procedure is often followed by an activation by UV/O_3 [3].

The immobilization process is then performed through direct adsorption or silanization, followed by an activation through carbodiimide (NHS/EDC) or glutaraldehyde. When the glass substrate is planar, many techniques may also involve intermediate probing membranes (nitrocellulose, polystyrene, specific nylon materials) using ultraviolet light to avoid the need for long binding procedures onto the

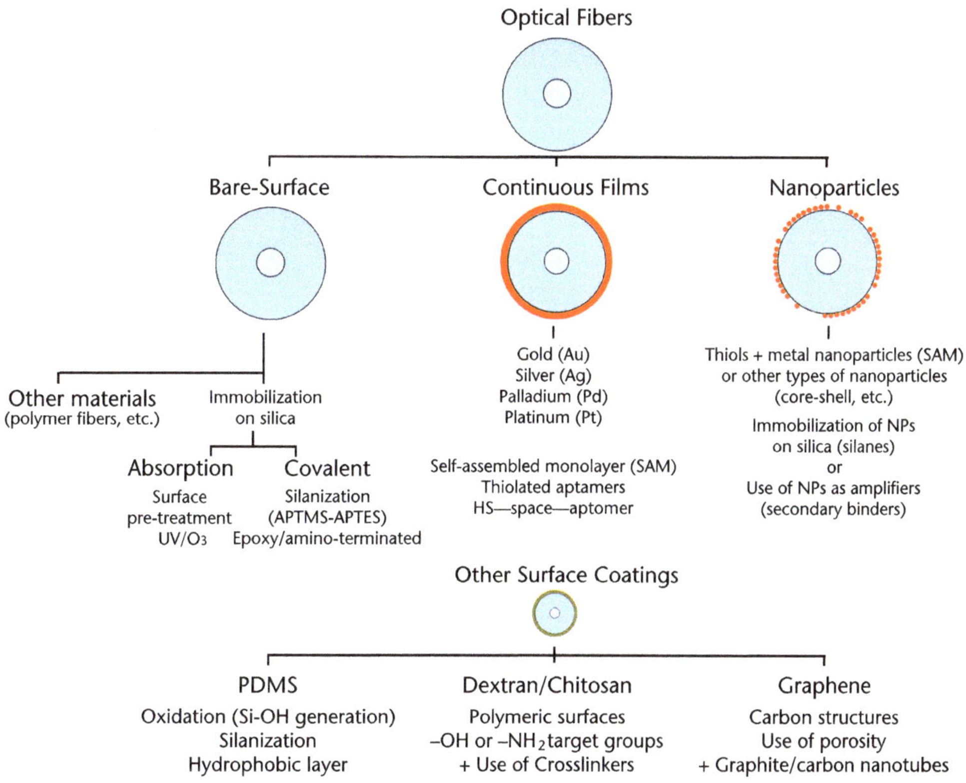

Figure 7.1 Diagram showing different optical fiber configurations for DNA immobilization. (*From:* [5]. (RightsLink License Number 5311201102204.))

transducer and therefore improve the process for large-scale production. These membranes are often avoided when it comes to cylindrical surface for practical reasons, so the development of such immobilization on optical fibers could be a great challenge in the following.

Another immobilization technique to be mentioned is a chemical modification of the 5′ or 3′ end of the DNA strand under controlled exposure through photolithography, usually chosen to produce high-end DNA arrays with a covering range of 10^6 receptors or more, per cm^2 [4]. The use of hydrogels on silica to locally increase the concentration of receptors may also be of interest in this particular case and easily adapted on optical fibers by probe-dipping.

7.1.2 Binding DNA Receptors on Plastic

Most disposable laboratory-based bioassays are made of plastic. Polymers such as polystyrene, polypropylene, or polycarbonate with some specific additive mixtures are the more common compositions of microtiter plates, such as those used for enzyme-linked immunosorbent assays (ELISA). It makes these substrates resistant to many constraints and perfectly compatible with bioassays. Direct adsorption of DNA can be exploited with such surfaces. Usually, the adsorption is conducted by electrostatic interactions between the negatively charged phosphate groups of DNA and the positively charged surface. Studies have also shown that chitosan (with free amine $-NH_3$ groups) or polymeric films provide good biocompatibility and high immobilization yields on these surfaces. This attraction can be amplified using a positive electrochemical potential (e.g., 0.2V to 0.5V for 5 minutes) to enhance and stabilize the phenomenon. The risk of this adsorption is again the possible desorption from the surface in case of nonadapted buffers (pH change, ionic strength, or temperature increase). Chemically modified surfaces for covalent binding of DNA are also commercially available. This chemistry usually involves amines ($-NH_2$) on 5′ or 3′ ends to provide a vertical orientation of the strands and, consequently, a higher binding strength.

7.1.3 Binding DNA Receptors on Metals

Covalent immobilization or chemisorption of DNA on metals make use of 5′ or 3′ amino (NH_2) or thiol (SH) modified sequences. The direct bonding of DNA through thiol chemistry allows the formation of a self-assembled monolayer (SAM) due to the high affinity between thiols (S) and gold (Au). The surface can also be passivated to prevent any nonspecific adsorption using truncated sequences or mercaptohexanol molecules and derivatives (e.g., 1 mM for 20 minutes). This direct binding can be further exploited on metal nanoparticles, which can, in turn, bind to secondary surfaces and therefore be used as an indirect binding method, improving the orientation of the receptors. Amino-modified DNA receptors can also be covalently linked to many reactive groups immobilized on metal surfaces such as carboxyl, aldehyde, sulfonic, or isothiocyanate groups by the help of the well-known carbodiimide (NHS-EDC) reaction. However, many commercial approaches do not immobilize bioreceptors by direct chemistry on the metal layer but make use of anchoring probes. For example, immobilizing an anchoring DNA sequence by thiol chemistry on a metal layer will be the first step to bind the complementary part of the bioreceptor sequence. In this case, the bioreceptor is not directly immobilized on the metal surface but is attached due to the anchoring DNA/RNA probe (also named capture-probe), which plays the role of a linker between the surface of the sensor and the active biological layer. This technique yields better detection dynamics and enhances the possibility to reuse the surface, in particular for conventional plasmonic equipment [6, 7].

7.1.4 Binding DNA Receptors on Polymeric Materials

This section covers both the immobilization of nucleic acids on polymer coatings deployed on silica fibers and on polymer optical fibers. As a part of biochemical strategies, polymer layers offer lots of possibilities and can lead to singular properties, especially on glass or metal-coated optical fibers. Polymer layers such as sol-gels or graphene (arranged carbon atoms in 2-D honeycomb lattice) are often chosen for their high stability or to create novel detection features. Matrices such as PDMS, dextran, or chitosan are also well-known to implement bioreceptors on many types of biosensing surfaces (Figure 7.2). The sol-gel process can be defined as a production of a solid material from a solution of small molecules. It is especially used for the fabrication of metal oxides such as for silicon or titanium oxides. This process can be represented by the following equation, where R is an alkyl group:

$$Si(OR)_4 + (4 - x)H_2O \rightarrow SiO_x(OH)_{4-2x} + 4\ ROH$$

Sol-gel chemistry takes place through hydrolysis and condensation of monomeric alkoxysilanes such as tetraethoxysilanes (TEOS) or tetramethoxysilanes (TMOS). The initial solution (sol) is made of water and cosolvents and the aforementioned reactions producing a gradual viscosity and rigidity (gel) can be accelerated by a catalyst. Most sol-gel based optical sensors are related with pH sensing, as many pH-sensitive dyes can diffuse though the porous sol-gel glass. Applications such as gas, solvent, or metal ions detection have also been explored. However, one of the main added values of sol-gels is their ability to encapsulate biomolecules such as aptamers and therefore lead to advanced biosensors. While conventional sol-gels do not enable that encapsulation, modified sol-gel processes in appropriate buffers have allowed the immobilization of many enzymes and proteins (acetylcholinesterase, concanavalin-A, lactate dehydrogenase, urease). Sol-gel hybrids are mainly studied to combine glasses with other polymeric matrices and can therefore exhibit flexible and highly stable properties [8, 9].

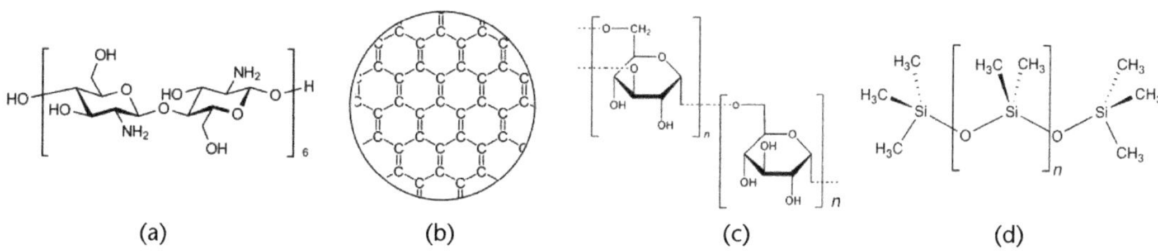

Figure 7.2 Molecular structures of different polymeric materials: (a) chitosan (CS), (b) graphene, (c) dextran, and (d) PDMS. (*From:* [5]. (RightsLink License Number 5311201102204.))

Polydimethylsiloxane (PDMS) is chemically inert and optically transparent at visible and near-infrared wavelength ranges, which makes its use compatible with optical sensing. It is cost-effective and easy to handle. PDMS is also investigated as an optical material itself due to its physical properties. Some studies are indeed developing unconventional optical fiber structures such as multimode, twisted, and looped fibers through this material [10]. However, when PDMS is used on top of a substrate as a sensing layer, it is known to present a high adsorption of non-specific agents. PDMS is therefore often chemically modified. PDMS first needs to be oxidized to generate silanols (Si-OH) and then allows a silanization strategy to further immobilize bioreceptors such as aptamers. As mentioned for the silanization process, PDMS deposition on optical fibers requires high precision to avoid granular and inhomogeneous layers. It is often performed by dip-coating strategies to avoid the apparition of droplets drying on the surface on one side of the optical fiber surface. Its hydrophobic nature may also cause issues when inserted in aqueous buffers, especially for its wettability [11]. As PDMS is also used to manufacture microfluidic channels, it could be interesting to investigate its use as a fluid carrier combined with its bio-functionalization properties (e.g., to trap proteins in a molecular flow or to optimize surface passivation by capturing some analytes in solution), preventing their interaction with the fiber surface [12].

Dextran is part of the carbohydrates family and is often presented as a polymeric gel. It is therefore considered, as for PDMS, as 3-D structures added on biosensor surfaces. Dextran matrices need to be activated by adding functional groups to enable covalent binding of bioreceptors, but the presence of hydroxyl groups increases non-specific adsorption. The activation of dextran matrices with aspartic acid (leading to the formation of aldehyde groups) can help to reduce these nonspecific interactions and adsorption. Dextran can then easily be attached to an aminated surface or self-contain amine residues (e.g., using periodate or enzymatic oxidation) to further react with aptamers. Considering its importance as amplifying the number of molecules around sensor surfaces in its 3-D configuration, this polysaccharide can be adapted to optical fibers and form hydrogels to reduce biofouling, especially for long-term measurements. Dextran plays a critical role in numerous applications including biomedical sensors [13, 14].

Chitosan (CS) is a well-known polysaccharide that is biocompatible, biodegradable, and perfectly suited for receptors immobilization, especially nucleic acids. Polymeric CS presents functional groups such as hydroxyl (–OH) or amino ($-NH_2$) that are crucial anchoring points for biochemical interactions. Chitosan is also known to have an important affinity for heavy metals such as Pb. It is a natural product derived from chitin (from exoskeletons of insects) and its structure is similar to cellulose. CS is produced by deacetylation of chitin through hydrolysis under alkaline conditions. CS is not often used as a standalone material because it provides few physicochemical properties by itself, but it is usually combined with other components by chemical and mechanical modification to allow the formation of a panel of different nanostructures. Blends with different polymers exist and provide singular properties with additional functional groups. Chitosan is water-soluble and can easily form gel or membranes. Many techniques can be adopted for bioreceptors immobilization on chitosan.

For instance, glutaraldehyde can be used as a crosslinker to improve chemical links between CS surface and bioreceptors. Polyaniline (PANI) can be grafted on $-OH$ or $-NH_2$ groups by ammonium persulfate in acidic solution and DNA can be detected through that process. Methylation, thiolation, N-succinylation, or copolymerization offers plenty of immobilization strategies [15]. Metal nanoparticles can be additionally implemented as intermediates with CS to anchor receptors and therefore improve the selectivity of the polymer layer. The use of such nanoparticles can improve the detection potential both by a higher RI change or by enhanced conductivity [16].

Many carbonaceous surfaces are used as solid supports for nucleic acids bonding where adsorption is mostly exploited. In those carbon structures such as graphite or carbon nanotubes usually generated through vapor deposition, the adsorption of bioreceptors is possible due to the porosity of the layer. Roughness, porosity, and permeability of the surfaces are therefore important parameters with which to deal. In most applications, porosity and thickness drive physical interactions, while the chemical structure drives polar or nonpolar interactions with target receptors. Graphene is a specific form of carbon material as it is a planar monolayer of carbon atoms, in a 2-D shape. It is highly stable, with important mechanical strength and optical properties. When graphene is immobilized on optical fibers, it provokes a unique atomic link between the fiber surface and the outer medium. It can be modified with polymers, metals, or other substances, which also provide new opportunities in terms of detection principles. Its study on optical fiber structures is still in its early stages despite a large number of publications on this topic. Graphene is considered as a plasmonic material itself such as gold of silver, but its deposition is much more complex, especially on cylindrical surfaces. It is also optically more lossy, which limits its use in practice. Immobilization of bioreceptors on graphene films also mainly relies on adsorption, as its hexagonal ring structure mainly allows adsorption of ions or molecules with high impermeability. Functional groups such as hydroxyl (–OH), carboxyl (COOH), and epoxy (–O) can be added to graphene structures to enable further covalent bonding as well. A large possibility of combinations with polymers and chemicals exists and is well described for specific chemistry [17–20].

While silica optical fibers are considered a gold standard, polymer optical fibers are emerging as an alternative to provide high sensing capabilities with improved biocompatibility of medical applications. Their actual low practical use is principally due to their high loss and the lack of commercial tools and instruments adapted to single-mode polymer fibers. Most polymer optical fibers are based on PMMA, but as the losses are high, the length of the sensors is limited to few centimeters at NIR and up to tens of centimeters at 500 to 800 nm. Alternatives exist, such as the use of cyclic transparent optical polymers (CYTOP, Asahi glass Co., Ltd.) where perfluorination improves their transmission properties. Structures such as fiber Bragg gratings have already been investigated in such CYTOP materials, especially for mechanical sensing. Optical fibers made of cyclic olefin copolymers (TOPAS and ZEONEX) can also be mentioned [21, 22]. All these polymer optical fibers are involved in expanding fields of research due to the numerous advantages brought by these various materials. Expectations for these new types of sensors are high [23, 24].

Polymer-based substrates such as PMMA are widely used, including for optical fibers sensing. The immobilization of DNA probes on this material has been studied for many reasons, especially to anchor receptors on chips and manufacture microarrays. The modification of PMMA surface to activate binding to aptamers can be done through different ways. The simpler method is the activation through hexamethylene-diamine to obtain animated PMMA with primary amino groups [25, 26]. Using crosslinkers such as glutaraldehyde, it is possible to bind DNA receptors onto PMMA. Thiolated DNA can also be attached onto animated PMMA surfaces reacting with NHS-ester groups such as the ones from N-maleimidocaproyl-oxy-sulfosuccinimide ester (sulfo-EMCS) to achieve a covalent immobilization. This method can be interesting if the same aptamers have to be immobilized on different substrates, for instance, both on metal layers and on other surfaces where a direct immobilization through the thiol end is not possible. The same receptors can therefore be used on different substrates following the needs for covalent immobilization. It has also been reported that ultraviolet exposure can help for direct binding oligonucleotides on many polymeric surfaces, including unmodified PMMA [26, 27].

Perfluorinated polymers such as CYTOP materials show higher transparency and lower losses than PMMA. CYTOP can be used to manufacture specific optical fibers but can also be present in microchips, as its absorption edge is lower than 190 nm, which is perfectly suited for DNA electrophoresis analysis. It is also more flexible and less fragile [11, 28]. Recent advances in optical fiber manufacturing have also opened a new era of research with bioresorbable optical fibers [29, 30]. While most attention was on polymer-type fibers for biocompatibility assets, resorbable materials become an inexorable trend all over the biosensing field [31, 32]. Using these materials, the implementation of fiber-optic sensors could therefore be deliberately degraded after a certain time of use, and this makes perfect sense if used *in vivo* or in the environment.

Bio-immobilization on metals or polymeric materials is important for many applications, so it can be inspired from the production of microtiter plates or ELISA wells. Among all the industrial strategies taking place to enhance molecular binding on substrates, the use of plasma treatment is increasingly investigated. Plasma treatment can be adapted through a large variety of gases such as argon, nitrogen, and carbon dioxide, which allows the introduction of specific groups for covalent immobilization of bioreceptors. The side effect of the plasma technique is the possible modification of surface morphology that can affect the optical properties of the sensor itself. The process should therefore be finely calibrated to obtain a fair balance between the molecular immobilization and the preservation of the surface and optical integrity, but is suited for many types of substrates [33, 34].

7.1.5 DNA Spot Arrays and Microstructures

Biochemical nanoprinting, dip-pen nanolithography, and DNA arrays are essential techniques to provide multiplexed detection on a same master chip. The duplication process of these spot chips is fast and mandatory for high-throughput genetic analyses. Both bottom-up or top-down chemistries can be used to build these arrays, but recent advances and improvements in nanoprinting technologies have boosted quality and quantity of the immobilized receptors.

By adjusting certain key parameters, it is possible to produce micrometric or even nanometric hotspots. For example, the viscosity of the immobilized solution can be tuned using glycerol or surfactants, and the surface wettability can be modified through UV/O_3 treatment. The incubation time, chamber humidity, and temperature are also playing an important role on the oligonucleotides' transfer and immobilization. All these parameters need to be finely studied before each application to ensure a good biosensor functionalization [35, 36].

7.1.6 Design and Synthesis of Specific DNA 2-D/3-D Structures

Among all nucleic receptors available, DNA nanotechnology relies on the information encoded within molecular strands to guide their own self-assembly and the hierarchical information to keep ultrahigh folding precision. The formation of DNA or RNA super-structures including 3-D or 2-D shapes, nanotubes, polyhedrons, nanocages, hydrogels, and crystals is a promising source for new bioreceptor edifices. A branch of this DNA nanotechnology called DNA origami is now very well handled to create 3-D objects at the nanoscale. Computer-assisted designing tools and automated fabrication equipment make this DNA-related fabrication an easy-to-play technology. Dynamic structures such as conformationally switchable domains or reconfigurable modules can be engineered for many applications, from target-responsive biosensing, bio-imaging, smart drug delivery, and many others. These structures are sometimes considered as nanorobots [37]. The assembly of such structures is determined by their size and related complexity. Most of the time, protocols used for the generation of DNA origami rely on pH 8 buffers containing approximately 12 mM Mg^{2+}, and up to 20 mM Mg^{2+} for the generation of 3-D structures. These origami are then folded by self-assembly. A key point is certainly the reduction of the aggregation to prevent precipitation or misfolding of the whole system.

Recent advances in DNA origami production led to their association with optical fibers to achieve specific biosensing. In 2018, the detection of tobacco mosaic virus (TMV) through chemiluminescence was performed on an optical fiber tip using a genosensor. The 3-D nanorobots made of DNA were immobilized using capture and anchor sequences. DNA origami were closed and opened once interacting with the target molecule. This conformational change provokes reaction yielding chemiluminescent signal, which was detected through a photomultiplier tube (PMT) and a detector/sensor module. The same year, DNA origami were tested on the surface of plasmonic optical fibers for thrombin detection. Their use was investigated to improve the orientation and the positioning of the receptors on the fiber surface [38, 39].

Origami-inspired folding of molecules or thin films provide unprecedented capabilities to design new biosystems. While enabling the creation of tunable 3-D systems at the nanoscale, the implementation of such complex and stimuli-responsive materials capable of molecular delivery is a major scientific advance, which will open the door to great progress in many areas. Their biocompatibility and biodegradability are probably their most critical assets, while their tunability to magnetic guidance and manipulation make them suited for *in vivo* exploration. Origami biorobotics coupling both the chemistry and integration of nanochips are sound to analyze and pilot single cells. Immobilizing such nanotech jewels on the surface of biosensors could make them much more efficient and reversible, with improved robustness against environmental disturbances.

7.2 RNA/miRNA Receptors

In the early days of molecular biology, it was thought that DNA was carrying all the genetic information (introns and exons) as a succession of genes expressed when necessary and that all the noncoding parts were useless sequences called junk DNA. Over the years, it was clear that many regulatory processes were involving these parts of DNA and that epigenetics including chemical modifications of DNA and histones modifying the way DNA is folded, was an essential key. Post-transcriptional control (pre-mRNA processing, mRNA actions and localization, translation, and stability) are also part of these complex regulatory stages.

In nature, oligonucleotides mainly exist as small pieces of noncoding RNA such as microRNA (miRNA) constituted of a maximum of approximately 25 nucleotides. Whereas the transcription of genes (from DNA to RNA) is mandatory for their expression, this process leads to the production of mRNA or noncoding RNA, depending on their nucleic acid contents and sequences. Promoters and enhancers enable the recruitment of RNA polymerases to begin the transcription, during which DNA is locally opened as single-strand and copied into one or more RNA transcripts. Then the RNA transcripts are processed and precursors RNA (pre-mRNA) are covalently modified on their 5′ end, essentially through the first base modifications called the trimethyl m^7G cap. This modified cap facilitates the nuclear export of the sequences and improve the mRNA translation and stability. The pre-mRNA is also cleaved at its 3′ end to be polyadenylated (addition of a tail made of A bases). Several molecular cascades are then activated, enabling alternative splicing, while exons are joined together until obtaining transcripts with coding potential. The matured mRNAs need to be delivered and translated in the cytoplasm through different steps: initiation, elongation, and termination. At the end of this process, which is highly regulated, proteins are produced and their expression is the direct result of a delicate balance between thousands of molecular cascades, happening in only few seconds to few minutes (Figure 7.3).

In parallel to this, noncoding RNA sequences (ncRNA) are not translated into proteins and take their origin from the transcription of genes called the RNA gene. These RNA types include transfer RNA (tRNA, a necessary component for the translation of mRNA to proteins), ribosomal RNA (rRNA, essential components to assemble the ribosome subunits), and a series of small RNAs. The latter is made of siRNA (short interfering RNA degrading mRNAs after transcription and aborting its translation), piRNA (piwi-interacting RNA, the largest class of small noncoding RNA mainly involved in epigenetic and post-transcriptional RNA silencing), snoRNA (small nucleolar RNA, guiding the chemical modifications of other RNA sequences), snRNA (small nuclear RNA, essentially working on the processing of pre-mRNA to RNA in the nucleus of eukaryote cells, but also help for telomere maintenance), and exRNA (extracellular RNA, carried within extracellular vesicles and lipoprotein complexes). These exRNA may play a role in intercellular

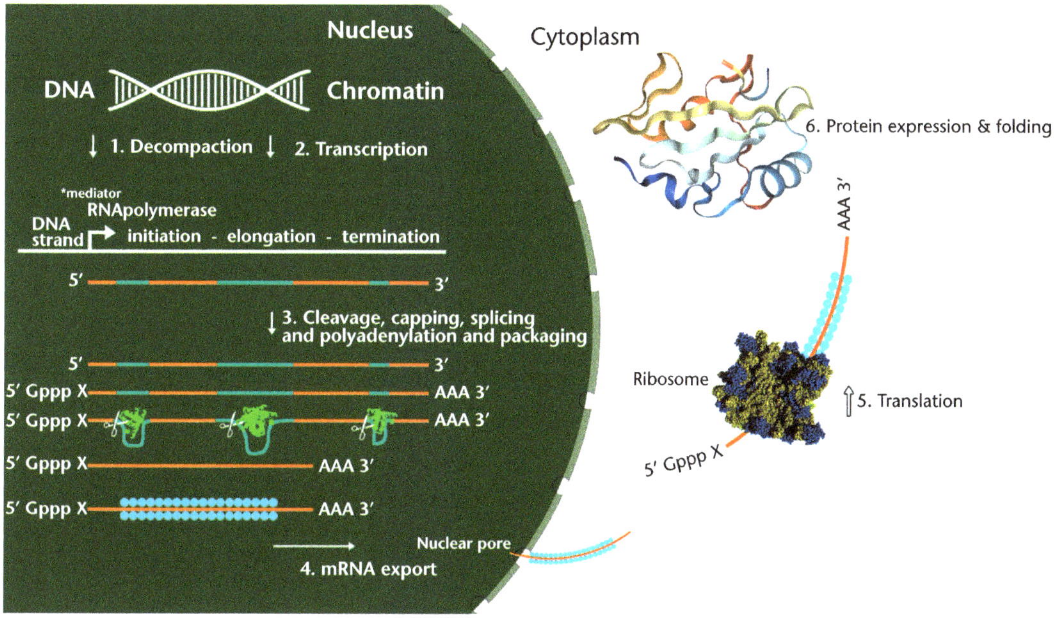

Figure 7.3 Summary of the transcription and translation process within an eukaryote cell.

communication and regulation but they are not well understood yet; and finally, scaRNA (small Cajal body RNA; located in specific nuclear organelles).

This being explained, we can summarize the natural RNA production by dividing it in two parts: coding mRNA are translated into proteins, whereas a lot of noncoding RNA types are regulating the translational process. All synthetic sequences produced *in vitro* by molecular biochemistry without purposes of protein production are therefore part of the noncoding family. In terms of biodetection, ingenuity comes from the adaptation of certain natural molecular behaviors in biomolecular engineering processes. The motivation behind this diversified oligonucleotides production and turnover was bolstered by a great number of biotech companies and an immense demand for bioassays production, therapeutics, and diagnostics. RNA-based probes and their analogs are therefore essential for many assays. They are considered templates that can be modified and adapted for very specific applications, depending on their scaffold and lifetime (Figure 7.4). RNA is less stable than DNA, but presents more diverse and intricate 3-D structures. The lifetime of RNA aptamers in plasma is often limited to few seconds to minutes, while DNA aptamers can persist for more than an hour. Their shorter lifetime can be compensated by chemical modifications or complexation with other molecules improving their stability. It can also be an asset for very fast measurements under specific conditions, especially when fast degradation of the sequences are required.

7.3 Aptamers

Aptamers are described as alternative immunoreceptors and bring novel detection perspectives compared to traditional antibodies [40, 41]. They are short, single-stranded DNA, RNA, or synthetic XNA molecules that are selected or designed to

	DNA	RNA / miRNA
Discovery	Starting library of ssDNA oligos	Starting library of dsDNA oligos Need for [retro]transcription
Structure	Larger structures	More diverse and versatile 3D shapes Smaller structures
Stability	Higher	Lower
	(can be improved by chemical modifications)	
Insights		Optimization and understanding of structural homogeneity
	Development as therapeutics, drug delivery systems, nanorobots, diagnostics	

Figure 7.4 Assets of DNA and RNA/miRNA-based bioreceptors.

target molecules of interest with high affinity and specificity [42]. These synthetic molecules are therefore made of nucleotides (or even peptides in some specific cases [43]) and are selected against a specific target through rounds of a method called SELEX (Systematic Evolution of Ligands by Exponential) enrichment. They have been studied since the 1990s and show practical assets as they present small sizes, high stability, and ease of production without the need for animal preproduction [44, 45]. While current aptamers are found from long trial-and-error experiments, the next generation of aptamers will surely rely on data learning and the possibility of generating computer-assisted sequences in silico by using the atomic properties of the strands to attach with a strong affinity to a definite binding site on the target. Many studies carried out recently also demonstrate the possibility to chemically improve the efficiency of existing aptamer-sequences and therefore avoid the need for further SELEX rounds of the candidate's isolation (Figure 7.5) [46, 47].

A myriad of variants have been developed so far, leading to changes in the structure and composition of these aptamers. Affimers consist of a subtype of peptide aptamer (12 to 14 kDa) derived from the cysteine protease inhibitor family of cystatins [48]. Aptamers can also be associated with ribozymes (small RNA molecules able to catalyze specific biochemical reactions) to self-cleave in the presence of targets. The latter can be useful to prepare riboswitches or to regulate the production of proteins while binding to mRNA in cells. Other discoveries such as X-aptamers can also be mentioned. The latter are made of a combination between both natural and chemically modified DNA (or RNA) sequences in order to improve the binding versatility of conventional aptamers [49, 50].

In a more general context, labeled aptamers are often used in solution to specifically bind to a target of interest. While immobilized as receptors, aptamer-based structures improve the selection and detection of biomarkers, and can translate binding events into signals (e.g., due to embedded quenchers and fluorophores or refractive index changes resulting from binding events or conformational modifications). These very few examples highlight the richness and diversity of these aptameric complexes [51, 52]. Aptamers have been highlighted for different means of sensing. Among the technologies considered, optical fiber-based aptasensors take place as an important part of the literature. They have been tested against proteins, cells, DNA, and metal ions' detection at ultralow concentration. The combination

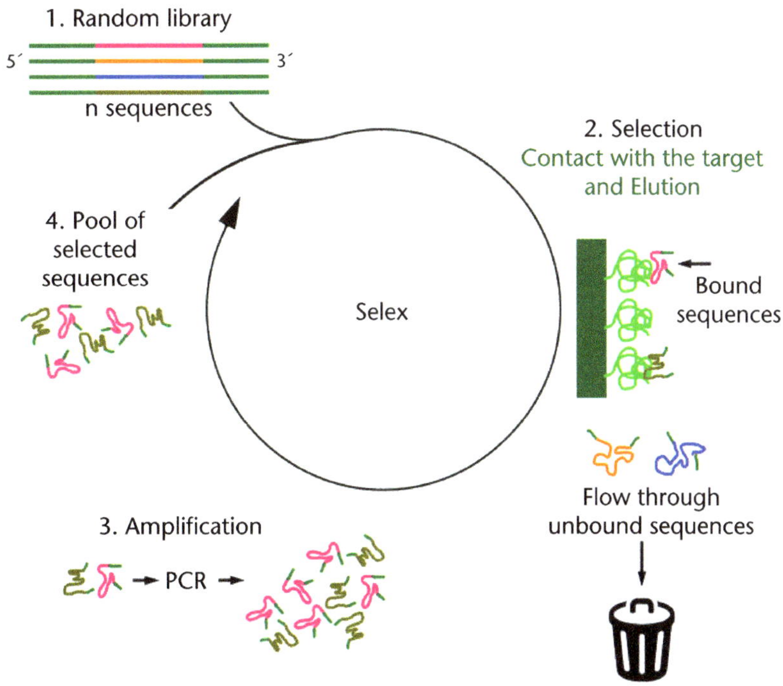

Figure 7.5 Diagram showing the SELEX method for aptamer selection.

of optical fibers with aptamers generates new challenges and prompts breakthrough sensing modalities for years to come [53, 54]. Their immobilization is often performed through thiol chemistry or using biotinylated sequences. Their binding on plasmonic optical fiber sensors can be tracked in real time and performed in traditional buffered conditions, which makes them particularly attractive.

A new branch of optimized aptamers called Optimers are also commercially available and tested against a wide range of targets, with the aims of improving their discovery and development and increasing tissue penetration for *in vivo* research.

7.4 Applications of Nucleic Acid-Based Biosensors

7.4.1 Hybridization/Complementary Strand Detection

The very first works on pioneering DNA detection using optical fibers were published in the early 1990s and were mainly focusing on fluorescent oligonucleotides [55, 56]. The use of plasmonics coupled with optical fiber biodetection has brought a plethora of approaches [57, 58]. Pollet et al., from the group led by Professor Jeroen Lammertyn, published in 2011 a work about DNA detection using an SPR optical fiber. This work was based on an unclad fiber of 400-μm core diameter, coated with 50 nm of gold. It showed the possibility of targeting DNA with a sensitivity of 140 pm/nM, in the range of 0 to 500 nM. Single-strand aptameric receptors were also exploited to target human immunoglobulin E (hIgE), and these initial contributions were the beginning of a long series, opening the doors to monitoring DNA amplification through PCR [59]. In terms of applications, it is possible to use these types of sensors to detect specific DNA or RNA sequences, such as some allergens or biomarkers [60]. Other works have followed in these footsteps and have demonstrated impressive achievements, especially using labeled DNA sequences, or, more recently, through specific enzymes cleaving DNA anchors, resulting in the release of gold nanoparticles away from the fiber surface [61]. Playing with the oligos' scaffold to bring nanoparticles closer or far away from the sensor surface and tune the plasmonic response depending on their binding are probably the most promising approach in this direction.

7.4.2 Protein, Toxin, and Organic Compound Detection

Protein detection using aptamers immobilized on a plasmonic optical fiber was profusely documented over the past few years. For instance, recent achievements have shown the possibility to bring D-shaped fiber aptasensors into the race for Covid-19 detection [62]. The detection of cancer biomarkers such as the human epidermal growth factor 2 (HER2) using unclad fiber aptasensors was also investigated. Gold-coated probes of 400-μm core diameter were covered with aptamers through thiol chemistry to detect HER2 in label-free conditions at different concentrations (Figure 7.6(a, b)), while the use of an amplificatory anti-HER2 antibody enabled the detection of lower concentrations, under the configuration of a sandwich assay [63]. Biosensors for the detection of thrombin [64–66], *Plasmodium falciparum* glutamate dehydrogenase [67], toxins, and organic compounds such as bisphenol A at ultralow concentrations can also be mentioned [68–70]. The performance reported for these types of biosensors bring them closer to the intended use, while many attractive works in signal analysis, automation, and encapsulation have been emerging [71]. More examples are shown in Table 7.1 [72–90], with associated binding strategy and surface sensitivity.

7.4.3 Cell Detection

Cellular detection with plasmonics could be considered at a first glance as a simple achievement because the imposing sizes of cellular structures (several micrometers to tens of micrometers) may lead to significant changes in surface refractive index. However, their large size is often a drawback in the context of a technique as fine as plasmonics, because it does not allow refined molecular screening as when the targets are freely circulating in buffer or the experimental biofluid [91, 92]. It is also known that the shift provoked by the target interaction is proportional to its size and surface density, so it may cause an important change in reflectivity at a certain incident light angle, and this setting is highly relevant for bulk plasmonics. Playing with these parameters may tune the plasmonic effect and bring new insights such as precise surface localization and imaging through SPR, bringing the perspective to detecting single-cell binding events (or even more located, cytoplasmic events) [93, 94]. When target molecules are trapped within a lipidic membrane or cell wall, nonspecific detections are numerous due to the interactions with other areas of the cellular surface. Moreover, the plasmon excitation field is often exceeded (the penetration depth of the plasmon excitation becomes smaller than the thickness of the biolayer) by the important thickness of the formed biolayer, which is not

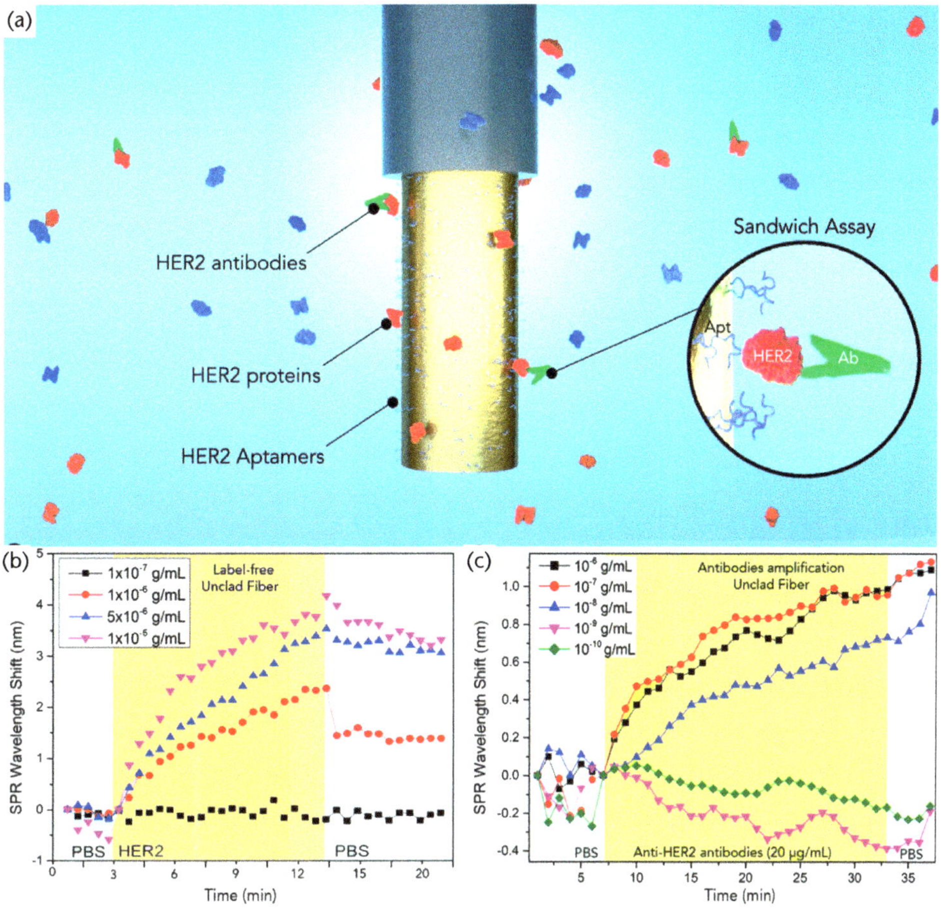

Figure 7.6 (a) Artistic representation of a gold-coated unclad optical fiber covered with thiolated aptamers for the detection of HER2 proteins. Anti-HER2 antibodies were then added to achieve a sandwich assay. (b) Evolution of the plasmonic response in growing HER2 concentrations. (c) Evolution of the plasmonic response after amplification using secondary antibodies as enhancers (sandwich assay). (RightsLink License Number 5311321318130.)

necessarily to the advantage of the technique compared to the detection of small molecular events, such as for single proteins or peptides. The latter are indeed localized in a range of a few tens of nanometers from the metal surface. Nevertheless, many studies have demonstrated that cell detection is possible and highly relevant with plasmonics, especially by taking care of the surface blocking and by refining the experimental conditions, for example, using a microfluidic system and by thoroughly rinsing the biosensor surface to sufficiently detach nonspecific cell interactions, often yielding false positive shifts with those sensors. In the example reported next, gold-coated TFBGs functionalized with anti-mammaglobin aptamers were tested against cancer cells (positive) and control cells (negative). Only the positive cells were expressing the target mammaglobin proteins on their outer membrane, allowing the interactions with the optical fiber surface (Figure 7.7). The plasmonic monitoring was assessed during both the functionalization (aptamers deposition by thiol chemistry and surface passivation using mercaptohexanol) and the detection of target cancer cells [78].

Table 7.1 Overview of Reported Optical Fiber Nucleic Acid-Based Receptors

Optical Fiber Configuration and Detection Method	*Receptor/Target*	*Bulk Refractive Index Sensitivity (nm/RIU)*	*LOD for Aptasensing*	*Reference*
600-μm core optical fiber with aptamer-modified gold nanorods (LSPR)	Aptamer/ochratoxin A (403.8 g/mol) in grape juice/10% methanol	601.05 nm/RIU (glycerol solutions between 1.333 and 1.442)	12 pM	[72]
600-μm core optical fiber with probe DNA immobilized on silica (fluorescence)	Fluorescent aptamer/Bisphenol A (~228.3 g/mol) in (waste)water	NA	1.86 nM	[73]
600-μm core optical fiber with aptamers immobilized on silica (luminescence)	Aptamer/Hg++ (~200.5 g/mol) in pure and tap water	NA	0.47 pM	[51]
600-μm core optical fiber with aptamers immobilized on silica (fluorescence)	Aptamer/melamine (126.12 g/mol) in PBS	NA	0.15 μM in buffer and 3.0 μM in milk	[74]
400-μm core optical fiber with aptamers immobilized on gold film (SPR) with antibody/AuNP amplification	Aptamer/Ara h 1 protein (~70.8 kDa) and Antibody AuNP amplification in PBS or TGK buffer with food matrices	NA	75 nM	[75]
400-μm core optical fiber with aptamers immobilized on gold film (SPR) label-free detection	Aptamer/hIgE (~190 kDa) in MES or TRIS buffer	~1,500 nm/RIU (sucrose concentrations between 1.333 and 1.354 and water-alcohol concentrations)	2 nM	[76]
400-μm core optical fiber with aptamers immobilized on gold film (SPR) with antibodies amplification	Aptamer/HER2 (115 kDa)/AuNPs in PBS	1,573.9 nm/RIU (LiCl dilutions between 1.337 and 1.363)	5.16 nM (label-free) 77.4 pM (AuNPs amp)	[63]
MMF with aptamers immobilized through NHS/EDC chemistry	Aptamer/ochratoxin A	NA	~0.9 nM	[70]
TFBGs with aptamers immobilized on gold film (SPR)	Aptamer/thrombin (72 kDa) in protein buffer	NA	22.6 nM	[77]
TFBGs with aptamers immobilized on gold film (SPR) and AuNPs amplification	Aptamers/cells (CTCs)/AuNPs in PBS	NA	49 cells (label-free) 10 cells (AuNPs amp)	[78]
TFBGs (MMF and SMF fibers) with aptamers immobilized on gold film (SPR)	Aptamer/HER2 (115 kDa) in PBS	102.03 nm/RIU (SMF) and 124.89 nm/RIU (MMF) (LiCl dilutions)	~8.6 fM	[79]
Etched FBG (from 125 μm to 13 μm) with aptamers immobilized on silica (SRI shift)	Aptamer/thrombin (72 kDa) in PBS	23.38 nm/RIU (sucrose dilutions between 1.342 and 1.442)	75–110 pM	[80]
Etched FBG with aptamers immobilized on silica (SRI shift)	Aptamer/Thrombin (72 kDa) in Tris buffer	17.4 nm/RIU (sucrose dilutions between 1.6 and 25%)	~10 nM	[64]

Table 7.1 *(Continued)*

TFBGs with aptamers immobilized on gold film (SPR) with AuNPs amplification	Aptamer/thrombin (72 kDa)/AuNPs in protein buffer	NA	1 nM	[81]
TFBGs with aptamers adsorbed onto graphene Au/Graphene layer (SPR)	Aptamer/dopamine (153.2 g/mol) in serum	NA	160 fM	[82]
LPFG with aptamers immobilized on TiO_2 layer through poly-L-lysine (SPR)	Aptamer/Thrombin (72 kDa) in PBS with 1% serum	3,100 nm/RIU (ethylene glycol concentrations in water between 1.335 and 1.355)	~10 nM	[65]
LPFG with aptamers immobilized on silica (SRI shift)	Aptamer/cocaine (303.3 g/mol) in PBS	~6,000 nm/RIU	~25 μM	[83]
Tapered optical fiber with aptamers immobilized on silica through poly-L-lysine (SRI shift)	Aptamer/dopamine (~153.2 g/mol) in PBS	NA	37 nM	[84]
Tapered optical fiber with aptamers immobilized on silica (fluorescence)	Aptamer sandwich/streptomycin (582.6 g/mol) in PBS and water samples	NA	33 nM	[85]
D-shaped plastic optical fiber with aptamers immobilized through PEG SAM on gold surface with streptavidin biotin affinity (SRI shift)	Aptamer/thrombin (72 kDa) in PBS	NA	1 nM	[86]
Dual-resonance LPFG (DR-LPFG) with aptamers immobilized on gold film (SPR)	Aptamer/microsystin-LR (toxin) (995.2 g/mol) in buffer	3,891.5 nm/RIU	~5 nM	[87]
D-shaped optical fiber with aptamers immobilized by thiol chemistry on gold nano antennae array (cLSPR)	Aptamer/Bisphenol A (~228.3 g/mol) in water	~1,200 nm/RIU	~330 aM	[69]
D-shaped plastic optical fiber with aptamers immobilized through biotin/streptavidin	Aptamer/SARS-CoV-2 spike protein	NA	37 nM	[62]
U-shaped MMF with aptamers immobilized on 3-D gold nanoparticles and graphene layers (LSPR)	DNA hybridization to immobilized aptamer in PBS	~1,251.4 nm/RIU	0.1 nM	[88]
U-shaped plastic MMF with aptamers immobilized on gold film, (L)SPR	Aptamer/PfGDH *Plasmodium falciparum* glutamate dehydrogenase	NA	264 pM	[67]
Microcavity in-line Mach-Zehnder interferometer with peptide aptamers	Peptide aptamers/cells (*E. coli* O157:H7), using culture dilutions in PBS	~17,000 nm/RIU (water glycerin solutions between 1.333 and 1.390)	10 cfu (colony forming unit)	[89]
Tapered optical fiber with fluorescent labeled aptamers (fluorescence FRET)	Aptamer/Aflatoxin M1 (328.3 g/mol) and ochratoxin A (403.8 g/mol) in buffer	NA	~64 pM AFM1 and ~818 pM OTA	[90]

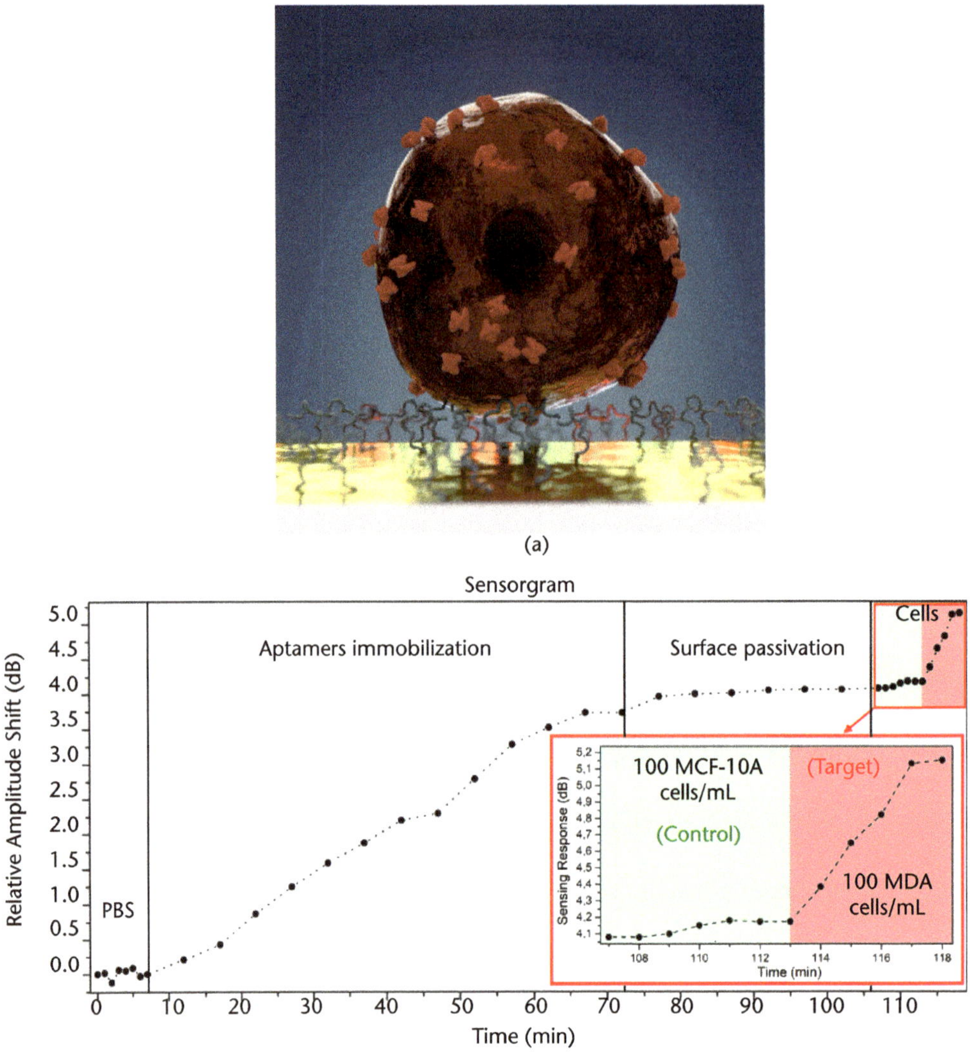

(b)

Figure 7.7 The 3-D scheme of a cell binding to aptamers immobilized on a gold surface and recognizing specific cell-surface proteins (a), and sensorgram showing the evolution of a plasmonic tilted fiber Bragg grating response during the aptamers deposition, passivation, and detection of target cancer cells (b). (Adapted with permission from [78]. American Chemical Society.)

Many optical fiber biosensors were also involved in bacteria detection, such as for *Escherichia coli*. This is highly relevant in water and food samples where pathogens need to be detected quickly and in the field. Sensors were tested with peptide aptamers to detect a pathogenic strain of *E. coli* (0157:H7) at a limit of detection of around 10 colony forming units (CFU)/mL, which was equal to or lower than reported for other optical fiber sensing platforms so far [43]. Their exploitation as fiber-tip probes could be highly relevant for *in situ* applications as well as for cellular detection leading to the diagnosis of diseases or cancers [95].

7.4.4 Ion Detection

Among all the available sets of targets, the detection of ions is probably the most challenging, as their small sizes and low molecular weights result in extremely small changes of the surface refractive index compared to their proteins or cells counterparts. However, the detection of ions, especially heavy metals, is a great concern for health and environmental considerations.

The most toxic ions are indeed Hg^{2+}, Cd^{2+}, Pb^{2+}, and As^{2+}. Current methods for such detection are based on laboratory devices and call on absorption spectroscopy, inductively coupled plasma atomic mass spectroscopy (ICP-MS), and piezoelectric or electrochemical measurements. Alternatives using fluorescence detection (chemosensors and chemodosimeters) are also very popular because they provide high sensitivity at a relatively low cost. The advantage of the optical fiber remains the ability to bring the sensor directly in the medium of interest without the need for sampling. It also brings perspectives of detection in a reversible way and therefore projects a significant interest in online measurements. Their detection in buffers already containing other ions is difficult, while their analysis in complex media is even more so. Therefore, the first fruits of these sensors currently require a great deal of work, on both reproducibility and protection of the sensor from the external environment (e.g., via filtration of the sample or encapsulation of the sensing area to be viable under field conditions). Recent achievements for ion sensing have been reported on optical fiber platforms, especially for lead detection (Pb^{2+}) using the thrombin aptamer (TBA, 5′-GGTTGGTGTGGTTGG-3′) that carries a stabilized G-quadruplex where Pb^{2+} ions can be complexed in the center of the two G-tetrads [96]. These aptamers initially designed for thrombin protein detection can therefore be diverted for ion detection. Using an SPR-TFBG sensor, it was possible to detect as low as 8.56 pM with a dynamic range between 10^{-11} and 10^{-6} M [97].

The first works about optical fiber-based biosensing using oligonucleotides as receptors clearly opened the race for sensing innovation for multiple applications. Nowadays, it is necessary to come back to the fundamentals and to better understand these strategies to improve their reliability and robustness (e.g., by playing on their smart integration via on-board micro-electronics, automated data analysis, or artificial intelligence to concretize and fully exploit their assets on field). Table 7.1 summarizes some recent approaches of optical fiber-based biosensing through DNA-based receptors. Depending on the optical fiber platform and surface chemistry, it is clear that the list of biosensors developed in recent years is still expanding and that sensing performance matters. It is also important to underline that biosensing methods that seem very similar on the paper can lead to very different experimental results, in particular because of variations in the experimental approach or the selected target, which cause very differentiated plasmonic responses. For all these reasons, it is currently almost impossible to clearly establish a performance list for each type of sensor and designate the most adapted configuration for a given application. Reproducibility tests and versatility tests against different targets and under the same experimental conditions are therefore essential data to be collected in the near future in order to validate the protocols presented in the literature so far and to order all these rich information disseminated to date.

7.5 Conclusion

Nucleic acid-based biosensors are essential tools for the development of fast and stable assays. Among the biosensors presented in this chapter, receptors such as DNA and RNA probes, aptamers, and certainly all the other X-NA molecules could have a great impact on environmental and *in situ* monitoring in the near future. This added value is particularly due to the many customizations and optimizations possible in terms of 3-D structure and their related detection strategy. They also lead to highly stable layers. For instance, DNA can be lyophilized and stored for years in dry conditions while traditional receptors such as antibodies usually need to be stabilized and frozen in particular buffers. Their development costs are also reduced compared to the selection and production of traditional bioreceptors. When looking at the evolution of aptamer prices, while first batches were extremely expensive, the current trend is a price drop, because of the increase in their diversification and production. Today, most traditional approaches and large-scale productions are still based on antibodies. The future will tell us whether aptamers will lead or not the market. Nevertheless, because these receptors can be synthesized and chemically modified with labels and tags at the whim of researchers and applications, they open up the path for new insights in terms of biosensing technologies.

References

[1] Vo-Dinh, T., et al., "Development of a Multiarray Biosensor for DNA Diagnostics," *Instrum. Sci. Technol.*, Vol. 26, No. 5, 1998, pp. 503–514.

[2] Sadana, A., and T. Vo-Dinh, "Single- and Dual-Fractal Analysis of Hybridization Binding Kinetics: Biosensor Applications," *Biotechnol. Prog.*, Vol. 14, No. 5, 1998, pp. 782–790.

[3] Luderer, F., and U. Walschus, "Immobilization of Oligonucleotides for Biochemical Sensing by Self-Assembled Monolayers: Thiol-Organic Bonding on Gold and Silanization on Silica Surfaces," Chapter 2 in *Immobilisation of DNA on Chips I*, C. Wittmann, (ed.), 2005, pp. 37–56.

[4] Zammatteo, N., et al., "Comparison Between Different Strategies of Covalent Attachment of DNA to Glass Surfaces to Build DNA Microarrays," *Anal. Biochem.*, Vol. 280, No. 1, 2000, pp. 143–150.

[5] Loyez, M., et al., "Overview and Emerging Trends in Optical Fiber Aptasensing," *Biosens. Bioelectron.*, Vol. 196, January 15, 2022.

[6] Ding, X., et al., "Surface Plasmon Resonance Biosensor for Highly Sensitive Detection of Microrna Based on DNA Super-Sandwich Assemblies and Streptavidin Signal Amplification," *Anal. Chim. Acta*, Vol. 874, 2015, pp. 59–65.

[7] Wu, Q., et al., "Review of Electrochemical DNA Biosensors for Detecting Food Borne Pathogens," *Sensors (Switzerland)*, Vol. 19, No. 22, 2019.

[8] Gill, I., "Bio-Doped Nanocomposite Polymers: Sol-Gel Bioencapsulates," *Chem. Mater.*, Vol. 13, No. 10, 2001, pp. 3404–3421.

[9] Jerónimo, P. C. A., A. N. Araújo, and M. Conceição, "Optical Sensors and Biosensors Based on Sol-Gel Films," *Talanta*, Vol. 72, No. 1, 2007, pp. 13–27.

[10] Martincek, I., D. Pudis, and M. Chalupova, "Technology for the Preparation of PDMS Optical Fibers and Some Fiber Structures," *IEEE Photonics Technol. Lett.*, Vol. 26, No. 14, 2014, pp. 1446–1449.

[11] Liu, Z., et al., "Multifunctional Smart Optical Fibers: Materials, Fabrication, and Sensing Applications," *Photonics*, Vol. 6, No. 2, 2019, pp. 1–24.

[12] Sui, G., et al., "Solution-Phase Surface Modification in Intact Poly(dimethylsiloxane) Microfluidic Channels," *Anal. Chem.*, Vol. 78, No. 15, 2006, pp. 5543–5551.

[13] Chandra, P., and R. Prakash, (eds.), *Nanobiomaterial Engineering: Concepts and Their Applications in Biomedicine and Diagnostics*, New York: Springer, 2020.

[14] Shynkarenko, O. V., and S. A. Kravchenko, "Surface Plasmon Resonance Sensors: Methods of Surface Functionalization and Sensitivity Enhancement," *Theor. Exp. Chem.*, Vol. 51, No. 5, 2015, pp. 273–292.

[15] Shukla, S. K., et al., "Chitosan-Based Nanomaterials: A State-of-the-Art Review," *Int. J. Biol. Macromol.*, Vol. 59, 2013, pp. 46–58.

[16] Mirzaie, A., M. Hasanzadeh, and A. Jouyban, "Cross-Linked Chitosan/Thiolated Graphene Quantum Dots as a Biocompatible Polysaccharide Towards Aptamer Immobilization," *Int. J. Biol. Macromol.*, Vol. 123, 2019, pp. 1091–1105.

[17] Kamaruddin, N. H., et al., "Binding Affinity of a Highly Sensitive Au/Ag/Au/Chitosan-Graphene Oxide Sensor Based on Direct Detection of Pb^{2+} and Hg^{2+} Ions," *Sensors (Switzerland)*, Vol. 17, No. 10, 2017.

[18] Luo, X., et al., "Plasmons in Graphene: Recent Progress and Applications," *Mater. Sci. Eng.*, Vol. R 74, No. 11, 2013, pp. 351–376.

[19] Kuila, T., et al., "Chemical Functionalization of Graphene and Its Applications," *Prog. Mater. Sci.*, Vol. 57, No. 7, 2012, pp. 1061–1105.

[20] Li, C., et al., "Optical Fiber SPR Biosensor Complying with a 3D Composite Hyperbolic Metamaterial and a Graphene Film," *Photon. Res.*, Vol. 9, No. 3, 2021, pp. 379–388.

[21] Theodosiou, A., et al., "Plane-by-Plane Femtosecond Laser Inscription Method for Single-Peak Bragg Gratings in Multimode CYTOP Polymer Optical Fiber," *J. Light. Technol.*, Vol. 35, No. 24, 2017, pp. 5404–5410.

[22] Theodosiou, A., and K. Kalli, "Recent Trends and Advances of Fibre Bragg Grating Sensors in CYTOP Polymer Optical Fibres," *Opt. Fiber Technol.*, Vol. 54, January 2020.

[23] Cennamo, N., M. Pesavento, and L. Zeni, "A Review on Simple and Highly Sensitive Plastic Optical Fiber Probes for Bio-Chemical Sensing," *Sensors Actuators B Chem.*, Vol. 331, 2021, p. 129393.

[24] Pasquardini, L., et al., "A Surface Plasmon Resonance Plastic Optical Fiber Biosensor for the Detection of Pancreatic Amylase in Surgically-Placed Drain Effluent," *Sensors*, Vol. 21, No. 10, 2021.

[25] Cennamo, N., F. Mattiello, and L. Zeni, "Slab Waveguide and Optical Fibers for Novel Plasmonic Sensor Configurations," *Sensors (Switzerland)*, Vol. 17, No. 7, 2017.

[26] Fixe, F., et al., "Functionalization of Poly(Methyl Methacrylate) (PMMA) as a Substrate for DNA Microarrays," *Nucleic Acids Res.*, Vol. 32, No. 1, 2004, pp. 1–8.

[27] Sabourin, D., et al., "Microfluidic DNA Microarrays in PMMA Chips: Streamlined Fabrication Via Simultaneous DNA Immobilization and Bonding Activation by Brief UV Exposure," *Biomed. Microdevices*, Vol. 12, No. 4, 2010, pp. 673–681.

[28] Obata, K., et al., "Fabrication of Microchip Based on UV Transparent Polymer for DNA Electrophoresis by F2 Laser Ablation," *Appl. Phys. A Mater. Sci. Process.*, Vol. 84, No. 3, 2006, pp. 251–255.

[29] Mohankumar, P., et al., "Recent Developments in Biosensors for Healthcare and Biomedical Applications: A Review," *Meas. J. Int. Meas. Confed.*, Vol. 167, 2021.

[30] Gallichi-Nottiani, D., et al., "Toward the Fabrication of Extruded Microstructured Bioresorbable Phosphate Glass Optical Fibers," *Int. J. Appl. Glas. Sci.*, Vol. 11, No. 4, 2020, pp. 632–640.

[31] Pugliese, D., et al., "Bioresorbable Optical Fiber Bragg Gratings," *Opt. Lett.*, Vol. 43, No. 4, 2018, pp. 671–674.

[32] Gierej, A., et al., "Poly(D,L-Lactic Acid) (PDLLA) Biodegradable and Biocompatible Polymer Optical Fiber," *J. Light. Technol.*, Vol. 37, No. 9, 2019, pp. 1916–1923.

[33] North, S. H., et al., "Plasma-Based Surface Modification of Polystyrene Microtiter Plates for Covalent Immobilization of Biomolecules," *ACS Appl. Mater. Interfaces*, Vol. 2, No. 10, 2010, pp. 2884–2891.

[34] Bilek, M. M., and D. R. McKenzie, "Plasma Modified Surfaces for Covalent Immobilization of Functional Biomolecules in the Absence of Chemical Linkers: Towards Better Biosensors and a New Generation of Medical Implants," *Biophys. Rev.*, Vol. 2, No. 2, 2010, pp. 55–65.

[35] Kumar, A., and Z. Liang, "Chemical Nanoprinting: A Novel Method for Fabricating DNA Microchips," *Nucleic Acids Res.*, Vol. 29, No. 2, 2001, p. 2.

[36] Liu, G., et al., "Evolution of Dip-Pen Nanolithography (DPN): From Molecular Patterning to Materials Discovery," *Chem. Rev.*, Vol. 120, No. 13, 2020, pp. 6009–6047.

[37] Dey, S., et al., "DNA Origami," *Nat. Rev. Methods Prim.*, Vol. 1, No. 1, 2021, pp. 1–24.

[38] Torelli, E., et al., "DNA Origami Nanorobot Fiber Optic Genosensor to TMV," *Biosens. Bioelectron.*, Vol. 99, January 15, 2018, pp. 209–215.

[39] Daems, D., et al., "3D DNA Origami as Programmable Anchoring Points for Bioreceptors in Fiber Optic Surface Plasmon Resonance Biosensing," *ACS Appl. Mater. Interfaces*, Vol. 10, No. 28, 2018, pp. 23539–23547.

[40] Pohanka, M., "Monoclonal and Polyclonal Antibodies Production—Preparation of Potent Biorecognition Element," *J. Appl. Biomed.*, Vol. 7, No. 3, 2009, pp. 115–121.

[41] Liu, M., et al., "In Vitro Selection of Circular DNA Aptamers for Biosensing Applications," *Angew. Chemie—Int. Ed.*, Vol. 58, No. 24, 2019, pp. 8013–8017.

[42] Tombelli, S., M. Minunni, and M. Mascini, "Analytical Applications of Aptamers," *Biosens. Bioelectron.*, Vol. 20, 2005, pp. 2424–2434.

[43] Janik, M., et al., "Optical Fiber Aptasensor for Label-Free Bacteria Detection in Small Volumes," *Sensors Actuators B Chem.*, Vol. 330, 2021, p. 129316.

[44] Kang, K. N., and Y. S. Lee, "RNA Aptamers: A Review of Recent Trends and Applications," *Adv. Biochem. Eng. Biotechnol.*, Vol. 131, 2013, pp. 153–169.

[45] Sun, H., et al., "Oligonucleotide Aptamers: New Tools for Targeted Cancer Therapy," *Mol. Ther.—Nucleic Acids*, Vol. 3, April 2014, p. e182.

[46] Im, J., B. Park, and K. Han, "A Generative Model for Constructing Nucleic Acid Sequences Binding to a Protein," *BMC Genomics*, Vol. 20, Suppl. 13, 2019, pp. 1–13.

[47] Knight, C. G., et al., "Array-Based Evolution of DNA Aptamers Allows Modelling of an Explicit Sequence-Fitness Landscape," *Nucleic Acids Res.*, Vol. 37, No. 1, 2009, pp. 1–10.

[48] Tans, R., et al., "Affimers as an Alternative to Antibodies for Protein Biomarker Enrichment," *Protein Expr. Purif.*, Vol. 174, October 2020.

[49] He, W., et al., "X-Aptamers: A Bead-Based Selection Method for Random Incorporation of Druglike Moieties onto Next-Generation Aptamers for Enhanced Binding," *Biochemistry*, Vol. 51, No. 42, 2012, pp. 8321–8323.

[50] Lokesh, G. L., et al., "X-Aptamer Selection and Validation," *Methods Mol. Biol.*, Vol. 1632, 2017, pp. 151–174.

[51] De Acha, N., C. Elosúa, and F. J. Arregui, "Development of an Aptamer Based Luminescent Optical Fiber Sensor for the Continuous Monitoring of HG^{2+} in Aqueous Media," *Sensors (Switzerland)*, Vol. 20, No. 8, 2020.

[52] Fang, X., et al., "Synthetic DNA Aptamers to Detect Protein Molecular Variants in a High-Throughput Fluorescence Quenching Assay," *ChemBioChem*, Vol. 4, No. 9, 2003, pp. 829–834.

[53] Chen, C., and J. Wang, "Optical Biosensors: An Exhaustive and Comprehensive Review," *Analyst*, Vol. 145, No. 5, 2020, pp. 1605–1628.

[54] Mowbray, S. E., and A. M. Amiri, "A Brief Overview of Medical Fiber Optic Biosensors and Techniques in the Modification for Enhanced Sensing Ability," *Diagnostics (Basel, Switzerland)*, Vol. 9, No. 1, 2019, p. 23.

[55] Piunno, P. A. E., et al., "Fiber Optic Biosensor for Fluorimetric Detection of DNA Hybridization," *Anal. Chim. Acta*, Vol. 288, No. 3, 1994, pp. 205–214.

[56] Ferguson, J. A., et al., "A Fiber-Optic DNA Biosensor Microarray for the Analysis of Gene Expression," *Nat. Biotechnol.*, Vol. 14, No. 13, 1996, pp. 1681–1684.

[57] Stewart, G., et al., "Surface Plasmon Resonances in Thin Metal Films for Optical Fibre Devices," *Optical Fiber Sensors*, 1988, paper ThEE2.

[58] Matsubara, K., S. Kawata, and S. Minami, "Optical Chemical Sensor Based on Surface Plasmon Measurement," *Appl. Opt.*, Vol. 27, No. 6, 1988, pp. 1160–1163.

[59] Pollet, J., et al., "Real-Time Monitoring of Solid-Phase PCR Using Fiber-Optic SPR," *Small*, Vol. 7, No. 8, 2011, pp. 1003–1006.

[60] Daems, D., et al., "Identification and Quantification of Celery Allergens Using Fiber Optic Surface Plasmon Resonance PCR," *Sensors (Switzerland)*, Vol. 17, No. 8, 2017.

[61] Peeters, B., et al., "Real-Time FO-SPR Monitoring of Solid-Phase DNAzyme Cleavage Activity for Cutting-Edge Biosensing," *ACS Appl. Mater. Interfaces*, Vol. 11, No. 7, 2019, pp. 6759–6768.

[62] Cennamo, N., et al., "SARS-CoV-2 Spike Protein Detection Through a Plasmonic D-Shaped Plastic Optical Fiber Aptasensor," *Talanta*, Vol. 233, October 1, 2021.

[63] Loyez, M., et al., "HER2 Breast Cancer Biomarker Detection Using a Sandwich Optical Fiber Assay," *Talanta*, Vol. 221, January 1, 2021.

[64] Bekmurzayeva, A., et al., "Etched Fiber Bragg Grating Biosensor Functionalized with Aptamers for Detection of Thrombin," *Sensors (Switzerland)*, Vol. 18, No. 12, 2018.

[65] Coelho, L., et al., "Aptamer-Based Fiber Sensor for Thrombin Detection," *J. Biomed. Opt.*, Vol. 21, No. 8, 2016, p. 087005.

[66] Shevchenko, Y., et al., "In Situ Biosensing with a Surface Plasmon Resonance Fiber Grating Aptasensor," *Anal. Chem.*, Vol. 83, No. 18, 2011, pp. 7027–7034.

[67] Sanjay, M., et al., "A Smartphone-Based Fiber-Optic Aptasensor for Label-Free Detection of Plasmodium Falciparum Glutamate Dehydrogenase," *Anal. Methods*, Vol. 12, No. 10, 2020, pp. 1333–1341.

[68] Alhamoud, Y., et al., "Advances in Biosensors for the Detection of Ochratoxin A: Bio-Receptors, Nanomaterials, and Their Applications," *Biosens. Bioelectron.*, Vol. 141, September 15, 2019, p. 111418.

[69] Allsop, T. D. P., et al., "An Ultra-Sensitive Aptasensor on Optical Fibre for the Direct Detection of Bisphenol A," *Biosens. Bioelectron.*, Vol. 135, June 15, 2019, pp. 102–110.

[70] Hua Liu, L., X. Hong Zhou, and H. Chang Shi, "Portable Optical Aptasensor for Rapid Detection of Mycotoxin with a Reversible Ligand-Grafted Biosensing Surface," *Biosens. Bioelectron.*, Vol. 72, 2015, pp. 300–305.

[71] Vidal, M., et al., "Relevance of the Spectral Analysis Method of Tilted Fiber Bragg-Grating-Based Biosensors: A Case-Study for Heart Failure Monitoring," *Sensors (Basel)*, Vol. 22, No. 6, March 10, 2022, p. 2141.

[72] Lee, B., et al., "An Optical Fiber-Based LSPR Aptasensor for Simple and Rapid *In-Situ* Detection of Ochratoxin A," *Biosens. Bioelectron.*, Vol. 102, April 15, 2018, pp. 504–509.

[73] Yildirim, N., et al., "A Portable Optic Fiber Aptasensor for Sensitive, Specific and Rapid Detection of Bisphenol-A in Water Samples," *Environ. Sci. Process. Impacts*, Vol. 16, No. 6, 2014, pp. 1379–1386.

[74] Qiu, Y., et al., "Aptamer-Based Detection of Melamine in Milk Using an Evanescent Wave Fiber Sensor," *Anal. Methods*, Vol. 10, No. 40, 2018, pp. 4871–4878.

[75] Tran, D. T., et al., "Selection of Aptamers Against Ara h 1 Protein for FO-SPR Biosensing of Peanut Allergens in Food Matrices," *Biosens. Bioelectron.*, Vol. 43, No. 1, 2013, pp. 245–251.

[76] Pollet, J., et al., "Fiber Optic SPR Biosensing of DNA Hybridization and DNA-Protein Interactions," *Biosens. Bioelectron.*, Vol. 25, No. 4, 2009, pp. 864–869.

[77] Shevchenko, Y., et al., "In Situ Biosensing with a Surface Plasmon Resonance Fiber Grating Aptasensor," *Anal. Chem.*, Vol. 83, 2011, pp. 7027–7034.

[78] Loyez, M., et al., "Rapid Detection of Circulating Breast Cancer Cells Using a Multiresonant Optical Fiber Aptasensor with Plasmonic Amplification," *ACS Sensors*, Vol. 5, No. 2, 2020.

[79] Lobry, M., et al., "Multimodal Plasmonic Optical Fiber Grating Aptasensor," *Opt. Express*, Vol. 28, No. 5, 2020, p. 7539.

[80] Sypabekova, M., et al., "Functionalized Etched Tilted Fiber Bragg Grating Aptasensor for Label-Free Protein Detection," *Biosens. Bioelectron.*, Vol. 146, 2019, p. 111765.

[81] Lao, J., et al., "Gold Nanoparticle-Functionalized Surface Plasmon Resonance Optical Fiber Biosensor: In Situ Detection of Thrombin with 1 nM Detection Limit," *J. Light. Technol.*, Vol. 37, No. 11, 2018, pp. 2748–2755.

[82] W. Hu, Y et al., "Highly Sensitive Detection of Dopamine Using a Graphene Functionalized Plasmonic Fiber-Optic Sensor with Aptamer Conformational Amplification," *Sensors Actuators B Chem.*, Vol. 264, 2018, pp. 440–447.

[83] Celebanska, A., et al., "Label-Free Cocaine Aptasensor Based on a Long-Period Fiber Grating," *Opt. Lett.*, Vol. 44, No. 10, 2019, p. 2482.

[84] Zibaii, M. I., et al., "Label Free Fiber Optic Apta-Biosensor for In-Vitro Detection of Dopamine," *J. Light. Technol.*, Vol. 34, No. 19, 2016, pp. 4516–4524.

[85] Zhu, Q., et al., "A Split Aptamer (SPA)-Based Sandwich-Type Biosensor for Facile and Rapid Detection of Streptomycin," *J. Hazard. Mater.*, Vol. 403, February 5, 2021.

[86] Cennamo, N., et al., "D-Shaped Plastic Optical Fibre Aptasensor for Fast Thrombin Detection in Nanomolar Range," *Sci. Rep.*, Vol. 9, No. 1, 2019, pp. 1–9.

[87] Tripathi, S. M., et al., "Gold Coated Dual-Resonance Long-Period Fiber Gratings (DR-LPFG) Based Aptasensor for Cyanobacterial Toxin Detection," *Sens. Bio-Sensing Res.*, Vol. 25, September 2019.

[88] Li, C., et al., "LSPR Optical Fiber Biosensor Based on a 3D Composite Structure of Gold Nanoparticles and Multilayer Graphene Films," *Opt. Express*, Vol. 28, No. 5, 2020, p. 6071.

[89] Janik, M., et al., "Optical Fiber Aptasensor for Label-Free Bacteria Detection in Small Volumes," *Sensors Actuators B Chem.*, Vol. 330. March 1, 2021.

[90] Song, D., et al., "A FRET-Based Dual-Color Evanescent Wave Optical Fiber Aptasensor for Simultaneous Fluorometric Determination of Aflatoxin M1 and Ochratoxin A," *Microchim. Acta*, Vol. 185, No. 11, 2018, pp. 1–10.

[91] Caucheteur, C., et al., "Cell Sensing with Near-Infrared Plasmonic Optical Fiber Sensors," *Opt. Laser Technol.*, Vol. 78, 2016, pp. 116–121.

[92] Yanase, Y., et al., "Development of an Optical Fiber SPR Sensor for Living Cell Activation," *Biosens. Bioelectron.*, Vol. 25, 2010, pp. 1244–1247.

[93] Abadian, P. N., C. P. Kelley, and E. D. Goluch, "Cellular Analysis and Detection Using Surface Plasmon Resonance Techniques," *Anal. Chem.*, Vol. 86, No. 6, 2014, pp. 2799–2812.

[94] Peterson, A. W., et al., "High Resolution Surface Plasmon Resonance Imaging for Single Cells," *BMC Cell Biol.*, Vol. 15, No. 1, 2014, pp. 1–14.

[95] Hong, P., W. Li, and J. Li, "Applications of Aptasensors in Clinical Diagnostics," *Sensors*, Vol. 12, No. 2, 2012, pp. 1181–1193.

[96] Liu, H., et al., "Structure-Guided Development of Pb^{2+}-Binding DNA Aptamers," *Sci. Rep.*, Vol. 12, No. 1, 2022, pp. 1–11.

[97] Wang, F., et al., "Near-Infrared Band Gold Nanoparticles-Au Film 'Hot Spot' Model Based Label-Free Ultratrace Lead (II) Ions Detection Via Fiber SPR DNAzyme Biosensor," *Sensors Actuators B Chem.*, Vol. 337, June 15, 2021, p. 129816.

CHAPTER 8

Other Bioreceptors for Plasmonic Biosensors

Chapters 6 and 7 were dedicated to antibodies and nucleic acid-based bioreceptors. However, a large number of biosensors uses other sensing strategies. This is typically the case for most glucose sensors relying on enzymes, such as the glucose oxidase. New technologies also lead to the development of mimicking agents and key-lock sets of molecules to retain the benefits of antibodies or DNA, while providing new physical or biochemical properties. Molecularly imprinted polymers (MIPs) are one of these synthetic recognition elements, but even more specific techniques aim to harness proteins or fragments of proteins, cells or cellular membranes, and microporous or nanoporous gels. All these molecular architectures are discussed in this chapter.

8.1 MIPs

MIPs are synthetic receptors built *in vitro* against a target molecule or cell. They are considered as analogs of natural antibody-antigen systems, mimicking their behavior through the lock-and-key principle. MIPs are popular due to their polymeric nature, so they can be produced at low cost and stored for long, as they do not require specific storage conditions. They can also be produced against almost any targets (from small molecules to cell surfaces). Many MIP production processes have been developed and optimized recently, but the main outline is that a polymer containing a template (or the target molecule) is synthesized. This template is then removed from the polymer, leaving a specific cavity, playing the role of a fingerprint that will enable further rebinding once the MIP will be exposed to the sample (Figure 8.1).

The first works published about imprinted polymers were released in the late 1970s and early 1980s by Andersson, Sellergren, and Mosbach [1] and Wulff et al. [2, 3]. The technique quickly gained momentum and a lot of chemical processes

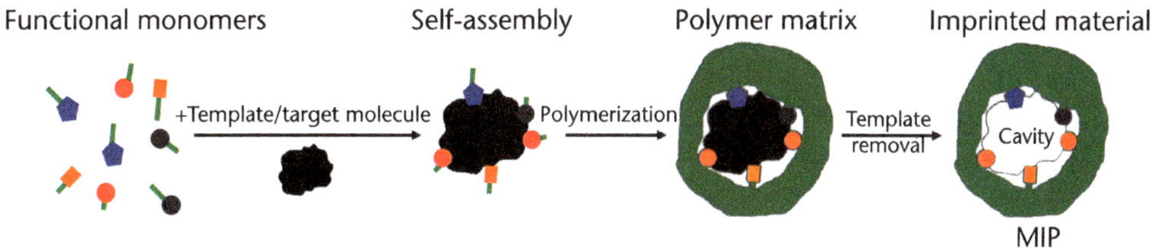

Figure 8.1 Typical principle of polymerization process for MIP generation using a target molecule or template.

were then developed in the 1990s [4]. They have been commercialized, and their main applications are about molecular separation and chromatography rather than for sensing purposes. This happens because the polymeric nature of the receptors often shows high adsorption properties and therefore can lead to low specificity. The production methods also need precise and restrictive extraction to remove the template from the polymer, which is usually the trickiest part. In sensing applications, the remain of unextracted targets may interfere with the ability to sense rebinding and lead to untimely signal changes. Nevertheless, MIP receptors are increasingly studied and show notable advantages and improvements. Some of them have demonstrated interesting responses for biosensing, especially on optical fiber-based platforms. Several companies are also emerging in this field, betting on this new technology to create high-stability diagnostic kits and high-performance sensing, with a strong tendency for their synthesis through green chemistry processes [5, 6].

Various techniques usually lead up to the production of MIPs. While the most known is the synthesis from monomers in the presence of the template or target, other methods such as phase inversion through polymer precipitation, lithography, and surface-stamping also exist (Figure 8.2). For example, a polymerization process can take place under ultraviolet radiation by mixing a cross-linking agent such as ethylene glycol dimethacrylate (EGDMA) and a polymerization initiator such as azobisisobutyronitrile (AIBN). This process can provide a powdered product but can also lead to thin films if the reaction takes place on a substrate [7]. It is also widely used with sol-gel particles. At the end of the process, the template/target molecule is removed using a specific solvent so the MIP becomes empty and available for rebinding. However, the phase inversion process does not require a polymerization step [8]. With that technique, a compatible solvent with both the template or target and the polymer needs to be selected in order to dissolve these two components. By mixing both dissolved polymers and targets, it becomes possible to precipitate them together by adding another solvent or to evaporate the initial solvent to obtain a film. The solvents used for this second technique are usually dimethylsulfoxide (DMSO) and water. Finally, lithographic processes enable the production of surface-imprinted polymers and are generally suited for sensing applications. They involve the production of a molecular stamp containing the template or target to put in contact with a polymeric surface and to carefully remove it to release the MIP binding sites. The polymeric surface can be made of modified epoxy resin and the stamp made of polydimethylsiloxane, as recently reported for the detection of *E. coli* bacteria [9, 10].

The choice of reagents, solvents, and the production process is driven by the target application. The polymerization method often leads to highly homogeneous but more rigid MIPs, meaning that the accessibility to the binding site can be affected. The phase-inversion technique provides faster production and lacks cross-linking agents, so the cavities are more susceptible to collapse during the extraction of the template. Regardless of the chosen technique, both productions need to be characterized by atomic force microscopy (AFM) and scanning electron microscopy (SEM) to determine the final morphology and porosity of the MIP products or membranes. Once produced and characterized, MIPs are usually immobilized on chemiresistors, capacitance sensors, chemical field effect transistors (Chemfet), and quartz microbalances. Most of the sensing applications reported so far are coupling

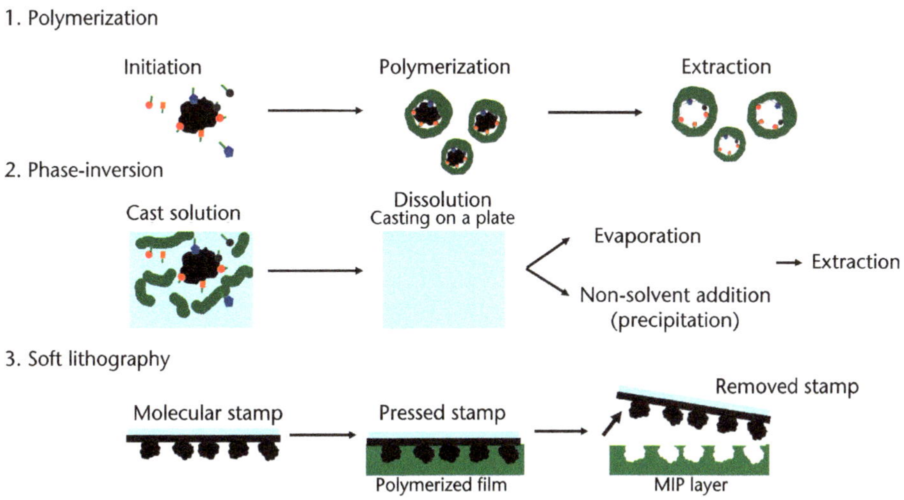

Figure 8.2 Scheme of: (1) polymerization, (2) phase inversion, and (3) soft lithography for MIP generation.

MIPs with electrochemical sensors, while several other nonsensing applications are about drug delivery [11], purification of biochemicals, analytical separations, safety, environment, and health [5, 12–14].

A paramount application of MIP sensing is the detection of gas and volatile organic compounds (VOCs). This category of molecules is indeed well targeted by these receptors and do not have any other counterparts in the traditional bioreceptors, although nucleic acid-based receptors are emerging for gaseous detections [15, 16]. MIP receptors also show high appeal for the detection of proteins or peptides and would certainly be an inexpensive alternative to current production of certain antibodies and aptamers. Recently, MIPs have also been studied to detect toxins, and pesticides (or residues), and to identify microorganisms [17, 18]. Their implementation on optical fibers as receptors is suited for several sensing principles such as SPR, Raman, intensity distribution, interferometry, fluorescence, and diffraction. The combination of such sensing strategies with MIPs therefore opens the door to new features that are being developed by several research groups and companies. Recent research papers and reviews have demonstrated this particular interest [19–21]. Over the past few years, many MIP optical fibers have been studied. Some of them are reported in Table 8.2 [22, 23]. The use of QD hybrids and MIP nanoparticles is also expanding fields of research. In addition to that, computational simulation for the synthesis of high-affinity synthetic receptors (*in silico*) is certainly a primary focus for drug recognition and detection with optical biosensors [24, 25].

To summarize, MIPs have been developed to satisfy the need for alternatives to traditional bioreceptors at low cost and with fast production rates. Their initial uses were limited due to poor reproducibility, permanent trapping of analytes, and long diffusion times. Today, the use of MIP nanoparticles and layers is still limited to specific applications, but the field is growing and increasingly exploits their advantages such as their long lifetime and stability, which are mandatory in harsh environments. Improvements in MIP generation, affinity, and reproducibility give optimistic insight about their future, especially when combined with biosafety detection purposes in either liquids or gases.

8.2 Enzymes

8.2.1 Enzymatic Biosensors

As this book deals with biosensors, it is impossible to bypass this section related to the enzymes. They are indeed at the origin of the whole biosensing concept and enzymatic-based platforms are the most widespread types of sensors worldwide, especially for glucose monitoring. The Clark and Lyons electrode released in 1962 with a glucose oxidase (GOx)-soaked membrane is considered as the first biosensor of any type. Its basic working principle is all its strength, and it gave birth to a series of breakthrough improvements and inspired subsequent sensors generations, progressively leading to ultralight and long-term monitoring systems [26].

Enzymes can be described as a group of molecules produced by living organisms and causing particular biochemical reactions. They are present in all cells and can be secreted, as it is the case in saliva, for example. Enzymes play essential roles in the living world and are the key to all metabolisms. Most of enzymes are proteins (amino acids), but they can be made of ribonucleic acids (RNA, called ribozymes) or involve other macromolecules. Enzymes are catalysts, so they enable reaction processes by increasing their speed (lowering the activation energy) without appearing in the final reaction balance. However, they differ by other chemical catalysts by their high efficiency and specificity. They are able to catalyze, chemically recognize, bind, cut, or modify substrate and are often extracted from cells to provide commercial processes. For instance, they are used in many cleaning and washing products, play the role of analytical devices, and have clinical and environmental applications. A typical enzymatic reaction only requires a small concentration of enzymes and provides the conversion of substrate molecules (A) into product molecules (B) (Figure 8.3). The selection of efficient enzymes is mandatory to obtain efficient biosensors.

Due to the enzymes, specific reactions can occur because they allow the reduction of the necessary activation energy. The system therefore moves to the side of the products and the specific reagent that undergoes catalysis is known as the substrate. The catalytic activity of enzymes selected for sensing purposes is usually high and can be expressed by a constant, k_{cat}, which is referred as the turnover rate (or turnover frequency). It represents the number of substrate molecules converted into products by a single enzyme per time unit. As an example, a single carbonic anhydrase is able to convert roughly 600,000 substrate molecules (CO_2 + H_2O) into bicarbonate (HCO_3^-) every second [27, 28]. The kinetic k_{cat} and K_M (Michaelis

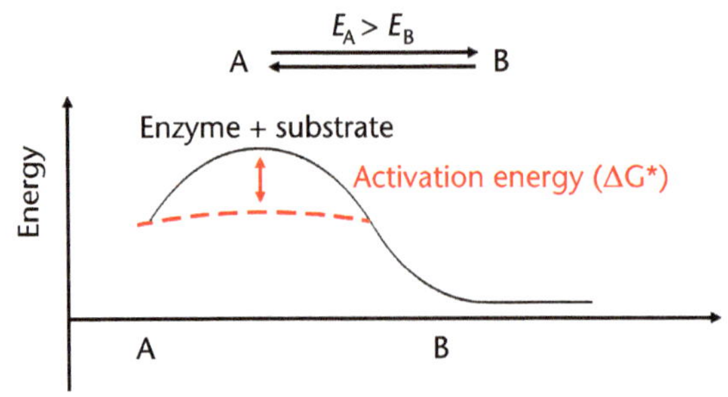

Figure 8.3 Scheme of a chemical reaction involving the reduction of its activation energy.

constant) are key parameters to determine the biosensing properties. Usually, biosensors possess high loadings of enzyme activity and, hence, a high k_{cat} (turnover). In the specific case of membrane sensors, a high k_{cat} would tend to result in a device limited by membrane transport and, thus, to one relatively insensitive to variations in enzyme loading or loss of enzyme activity [29]. However, the Michaelis constant K_m shows its effect on the linear range of the sensor. Ideally, a linear relationship between the output signal and the analyte concentration should be observed, and this is reflected in the K_M. The ratio k_{cat}/K_M is defined as the specificity constant and is expected to be high for the analyte of interest compared to other nonspecific substrates. In other words, it means that the enzyme operates at a maximum rate when all the active sites are filled, so this is true when the substrate concentration is very high. This case is typically when K_M is extremely lower than the substrate concentration (Figure 8.4). Under normal physiological conditions, the substrate concentration is relatively low, so the enzyme activity is never at its maximum [30].

The substrate concentration is usually smaller than the K_M value in normal physiological conditions and for the biosensors, aiming at targeting small concentrations of analytes. In these conditions, the substrate concentration is usually located somewhere between 0 and K_M. When the enzyme and substrate combine in the active site, the enzyme/substrate complex (ES) is formed. If the substrate is bound strongly, it can be transformed into the product. If not, it dissociates and the reaction goes back to the left side. When (8.1) is leading to products, the rate law is given by (8.2).

$$E + S \rightleftharpoons ES \rightarrow E + P \tag{8.1}$$

$$V_0 = k_{cat}\,[ES] \tag{8.2}$$

Under steady state conditions, the rate of the enzyme/substrate complex formation is equal to the rate of dissociate. In this case:

$$k_1[E][S] = k_{-1}[ES] + k_{cat}[ES] \rightarrow \frac{k_{-1} + k_{cat}}{k_1} = \frac{[E][S]}{[ES]} \tag{8.3}$$

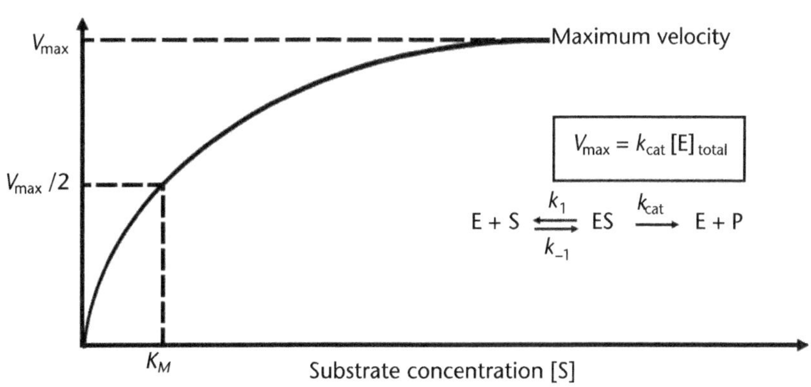

Figure 8.4 Michaelis-Menten kinetics. Reaction rate versus substrate concentration (E = enzyme, S = substrate, and P = products).

The Michaelis constant is equal to this ratio, so we can obtain the following equation:

$$K_M = \frac{[E][S]}{[ES]} \quad \text{so} \quad [ES] = \frac{[E][S]}{K_M} \tag{8.4}$$

This relation means that we can replace the concentration of ES in (8.2) to obtain (8.5):

$$V_0 = \frac{k_{cat}}{K_M}[E][S] \tag{8.5}$$

Equation (8.5) can be used to define how enzymes operate under normal conditions, when the substrate concentration is lower than the K_M value. This means that the total amount of enzymes is approximately equal to the sum of free enzymes with [ES] complexes.

$$[E]_{\text{total}} = [ES] + [E] \approx [E] \tag{8.6}$$

$$V_0 = \frac{k_{cat}}{K_M}[E]_{\text{total}}[S] \tag{8.7}$$

Equation (8.7) can therefore be used to describe how enzymes catalyze reactions under normal physiological conditions (bimolecular second-order chemical reaction). According to this equation, we can assume that the catalysis depends on three factors: the concentration of the substrate, the total amount of available enzymes, and the rate constant. The latter is used to measure the efficiency of enzymes (8.8). As k_{cat} measures the turnover and K_M describes the attraction of the substrate to the active site, the ratio of k_{cat}/K_M is used to quantify the enzymatic efficiency.

$$\frac{k_{cat}}{K_M} = \frac{k_1}{k_{-1} + k_{cat}} \times k_{cat} = \left(\frac{k_{cat}}{k_{-1} + k_{cat}}\right) \times k_1 < k_1 \tag{8.8}$$

When the k_{-1} value is low and reaches 0, the ratio tends to 1, which is its maximum value (8.9). The total ratio is then equal to k_1, the maximum value expressed for the rate constant. This means that the key factor is the formation of the ES complex, driving the rest of the reaction. Determining the parameters of the selected enzyme therefore drives the efficiency of the reaction and, consequently, the biosensing rate and sensitivity.

$$\lim_{k_{-1}\to 0}\left(\frac{k_{cat}}{K_M}\right) = k_1 \tag{8.9}$$

Unfortunately, a steady state kinetic measurement gives a little insight as to the nature of the enzyme/substrate complex. An enzymatic reaction may in reality pass through several intermediate states or to complex paths where different enzymes and substrates can be present. The ES state therefore needs to be considered as a black box that may content many other states and reactions. Enzymes are also classified into six categories, depending on their own function and activity (Table 8.1). Any individual enzyme can use the international nomenclature database, which is available at http://enzyme.expasy.org. The related enzymatic activity is expressed in international units (IU) where 1 IU represents 1 μmole of substrate reacting per minute.

Most enzymes suited for electrochemistry are named oxidoreductases because they catalyze specific oxidation or reduction reactions enabling electrochemical activity. Their activity can be tracked through direct methods (DET) or mediated electron transfer (MET). The detection of byproducts, cosubstrates, or coproducts is also possible such as for the glucose oxidase described hereafter, where the glucose level is determined through the O_2 or H_2O_2 levels. It should also be specified that in natural environments, enzymes do not operate constantly. Regulatory units (inhibitors or activators) are therefore present to ensure controlled reaction rates and to maintain a balanced system. For this, enzymes usually require the presence of cofactors in their active sites (chemical compounds or metal ions) to assist reactions.

Well-known cofactors such as flavin adenine dinucleotide (FAD), Mg^{2+} ion, cosubstrate nicotinamide adenine dinucleotide (NAD+) or coenzyme A (CoA) can be cited. In order to pursue its activity, the system must be synchronized with other enzymes, to regenerate the cofactors (oxidation/reduction) and the electronic transmission. Cellular metabolic pathways are therefore extremely regulated and their study is complex due to the many reactions that compose them [31].

8.2.2 Glucose Biosensors

Biosensors, meaning the ones including biological materials, were first described as electrochemical sensors for glucose detection. The combination of GOx with a Clark oxygen electrode was a genius idea to measure glucose by detecting the

Table 8.1 Enzyme Classification

Category	*Function*	*Examples*
Oxidoreductase	Electron transfer, changing the oxidation state	Dehydrogenase, glucose oxidase (GOx)
Transferase	Transfer of functional groups from one molecule to another	Kinase, phosphorylase, acetyltransferase, glycotransferase
Hydrolase	Breakdown of covalent bonds using water molecules	Protease, phosphatase, glycosidase, nucleosidase, lipase
Lyase	Breakdown of covalent bonds without need for water molecules or oxidation	Decarboxylase, citrate lyase, pectin lyase
Isomerase	Rearrangement of atoms within a molecule (intra-molecular group transfer)	Mutase, phosphoisomerase
Ligase	Formation of covalent bond between two molecules	DNA ligase

drop of oxygen level when the latter was converted to gluconic acid and hydrogen peroxide (8.10).

$$\text{Glucose} + \text{oxygen} \rightarrow \text{gluconic acid} + \text{hydrogen peroxide} \tag{8.10}$$

In the first part, the analyte consumes oxygen in the presence of an enzyme and the detection depends on a change in the oxygen tension. In the second part, the enzyme converts the analyte to a substance to which the sensor is sensitive. The first design of the Clark and Lyons electrode provides a combination of Ag/AgCl and Pt electrodes to measure oxygen concentration increasing as a function of the GOx activity, in turn, depending on the glucose concentration. When negative DC voltages are applied to the Pt electrode, the Ag/AgCl electrode acts as a capacitor and a reductive detection of the oxygen consumption occurs. At around –0.6 to –0.8, the current increases up to reach a plateau around –1V. In that particular voltage range, the following chemical reaction takes place (8.11):

$$O_2 + 4\ e^- + 4\ H^+ \rightleftharpoons 2\ H_2O \tag{8.11}$$

This means that, for each dioxygen molecule reaching the electrode, four electrons can be measured as current. Moreover, the latter is directly proportional to the amount of oxygen present in solution. The system is simple and stable, but the small currents limit its lifetime and so, its use on a long time scale. A buffer also needs to be used at the Pt electrode in order to stabilize the pH in that area, because the chemical reaction (8.10) stops in acidic conditions when:

$$2H^+ + O_2 + 2e^- \rightleftharpoons H_2O_2 \tag{8.12}$$

In its initial state, it means that the current stops before the end of reaction and leads to few reliable measurements. However, oxygen is removed from the layer close to the Pt electrode through (8.11) and only diffusion can take place. As a result, an O_2 concentration gradient is established steadily when the O_2 reaction rate at the electrode equals the diffusive flow from the bulk concentration. A convective flow is therefore mandatory to provide fresh solution with the bulk oxygen concentration by stirring the test solution. In short, each molecule of glucose is oxidized by the glucose oxidase enzyme, reducing its cofactor. This reduced cofactor is immediately reoxidized by oxygen dissolved in the medium so the enzyme becomes ready again to oxidize another glucose molecule, meaning that the enzymatic activity provokes an oxygen depletion in the medium (Figure 8.5).

Updike and Hicks [32, 33] further developed the working principle using two oxygen electrodes. One of them was covered by the GOx. Both electrodes were used to correct the oxygen background variation in samples. Some years later, another concept was based on the detection of blood glucose using the amperometric monitoring (anode) of the hydrogen peroxide.

$$H_2O_2 \rightarrow O_2 + 2\ H^+ + 2\ e^- \tag{8.13}$$

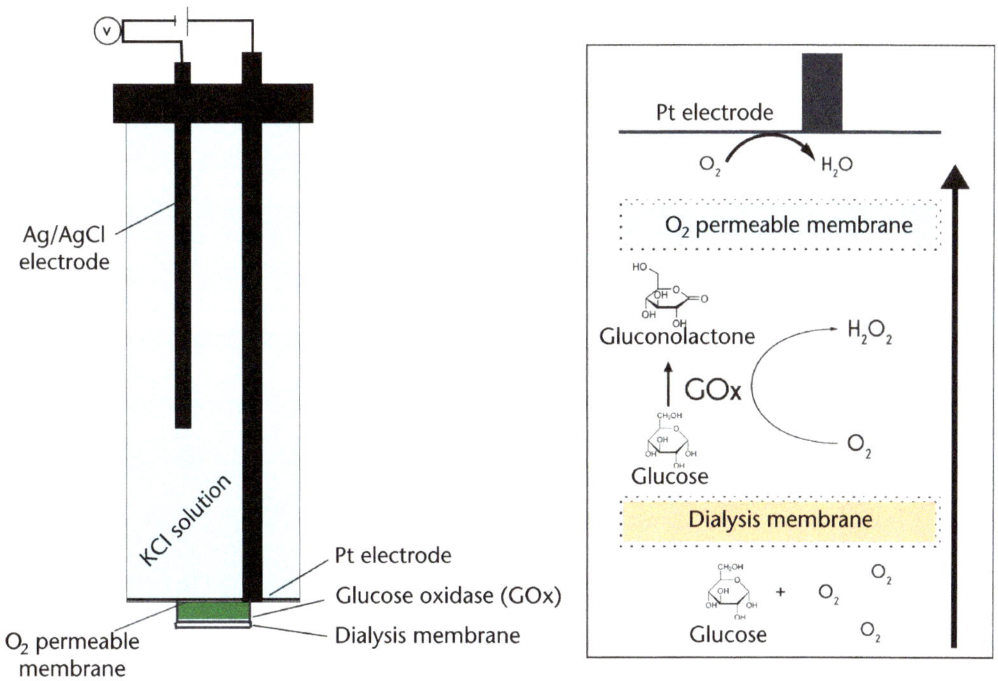

Figure 8.5 Working principle of the Clark and Lyons electrode.

First-generation devices can be described as electrode surfaces where a reduction reaction is monitored and where the detection is driven by the reaction. The detection of hydrogen peroxide produced by reoxidation of the enzymatic active site was indeed an interesting option. Then second-generation devices rely on artificial final electron acceptors (or donors) that are different from the ones of oxygen in the first generation. Large amounts of mediators can be implemented and lead to high versatility in comparison to the first generation. In third-generation biosensors, electron shuttling via mediators is omitted and the electrochemical behavior of the sensor is only driven by the enzymatic activity, so the mediator absence is an asset from both fabrication and thermodynamic points of view. However, it is conditioned by a physical contact between the enzyme and the electrode surface. GOx with a flavin adenine dinucleotide (FAD) cofactor is one of the most frequently chosen configurations for the fabrication of enzymatic biosensors [29, 34].

In very similar configurations, other oxidases and enzymes were exploited over the years. This is essentially true concerning the food industry and environmental analyses to detect fructose, lactose, ethanol, glycerol, aromatic compounds, cholesterol, and so many others. More detailed books are available on this fascinating topic, [35] in particular providing a general approach. Nowadays, glucose sensors are still at the heart of industrial concerns due to their incredible impact and the immense market. However, the progress made over the last decade is impressive and miniaturization now makes it possible to monitor glucose concentrations for days or weeks remotely. Many concepts such as connected watches and onboard tools are eager to include this kind of biosensor. It is therefore up to the smartest companies to launch new technological trends in this field of research, by implementing new sensors among the many ones already on the market.

As an example, under the GlucoWatch concept is based a glucose extraction principle named reverse iontophoresis (Figure 8.6). It happens in close contact with the skin, where glucose can be extracted by electro-osmotic flow. When a low-level current is applied (e.g., through a remote sensor for continuous glucose monitoring (CGM)), the migration of Na^+ ions cause a convective flow (electro-osmotic flow) of the interstitial fluid (ISF), carrying the glucose molecules towards the cathode. At the cathode, a standard enzymatic glucose sensor (GOx) is present to measure its concentration [36]. Most of these CGMs are almost noninvasive but need a renew of the contact patch after some days or weeks. It is able to monitor the glucose level in real time and launch alerts when glucose blood levels become critical. This is particularly interesting for patients with diabetes who practice sports or intense physical activity. It is one of the recent wearable biosensors. Their use is still highly controversial and has suffered many pitfalls, but many wearable biosensors are still on track and we will see if the market tends towards these new technological progresses.

In vivo implantable biosensors are a new variety of sensors that go one step further than wearable sensors. They are promising for many clinical applications. Although they were considered as too intrusive for a long time, they are finally starting to hit the market and their study and fields of applications are sharply expanding [37, 38].

8.2.3 Snapshot of Other Enzymatic Biosensors

The use of enzymes for biosensing is not only relying on the production of electrons to measure a current as it is the case for most CGM systems. The majority of enzymatic assays are actually based on light emission, absorption, reflectance, and heat emission or by measuring the products generated by the reaction itself. Enzymes are therefore suitable for colorimetric assays and mainly implemented on paper-based tests, where they excel and are the go-to items. Nowadays, Abbott, Roche Diagnostics, Bayer, and LifeScan dominate together the world market for biosensors, mostly due to enzyme-based platforms.

Despite the vast majority of commercial enzymatic sensors and the important interest for glucose sensors, these types of bioassays are less widespread than those involving antibodies and nucleic acid-based receptors on optical devices. This is

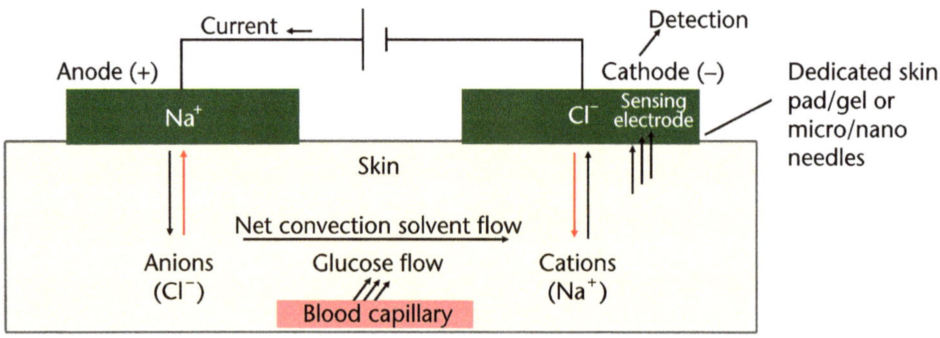

Figure 8.6 Working principle of the reverse-iontophoresis detection of glucose using pads on the skin.

probably due to the fact that these devices exploit optical phenomena rather than electrochemical properties usually brought by the enzymes. Enzymes have nevertheless led to many original contributions because they can also be involved in the production of light, fluorescence, cut proteins or DNA, replicate DNA, or RNA strands such as for polymerases. For all these reasons, enzymes still offer possibilities unmatched with other types of receptors and remain at the forefront of detection strategies, both from commercial and academic points of view.

8.2.4 The Case of ELISA

Immunoassays involve tests using antibodies as reagents, while enzyme immunoassays make use of enzymes attached to one or more molecules involved in an immunoassay, to allow quantification through the apparition of color after the addition of the suitable substrate or chromogen. ELISA is an analytical method taking part of this category. This technique is not directly a biosensor as such. It is indeed considered a gold standard in biology and biochemistry, but it is a routine laboratory assay. Biosensors, on their end, make extensive use of detection techniques implemented and optimized by reference ELISA tests. This is the main reason why they are described here in this enzymatic section, as ELISA tests are good validation and comparison tools for detection techniques such as for the development of novel biosensors.

ELISA tests involve reaction reagents to a solid phase-bound substance through incubation and washing steps. An enzymatic reaction is then used to provoke a color change often visible to the naked eye and quantified by a spectrometer (or fluorescence/luminescence). Usually, ELISA tests take place on a microtiter plate (96 wells) where antibodies or antigens of interest are attached. These molecules are often immobilized by adsorption on the solid phase. Then different assays can take place, depending on its nature. We can categorize these assays as direct ELISA, indirect ELISA, and sandwich ELISA (Figure 8.7) [39].

In traditional ELISA, antigens from a test sample are placed in a microtiter plate and bound on the surface of the wells. Then a matching antibody (primary) is applied over the surface so it can bind if the antigen of interest is present in the

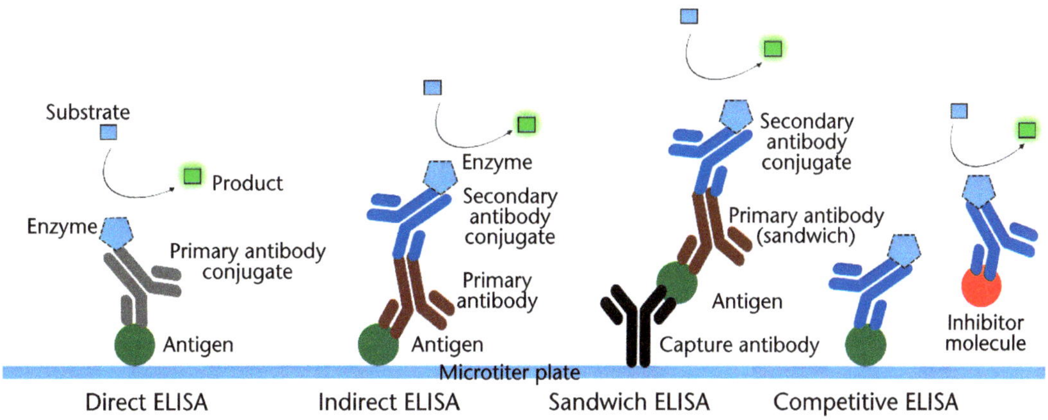

Figure 8.7 Scheme of different types of ELISA. From left to right, direct, indirect, sandwich, and competitive ELISA.

sample. This antibody is bound to an enzyme and the ones that are not tightly linked to the target analyte are washed away. At the end of the assay, a substrate is added so the enzymatic reaction can occur and lead to detection. Depending on the intensity of the signal, the quantification can be performed. This is usually done by a calibration curve using a batch of antigens at a known concentration (Figure 8.8).

Indirect ELISA and sandwich ELISA are often used to exploit secondary antibody conjugates as signal enhancers but specially to avoid the need for labeled antibodies against the target antigen. For instance, you can target different antigens using mouse antibodies (primary) but then use the same anti-mouse antibody carrying the enzyme to build your assay. It is therefore possible to save the same batch of molecules to create several assays. Most plasmonic assays using these configurations (direct, indirect) are actually transpositions of ELISA on other platforms.

8.2.5 Immobilization of Enzymes on Different Surfaces

With regard to immobilization on surfaces, the simplest method is an adsorption of the enzymes using hydrophobic, electrostatic, and other weak bonds. This technique is interesting to keep the integrity and structure of the enzymes and ensure highest activity. However, exactly as for nucleic acid receptors or antibodies, covalent immobilization provides longer stability. This covalent bonding takes advantage of the same working principle as for protein immobilization, using exposed amines and carboxyl groups to be coupled on activated surfaces. Glutaraldehyde or other cross-linkers can be implemented and provide nice binding kinetics.

As with any assay, surface development with active site concentration, blocking agent, and binding technique should be refined experimentally to the desired optimization. It is therefore impossible to provide a miracle recipe for perfect immobilization, but it must be tested and refined under the given experimental conditions. In some cases, the presence of too many active sites (or binding sites) can be detrimental to the bioassay because it prevents proper binding of targets to the

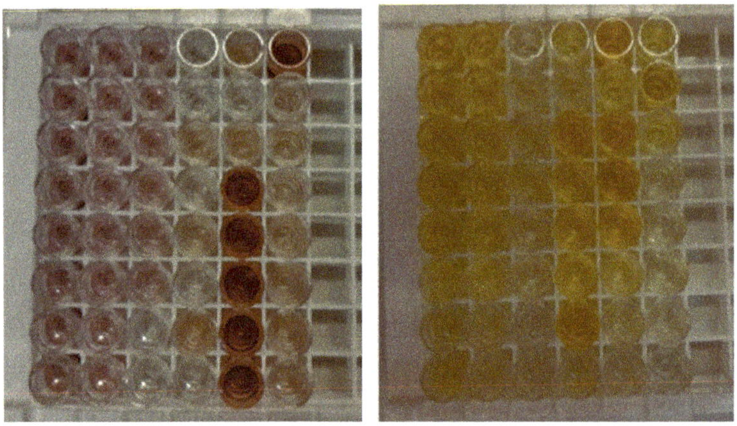

Figure 8.8 Picture of an ELISA performed on *Plasmodium falciparum* cultures as a validation for the detection of HRP2 proteins using optical fiber biosensor measurements. The left picture shows culture media and blood at different dilutions, while the right picture shows the color after washing and after the enzymatic reaction. The yellow intensity determines the concentration of HRP2 protein in wells [40].

ligand. It is therefore sometimes required to reduce the concentrations used and make use of spacer molecules. It is also important to consider that enzymes do not necessarily require being purified and immobilized on a substrate but can be self-contained on or within a biological matrix made of cells, vesicles, or others. The byproducts of the reaction can also be exploited for the assay and lead to direct or indirect indications of the target activity. More versatile approaches to avoid ligand immobilization are therefore possible due to enzymes.

8.2.6 Optical Fiber-Based Enzymatic Biosensors

In addition to the classical use of enzymes to carry out an enzyme-substrate type reaction, other original processes can be generated and provoke optical responses. One of these original processes that can be exploited with optical transducers is bioluminescence. It is a particular phenomenon exploited by specific symbiotic organisms. The chemical compound causing luminescence is luciferin, emitting light by oxidizing due to the intervention of an enzyme, the luciferase (Figure 8.9).

The generated signal can be flashing or glowing, depending on the kinetics of the reaction. Playing with absorbance and fluorescence parameters can therefore yield specific biosensing while tuning the reagents with these components.

Regarding the catalytic activities of enzymes and their ability to cleave target molecules, they can also be studied and used on optical fibers. This is particularly interesting in plasmonics, as the surface is highly sensitive to the molecular weight of both ligands and analytes. Using DNAses or specific proteases, it is possible to cut molecules present on the surface of a refractometer or plasmonic sensor to carry out a mass-sensitive reverse detection. This approach is often used via nanoparticles, further amplifying this surface mass depletion (Figure 8.10).

However, the use of a ligase or polymerase can play the role of increasing the signal to be detected by increasing the surface organic mass. This principle was used to demonstrate the live PCR monitoring using a plasmonic optical fiber [39]. By immobilizing other types of enzymes such as those causing conformational changes and chemiluminescence or fluorescence emission, the reversibility of the system can also be further investigated.

8.3 Proteins

8.3.1 Anchor Proteins (A, G, L)

The ideal proteins to immobilize in order to create a strong intermediary layer between antibody ligands and the surface of a transducer are certainly the ones from the Fc-binding family. Antibodies present a constant Fc region (corresponding

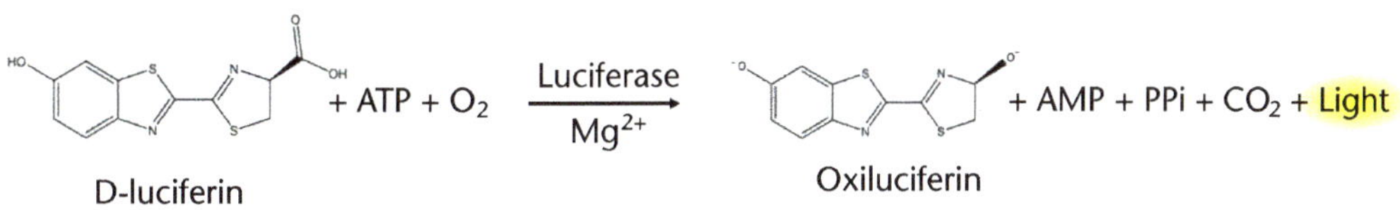

Figure 8.9 Chemical reaction of light production through luciferase enzymatic activity.

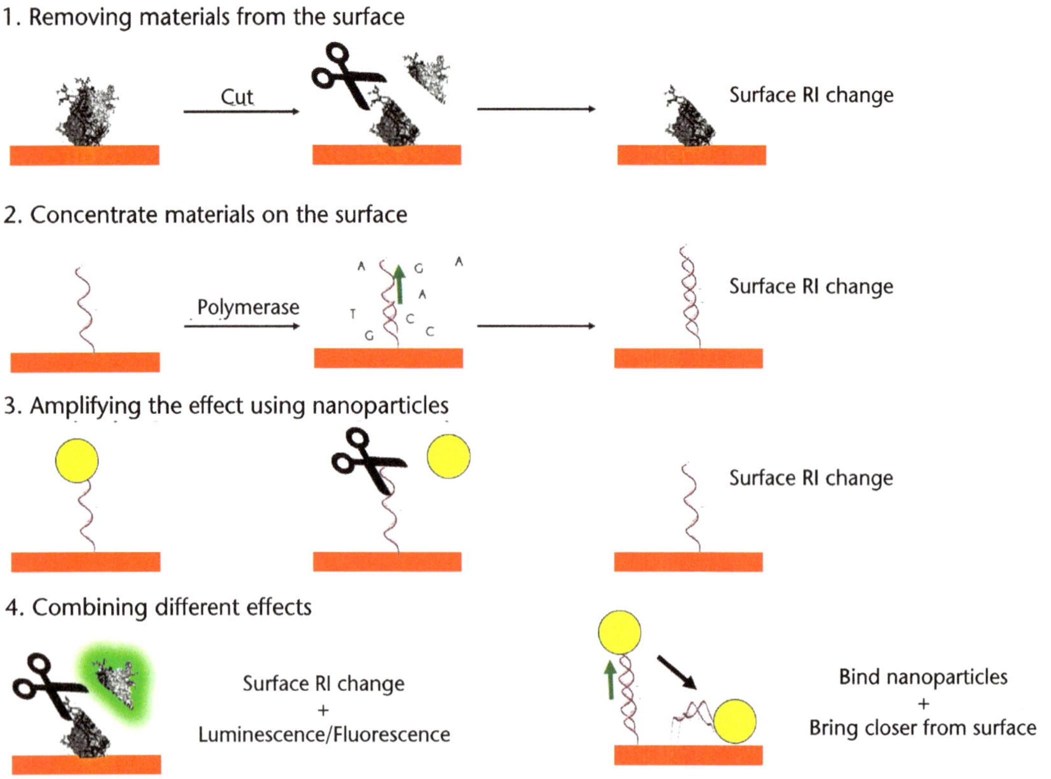

Figure 8.10 Schematic overview of different biosensing principles' integrating enzymes.

to the fragment crystallizable region) on their tail, which is the primary binding site for secondary antibodies (e.g., in the case of a sandwich assay) or certain surface proteins expressed at the surface of pathogenic bacteria such as *Staphylococcus aureus*. Among these well-known proteins, proteins A, G, and L are largely exploited in antibody purification. At the origin of their discovery, bacteria succeeded in thwarting our immune system by directing the antibodies in the wrong direction, inducing nonrecognition. This ability has been reversed to our advantage by using these proteins in biosensing in order to well-orient antibodies. They are therefore commercially available and are able to bind both monoclonal or polyclonal forms. Depending on the nature of the antibodies (especially their species) and the target application conditions, optimization or the binding is needed to select the best-suited anchor. Compatibility lists can often be downloaded directly from supplier websites to cross the type of antibody with the selected anchoring protein, but the offer is changing very quickly as many variants exist and new selection proteins are always developed. Binding proteins are therefore areas whose evolution must be followed regularly to check the availability of products and new molecular improvements [42, 43].

8.3.2 Protein/Antibodies or Protein/Protein Interactions

While many types of bioreceptors have been discussed over the last chapters, it is also possible to find studies that immobilize a target rather than a receptor (that become their opposite, by definition) and is similar to the strategy of ELISA. Circulating receptors therefore need to play the role of binders and this often happens in competitive assays, where the detection occurs only if the target is immobilized among a host of other analytes, also present on the sensor surface. This strategy of immobilizing the sampling matrix rather than the receptors requires great control of the detection because the amount of nonspecific binders directly available on the transducer can easily lead to false-positive responses. In the same philosophy, studying protein-protein interactions is also possible.

Protein engineering in its broad sense to achieve modifications of the native structure by addition, substitution, or deletion of chemical groups can provide completely different reactions through a change in substrate specificity or increase in the enzyme lifetime, under specific conditions (pH, salt concentration, temperature). This protein engineering can be implemented from the nucleotide sequence of the DNA coding for that protein to its anchoring on the final biosensor. The target improvements are to enhance the turnover number (k_{cat}), shift in or remove the pH dependence, improve the storage stability, and change the cofactor requirement. It can also lead to the widening or narrowing of the substrate specificity. All this is possible due to high-resolution 3-D structure evidence to locate the individual residues and binding sites and study the conformational changes of the protein in specific environments [44, 45].

8.4 Cells

Although most of the molecular bioreceptors have been mentioned over the last chapters and in this chapter, we still need to discuss cellular use. Uncommon cells are nevertheless very interesting because it is possible to use their integrity as a whole detection machinery. The first microbial biosensor was described in 1975 by Diviés [44] and was based on *Acetobacter xylinum* with an oxygen electrode. As first reported in the 1990s, cell receptors were roughly considered bags of enzymes. When analytes enter the cells, they are converted by intracellular enzymes. Cosubstrates are consumed and the reaction products are generated, yielding electrochemical signal. Oxygen levels and ionic composition, among others, can therefore be tracked as biologically active compounds to determine the metabolic state of the cells.

The use of nanostructures and nanowells designed at the surface of optical sensors are also of interest to specifically anchor cells on the transducer. The use of bacteria was clearly prioritized because of their ease of culture and immobilization, and the literature quickly gave rise to the term "microbial receptors" rather than the more generic "cellular receptors." Nonetheless, many different cell types have been used in recent years [46–48].

Among these nonmicrobial cells suited for sensing purposes, taste cells are certainly the most promising. Cells and receptors present in the nose or on the tongue are indeed able to naturally detect various chemical signals and are extensively studied for food and taste research, drug discovery, and environment monitoring [49, 50]. Taste cells are mainly located on the lingual epithelium of the tongue and can detect the five basic tastes including salty, sweet, bitter, sour, and umami. The receptors themselves are present in the membrane of taste cell microvilli, which are extended to the apex of taste buds. Different receptors are expressed in different subsets of taste cells, which make different types of cells showing different functions and properties. Once detected, the target signal is often transduced and amplified within the cell, resulting in an enhanced SNR. They have been studied on different transducers such as quartz microbalances and potentiometric sensors, and diverting their cellular functions within the framework of surface biosensing would certainly be a great challenge to tackle [51, 52].

Although cells are exploited as receptors, it is mostly their lipidic membranes that contain the receptors, yielding signal transduction and/or transport of molecules or organelles [53]. By inspiration from biological models, artificial membranes were studied to play the role of sensing surfaces. The phospholipids bilayer structure creates a hydrophobic barrier to soluble and ionic substances, while membrane proteins enable the transportation of signals and substances across the membrane through conformational changes. As they are the entry point to every cell, they became a significant focus of efforts to identify new pharmaceutical drug targets and new bioreceptors. In specific applications, artificial bilipid layers with both natural or synthetic molecules and protein receptors are created. Lipid vesicles that encapsulate an aqueous phase can also be synthesized. Other forms of lipid membranes such as suspended bilayers or nanodiscs also exist. These structures, including phospholipid vesicles, play an important role in artificial olfaction and the detection of volatile organic compounds [54–56].

8.5 Additional Layers and Matrices

8.5.1 Hydrogels

Hydrogels are not receptors as we understand them. Indeed, hydrogels are materials often used in biosensors to create protecting layers and to control molecular diffusion, enhancing biocompatibility and/or specificity. Usually, hydrogels are used as a 3-D matrix to encapsulate enzymes or other bioreceptors on a transducer. They present specific properties as they behave at the same time as solid and liquid materials. They are composed of polymer (and cross-linker) networks and are superabsorbent material, so they can be composed of more than 99% of the used solvent (often water or physiological buffer). The polymerization can be initiated after mixing enzymes or receptors so they can be homogeneously distributed in their volume. Moreover, depending on the needed properties, different cross-linkers and cross-linker concentrations can be chosen to tune the properties of the layers. An example of well-known composition is the use of acrylamide/bisacrylamide with water. The biggest challenge is to control the deposition and thickness of such layers of hydrogels on a surface while also controlling its anchoring over time on the latter. An abundant literature on the topic is available, so we would recommend more specific books and articles for more information [57–60].

8.5.2 Dextran Matrices

A second example of a 3-D layer is certainly the use of dextran, which is widespread and well-known with plasmonic biosensors. This is particularly the case while targeting low molecular weight targets to improve the distribution and binding along the surface. Using a sufficient amount of dextran layers on top of plasmonic materials can lead to signal enhancements roughly 10 times higher than the ones obtained with classical monolayers. However, they can be general artifacts due to the higher diffusion within the matrix. Hydrogel-like dextran matrix are manufactured and are commercially available for most plasmonic devices. The surface density and height can be modified according to the target application, leading to plethora of combinations when it comes to surface structures. Specific chemical groups such as carboxyl groups can also be grafted in order to apply traditional chemistries and consequently covalently immobilize bioreceptors. It was also reported that such 3-D matrices preserve the native structures of receptors and their binding efficiency. A smooth control of all these parameters need to be performed to certify the effectiveness of the method without impacting the diffusion and the selectivity negatively [61, 62].

8.6 Optical Fiber-Based Applications

After this brief overview about the other receptors, this section highlights recent original publications about optical fiber-based biosensors. This nonexhaustive list of contributions depicts the diversity and richness of this research area, which contributes to the development of this technology. It is intriguing to realize that technologies such as MIPs and enzymatic receptors are very popular and have shown their assets for the detection of many molecular targets while certain categories of receptors such as cells, hydrogels, or further molecular engineering are still weakly represented regarding optical fiber applications (Table 8.2).

Table 8.2 Biosensing Applications Using Optical Fibers and Other Bioreceptors

Optical Fiber Type	*Receptor/Target*	*LOD*	*Reference*
D-shaped plastic optical fiber (SPR)	MIP film/trinitrotoluene (TNT)	11 μg/mL (50 μM)	[63]
D-shaped plastic optical fiber (SPR)	MIP/L-nicotine and D-nicotine	1.86×10^{-4} M	[23]
D-shaped plastic optical fiber (SPR)	MIP/Furfural (2-furaldheide) in wine	0.004 mg/L	[20]
D-shaped plastic optical fiber (SPR)	MIP/perfluorinated alkylated substances (PFAs) in water	~0.5 ppb	[64]
D-shaped plastic optical fiber (SPR)	MIP/SARS-CoV-2 (subunit 1 spike protein)	0.058 μM	[65]
Two D-shaped plastic optical fiber platform (SPR)	MIP/dibenzyl disulfide (DBDS) in transformer oil	0.013 ppm (5.3×10^{-8} M)	[66]
Unclad polymer fiber (plastic optical fiber) (SPR)	MIP/Profenofos	2.5×10^{-6} μg/L (= 2.5 pg/L)	[67]
Fluorescent plastic optical fiber	MIP/ciprofloxacin	6.86 μM	[68]
Fluorescent plastic optical fiber	MIP/red dye detection in water	μM range	[69]
Fluorescent plastic optical fiber	MIP/herbicide (2,4-dichlorophenoxyacetic acid) and mycotoxin citrinin	nM and μM range	[70]

Table 8.2 *(Continued)*

Optical Fiber Type	*Receptor/Target*	*LOD*	*Reference*
Evanescent wave optical fiber sensor (FOEWS)	MIP/Bisphenol A	1.7 ng/mL	[71]
Polished plastic optical fiber	MIP—Graphene sensitization of glucose-imprinted polymer (G-IP)/ glucose	2.54 nM	[72]
Lossy mode resonance optical fiber (ZnO/MoS_2)	MIP/p-cresol in artificial urine	28 nM	[73]
Lossy mode resonance optical fiber (MoS_2/SnO_2)	MIP/creatinine	1.86 μg/mL	[74]
Silica light-diffusing fibers (SPR)	MIP/human serum transferrin (HTR)	4 fM	[75]
TFBG (SPR)	MIP coating/formaldehyde	More than a few ppm	[22]
TFBG (SPR)	Concanavalin A (protein) using polydopamine/glucose	~10^{-7} M	[76]
TFBG with graphene oxide (GO)	Glucose oxidase/glucose	NA, sensitivity of ~0.25 nm/mM	[77]
Tapered microfiber SMF	MIP/urea	17.1 fM	[78]
Unclad and polished polymer multimode fibers (lead and tip)	Haloalkane dehalogenase/1,2-dichloroethane (DCA) in aqueous solution	Starting at few mg/L	[79]
Fluorescence pH indicator polymer multimode fibers (lead and tip)	Haloalkane dehalogenase/ethylene bromide (1,2-dibromoethane)	1 μg/L	[80]
Fluorescence pH indicator polymer multimode fiber	Haloalkane dehalogenase/halogenated hydrocarbons	0.014 to 0.133 mM	[81]
MMF/SMF/MMF optical fiber sensor (SPR)	Glucose oxidase (GOx)/glucose	NA, sensitivity of 3.1 pm/(mg/dL)	[82]
U-bent fiber with polyaniline	Beta-lactamase/Ceftazidime (β-lactam antibiotics)	0.18 nM in milk, 9 nM in chicken, 0.18 nM in water	[83]
Single-mode hollow core tip fiber sensor (HCF) (LSPR)	Cholesterol oxidase/cholesterol	25.5 nM	[84]
Unclad fiber (SPR)	Glucose oxidase (GOx)/glucose	~0.01 mg/mL	[85]
Tapered optical fiber coated with Py/PVA	Glucose oxidase (GOx)/glucose	NA, sensitivity of 8.7×10^{-3} μW/ mM	[86]
Optical fiber with oxygen-sensitive ruthenium-based phosphorescent dye	Ortho-monooxygenase (TOM)/toluene	3 μM	[87]
Micro-ball fiber with gold nanoparticles (AuNPs) and graphene oxide (GO)	Uricase/uric acid (UA) in human serum	65.60 μM	[88]
200-μm core optical fiber tip sensor	Glutathione S-transferase and sol-gel entrapped bromocresol green/atrazine	0.84 μM	[89]
200-μm core optical fiber tip sensor	Flavobacterium species whole cells/ methyl parathion (pesticide)	0.3 μM	[90]
Quartz optical fiber bundle (electrochemiluminescence)	Glucose oxidase (GOx)/glucose	26 μM	[91]

Structures such as MIP layers and 3-D matrices require precise immobilization on cylindrical surfaces. The study of such fiber deposition processes is therefore a major concern to better understand these phenomena and improve the homogeneity and reproducibility of these biosensors. The works presented here contribute to this understanding and move this research forward. The integration of MIPs in 3-D layers or lipidic layers, enzymatic chain reactions coupling a cascade of transduction mechanisms, or integrating enzymes inside MIP layers are all promising challenges that could be addressed in future works on the subject.

References

[1] Andersson, L., B. Sellergren, and K. Mosbach, "Imprinting of Amino Acid Derivatives in Macroporous Polymers," *Tetrahedron Lett.*, Vol. 25, No. 45, 1984, pp. 5211–5214.

[2] Wulff, G., H.-G. Poll, and M. Minárik, "Enzyme-Analogue Built Polymers. XIX. Racemic Resolution on Polymers Containing Chiral Cavities," *J. Liq. Chromatogr.*, Vol. 9, No. 2–3, 1986, pp. 385–405.

[3] Wulff, G., A. Sarhan, and K. Zabrocki, "Enzyme-Analogue Built Polymers and Their Use for the Resolution of Racemates," *Tetrahedron Lett.*, Vol. 14, No. 44, 1973, pp. 4329–4332.

[4] Mosbach, K., and K. Haupt, "Some New Developments and Challenges in Non-Covalent Molecular Imprinting Technology," *J. Mol. Recognit.*, Vol. 11, No. 1–6, 1998, pp. 62–68.

[5] Wackerlig, J., and R. Schirhagl, "Applications of Molecularly Imprinted Polymer Nanoparticles and Their Advances Toward Industrial Use: A Review," *Anal. Chem.*, Vol. 88, No. 1, 2016, pp. 250–261.

[6] Viveiros, R., S. Rebocho, and T. Casimiro, "Green Strategies for Molecularly Imprinted Polymer Development," *Polymers (Basel)*, Vol. 10, No. 3, 2018.

[7] Yoshimi, Y., "Molecularly Imprinted Polymers Applicable for Biomimetic Catalysts in Sensors," Chapter 12 in *Molecularly Imprinted Catalysts: Principles, Syntheses, and Applications*, S. Li, et al. (eds.), Boston, MA: Elsevier, 2016, pp. 241–252.

[8] Barbani, N., et al., "Molecularly Imprinted Polymers by Phase Inversion Technique for the Selective Recognition of Saccharides of Biomedical Interest in Aqueous Solutions," *Polym. Int.*, Vol. 66, No. 6, 2017, pp. 900–907.

[9] Poller, A. M., et al., "Surface Imprints: Advantageous Application of Ready2use Materials for Bacterial Quartz-Crystal Microbalance Sensors," *ACS Appl. Mater. Interfaces*, Vol. 9, No. 1, 2017, pp. 1129–1135.

[10] Piletsky, S., et al., "Molecularly Imprinted Polymers for Cell Recognition," *Trends Biotechnol.*, Vol. 38, No. 4, 2020, pp. 368–387.

[11] He, S., et al., "Advances of Molecularly Imprinted Polymers (MIP) and the Application in Drug Delivery," *Eur. Polym. J.*, Vol. 143, 2021, p. 110179.

[12] Belbruno, J. J., "Molecularly Imprinted Polymers," *Chem. Rev.*, Vol. 119, No. 1, 2019, pp. 94–119.

[13] Ramanavicius, S., A. Jagminas, and A. Ramanavicius, "Advances in Molecularly Imprinted Polymers Based Affinity Sensors (Review)," *Polymers (Basel)*, Vol. 13, No. 6, March 22, 2021, p. 974.

[14] Piletsky, S. A., N. W. Turner, and P. Laitenberger, "Molecularly Imprinted Polymers in Clinical Diagnostics—Future Potential and Existing Problems," *Med. Eng. Phys.*, Vol. 28, No. 10, 2006, pp. 971–977.

[15] Zhang, Y., J. Zhang, and Q. Liu, "Gas Sensors Based on Molecular Imprinting Technology," *Sensors (Basel)*, Vol. 17, No. 7, 2017, p. 1567.

[16] Gaggiotti, S., et al., "Peptides, DNA and MIPs in Gas Sensing: From the Realization of the Sensors to Sample Analysis," *Sensors (Basel)*, Vol. 20, No. 16, August 8, 2020, p. 4433.

[17] Sroysee, W., et al., "Molecularly Imprinted Polymers to Detect Profenofos and Carbofuran Selectively with QCM Sensors," *Phys. Med.*, Vol. 7, June 2019.

[18] Dar, K. K., et al., "Molecularly Imprinted Polymers for the Selective Recognition of Microorganisms," *Biotechnol. Adv.*, Vol. 45, 2020, p. 107640.

[19] Yang, W., et al., "Molecularly Imprinted Polymers Based Optical Fiber Sensors: A Review," *TrAC Trends Anal. Chem.*, Vol. 152, 2022, p. 116608.

[20] Pesavento, M., et al., "SPR-Optical Fiber-Molecularly Imprinted Polymer Sensor for the Detection of Furfural in Wine," *Biosensors*, Vol. 11, No. 3, 2021.

[21] Rico-Yuste, A., and S. Carrasco, "Molecularly Imprinted Polymer-Based Hybrid Materials for the Development of Optical Sensors," *Polymers (Basel)*, Vol. 11, No. 7, 2019.

[22] González-Vila, Á., et al., "Molecularly Imprinted Electropolymerization on a Metal-Coated Optical Fiber for Gas Sensing Applications," *Sensors Actuators B Chem.*, Vol. 244, 2017, pp. 1145–1151.

[23] Cennamo, N., et al., "High Selectivity and Sensitivity Sensor Based on MIP and SPR in Tapered Plastic Optical Fibers for the Detection of L-Nicotine," *Sensors Actuators B Chem.*, Vol. 191, 2014, pp. 529–536.

[24] Altintas, Z., et al., "Computationally Modelled Receptors for Drug Monitoring Using an Optical Based Biomimetic SPR Sensor," *Sensors Actuators B Chem.*, Vol. 224, 2016, pp. 726–737.

[25] Abdin, M. J., Z. Altintas, and I. E. Tothill, "In Silico Designed NanoMIP Based Optical Sensor For Endotoxins Monitoring," *Biosens. Bioelectron.*, Vol. 67, 2015, pp. 177–183.

[26] Clark, L. C., and C. Lyons, "Electrode Systems for Continuous Monitoring in Cardiovascular Surgery," *Ann. N. Y. Acad. Sci.*, Vol. 102, No. 1, 1962, pp. 29–45.

[27] Newman, J. D., and S. J. Setford, "Enzymatic Biosensors," *Mol. Biotechnol.*, Vol. 32, No. 3, 2006, pp. 249–268.

[28] Imtaiyaz Hassan, M., et al., "Structure, Function and Applications of Carbonic Anhydrase Isozymes," *Bioorg. Med. Chem.*, Vol. 21, No. 6, 2013, pp. 1570–1582.

[29] Turner, A. P. F., I. Karube, and G. S. Wilson, *Biosensors: Fundamentals and Applications*, Oxford, U.K.: Oxford University Press, 1987.

[30] Cornish-Bowden, A., "The Origins of Enzyme Kinetics," *FEBS Lett.*, Vol. 587, No. 17, 2013, pp. 2725–2730.

[31] Richter, M., "Functional Diversity of Organic Molecule Enzyme Cofactors," *Nat. Prod. Rep.*, Vol. 30, No. 10, 2013, pp. 1324–1345.

[32] https://www.nature.com/articles/214986a0.

[33] https://www.science.org/doi/abs/10.1126/science.158.3798.270.

[34] Wang, J., "Electrochemical Glucose Biosensors," *Chem. Rev.*, Vol. 108, No. 2, 2008, pp. 814–825.

[35] Kessel, A., and N. Ben-Tal, *Introduction to Proteins, Structure, Function, and Motion*, 2nd ed., Boca Raton, FL: CRC Press, 2018.

[36] Potts, R. O., J. A. Tamada, and M. J. Tierney, "Glucose Monitoring by Reverse Iontophoresis," *Diabetes. Metab. Res. Rev.*, Vol. 18, No. S1, 2002, pp. S49–S53.

[37] Gray, M., et al., "Implantable Biosensors and Their Contribution to the Future of Precision Medicine," *Vet. J.*, Vol. 239, 2018, pp. 21–29.

[38] Heo, Y. J., and S.-H. Kim, "Toward Long-Term Implantable Glucose Biosensors for Clinical Use," *Appl. Sci.*, Vol. 9, No. 10, 2019.

[39] Crowther, J. R., *The ELISA Guidebook*, 2nd ed., New York: Springer, 2009.

[40] Loyez, M., et al., "PfHRP2 Detection Using Plasmonic Optrodes: Performance Analysis," *Malar. J.*, Vol. 20, No. 1, 2021, pp. 1–9.

[41] Pollet, J., et al., "Real-Time Monitoring of Solid-Phase PCR Using Fiber-Optic SPR," *Small*, Vol. 7, No. 8, 2011, pp. 1003–1006.

[42] Rigi, G., S. Ghaedmohammadi, and G. Ahmadian, "A Comprehensive Review on Staphylococcal Protein A (SpA): Its Production and Applications," *Biotechnol. Appl. Biochem.*, Vol. 66, No. 3, 2019, pp. 454–464.

[43] Choe, W., T. A. Durgannavar, and S. J. Chung, "Fc-Binding Ligands of Immunoglobulin G: An Overview of High Affinity Proteins and Peptides," *Mater. (Basel, Switzerland)*, Vol. 9, No. 12, 2016.

[44] Lutz, S., and S. M. Iamurri, "Protein Engineering: Past, Present, and Future," *Methods Mol. Biol.*, Vol. 1685, 2018, pp. 1–12.

[45] Woodley, J. M., "Protein Engineering of Enzymes for Process Applications," *Curr. Opin. Chem. Biol.*, Vol. 17, No. 2, 2013, pp. 310–316.

[46] Diviés, C., "[Remarks on ethanol oxidation by an 'Acetobacter xylinum' microbial electrode (author's transl)]," *Ann. Microbiol. (Paris)*, Vol. 126, No. 2, 1975, pp. 175–186. French.

[47] Reshetilov, A. N., P. V. Iliasov, and T. A. Reshetilov, "The Microbial Cell Based Biosensors," Chapter 15 in *Intelligent and Biosensors*, V. S. Somerset, (ed.), London, U.K.: InTechOpen, 2010.

[48] Lim, J. W., et al., "Review of Micro/Nanotechnologies for Microbial Biosensors," *Front. Bioeng. Biotechnol.*, Vol. 3, 2015.

[49] Wu, C., et al., "Recent Advances in Taste Cell- and Receptor-Based Biosensors," *Sensors Actuators B Chem.*, Vol. 201, 2014, pp. 75–85.

[50] Lindemann, B., "Receptors and Transduction in Taste," *Nature*, Vol. 413, No. 6852, 2001, pp. 219–225.

[51] Chen, P., et al., "Taste Receptor Cell-Based Biosensor for Taste Specific Recognition Based on Temporal Firing," *Biosens. Bioelectron.*, Vol. 25, No. 1, 2009, pp. 228–233.

[52] Zhang, N., et al., "Recent Advances in Development of Biosensors for Taste-Related Analyses," *TrAC Trends Anal. Chem.*, Vol. 129, 2020, p. 115925.

[53] Osaki, T., and S. Takeuchi, "Artificial Cell Membrane Systems for Biosensing Applications," *Anal. Chem.*, Vol. 89, No. 1, 2017, pp. 216–231.

[54] Steller, L., M. Kreir, and R. Salzer, "Natural and Artificial Ion Channels for Biosensing Platforms," *Anal. Bioanal. Chem.*, Vol. 402, No. 1, 2012, pp. 209–230.

[55] Barbosa, A. J. M., A. R. Oliveira, and A. C. A. Roque, "Protein- and Peptide-Based Biosensors in Artificial Olfaction," *Trends Biotechnol.*, Vol. 36, No. 12, 2018, pp. 1244–1258.

[56] Vörös, J., et al., "Optical Grating Coupler Biosensors," *Biomaterials*, Vol. 23, No. 17, 2002, pp. 3699–3710.

[57] Wolfbeis, O. S., "Fiber-Optic Chemical Sensors and Biosensors," *Anal. Chem.*, Vol. 78, 2006, pp. 3859–3874.

[58] Zhenglan, B., et al., "Glucose Biosensing Based on a Hydrogel Optical Fiber Immobilized with Glucose Oxidase," *Optik (Stuttg.)*, Vol. 255, 2022, p. 168655.

[59] Mateescu, A., et al., "Thin Hydrogel Films for Optical Biosensor Applications," *Membranes (Basel)*, Vol. 2, No. 1, 2012, pp. 49–69.

[60] Herrmann, A., R. Haag, and U. Schedler, "Hydrogels and Their Role in Biosensing Applications," *Adv. Healthc. Mater.*, Vol. 10, No. 11, 2021, pp. 1–25.

[61] Chiodi, E., et al., "The Effects of Three-Dimensional Ligand Immobilization on Kinetic Measurements in Biosensors," *Polymers (Basel)*, Vol. 14, No. 2, 2022, pp. 1–12.

[62] Ambrosetti, E., et al., "Patterned Carboxymethyl-Dextran Functionalized Surfaces Using Organic Mixed Monolayers for Biosensing Applications," *ACS Appl. Bio Mater.*, 2022.

[63] Cennamo, N., et al., "Sensors Based on Surface Plasmon Resonance in a Plastic Optical Fiber for the Detection of Trinitrotoluene," *Sensors Actuators B Chem.*, Vol. 188, 2013, pp. 221–226.

[64] Cennamo, N., et al., "A Simple and Low-Cost Optical Fiber Intensity-Based Configuration for Perfluorinated Compounds in Water Solution," *Sensors*, Vol. 18, No. 9, 2018.

[65] Cennamo, N., et al., "SARS-CoV-2 Spike Protein Detection Through a Plasmonic D-Shaped Plastic Optical Fiber Aptasensor," *Talanta*, Vol. 233, October 1, 2021.

[66] Cennamo, N., et al., "Intensity-Based Plastic Optical Fiber Sensor with Molecularly Imprinted Polymer Sensitive Layer," *Sensors Actuators B Chem.*, Vol. 241, 2017, pp. 534–540.

[67] Shrivastav, A. M., S. P. Usha, and B. D. Gupta, "Fiber Optic Profenofos Sensor Based on Surface Plasmon Resonance Technique and Molecular Imprinting," *Biosens. Bioelectron.*, Vol. 79, 2016, pp. 150–157.

[68] Huang, Q.-D., et al., "A Novel Fluorescent Optical Fiber Sensor for Highly Selective Detection of Antibiotic Ciprofloxacin Based on Replaceable Molecularly Imprinted Nanoparticles Composite Hydrogel Detector," *Sensors Actuators B Chem.*, Vol. 328, 2021, p. 129000.

[69] Foguel, M. V., et al., "A Molecularly Imprinted Polymer-Based Evanescent Wave Fiber Optic Sensor for the Detection of Basic Red 9 Dye," *Sensors Actuators B Chem.*, Vol. 218, 2015, pp. 222–228.

[70] Ton, X.-A., et al., "A Disposable Evanescent Wave Fiber Optic Sensor Coated with a Molecularly Imprinted Polymer as a Selective Fluorescence Probe," *Biosens. Bioelectron.*, Vol. 64, 2015, pp. 359–366.

[71] Xiong, Y., et al., "A Microvolume Molecularly Imprinted Polymer Modified Fiber-Optic Evanescent Wave Sensor for Bisphenol A Determination," *Anal. Bioanal. Chem.*, Vol. 406, No. 9, 2014, pp. 2411–2420.

[72] Azargoshasb, T., et al., "Evanescent Wave Optical Trapping and Sensing on Polymer Optical Fibers for Ultra-Trace Detection of Glucose," *ACS Omega*, Vol. 5, No. 35, 2020, pp. 22046–22056.

[73] Usha, S. P., and B. D. Gupta, "Urinary P-Cresol Diagnosis Using Nanocomposite of ZnO/MoS2 and Molecular Imprinted Polymer on Optical Fiber Based Lossy Mode Resonance Sensor," *Biosens. Bioelectron.*, Vol. 101, 2018, pp. 135–145.

[74] Sharma, S., A. M. Shrivastav, and B. D. Gupta, "Lossy Mode Resonance Based Fiber Optic Creatinine Sensor Fabricated Using Molecular Imprinting over Nanocomposite of MoS_2/SnO_2," *IEEE Sensors Journal*, Vol. 20, No. 8, 2020, pp. 4251–4259.

[75] Arcadio, F., et al., "A Plasmonic Biosensor Based on Light-Diffusing Fibers Functionalized with Molecularly Imprinted Nanoparticles for Ultralow Sensing of Proteins," *Nanomaterials*, Vol. 12, No. 9, 2022.

[76] Lobry, M., et al., "Non-Enzymatic D-Glucose Plasmonic Optical Fiber Grating Biosensor," *Biosens. Bioelectron.*, Vol. 142, 2019, p. 111506.

[77] Jiang, B., et al., "Label-Free Glucose Biosensor Based on Enzymatic Graphene Oxide-Functionalized Tilted Fiber Grating," *Sensors Actuators B Chem.*, Vol. 254, 2018, pp. 1033–1039.

[78] Gorai, P., and R. Jha, "Artificial Receptor-Based Optical Sensors (AROS): Ultra-Sensitive Detection of Urea," *Adv. Photonics Res.*, Vol. 2, No. 8, 2021, p. 2100044.

[79] Campbell, D. W., C. Müller, and K. F. Reardon, "Development of a Fiber Optic Enzymatic Biosensor for 1,2-Dichloroethane," *Biotechnol. Lett.*, Vol. 28, No. 12, 2006, pp. 883–887.

[80] Reardon, K. F., D. W. Campbell, and C. Müller, "Optical Fiber Enzymatic Biosensor for Reagentless Measurement of Ethylene Dibromide," *Eng. Life Sci.*, Vol. 9, No. 4, 2009, pp. 291–297.

[81] Bidmanova, S., et al., "Development of an Enzymatic Fiber-Optic Biosensor for Detection of Halogenated Hydrocarbons," *Anal. Bioanal. Chem.*, Vol. 398, No. 5, 2010, pp. 1891–1898.

[82] Zhang, J., et al., "Optical Fiber SPR Biosensor with a Solid-Phase Enzymatic Reaction Device for Glucose Detection," *Sensors Actuators B Chem.*, Vol. 366, 2022, p. 131984.

[83] Nag, P., et al., "Evanescent Wave Optical Fiber Sensors Using Enzymatic Hydrolysis on Nanostructured Polyaniline for Detection of β-Lactam Antibiotics in Food and Environment," *Anal. Chem.*, Vol. 93, No. 4, 2021, pp. 2299–2308.

[84] Kumar, S., et al., "LSPR-Based Cholesterol Biosensor Using Hollow Core Fiber Structure," *IEEE Sens. J.*, Vol. 19, No. 17, 2019, pp. 7399–7406.

[85] Zheng, W., et al., "Highly-Sensitive and Reflective Glucose Sensor Based on Optical Fiber Surface Plasmon Resonance," *Microchem. J.*, Vol. 157, 2020, p. 105010.

[86] Idris, S., et al., "Gamma Irradiated Py/PVA for GOx Immobilization on Tapered Optical Fiber for Glucose Biosensing," *Sensors Actuators B Chem.*, Vol. 273, 2018, pp. 1404–1412.

[87] Zhong, Z., et al., "Fiber Optic Monooxygenase Biosensor for Toluene Concentration Measurement in Aqueous Samples," *Biosens. Bioelectron.*, Vol. 26, No. 5, 2011, pp. 2407–2412.

[88] Kumar, S., et al., "Development of Uric Acid Biosensor Using Gold Nanoparticles and Graphene Oxide Functionalized Micro-Ball Fiber Sensor Probe," *IEEE Transactions on Nanobioscience*, Vol. 19, No. 2, 2020, pp. 173–182.

[89] Andreou, V. G., and Y. D. Clonis, "Novel Fiber-Optic Biosensor Based on Immobilized Glutathione S-Transferase and Sol–Gel Entrapped Bromcresol Green for the Determination of Atrazine," *Anal. Chim. Acta*, Vol. 460, No. 2, 2002, pp. 151–161.

[90] Kumar, J., S. K. Jha, and S. F. D'Souza, "Optical Microbial Biosensor for Detection of Methyl Parathion Pesticide Using Flavobacterium Sp. Whole Cells Adsorbed on Glass Fiber Filters as Disposable Biocomponent," *Biosens. Bioelectron.*, Vol. 21, No. 11, 2006, pp. 2100–2105.

[91] Zhu, L., Y. Li, and G. Zhu, "A Novel Flow Through Optical Fiber Biosensor for Glucose Based on Luminol Electrochemiluminescence," *Sensors Actuators B Chem.*, Vol. 86, No. 2, 2002, pp. 209–214.

CHAPTER 9

Combined Plasmonic Sensors

While plasmonic optical fiber biosensors bring their share of originality and versatility as demonstrated in previous chapters, other technologies can nonetheless be mingled to enhance the performance of detection or widen the sensing modalities. This is particularly the case via the use of fluorescence, electrical current, magnetic particles, or ultrasound or through the Raman effect. The great diversity of these technologies leads to many relevant associations resulting in new detection perspectives and brings a new energy to plasmonic-based research. The phenomena observed with these sensors help to improve knowledge of surface interactions taking place on thin layers and so to better understand the ins and outs to inspire future advanced sensors. Hence, this chapter will concentrate on such combined plasmonic platforms and present their inherent sensing mechanisms together with their relevant applications. For each technology, basic considerations about their operating principle and practical implementation will first be provided. Then a review of their combination with plasmonic optical fibers will be made.

9.1 Electro-Plasmonics

Electrochemistry is a very old practice, as it dates back to the end of the eighteenth century and the work of Volta. It corresponds to the study of chemical processes that force electrons' movements, electricity. The latter can be generated by the movements of electrons from one element to another in a chemical reaction called oxidation-reduction (most often contracted as redox) reaction. A redox reaction involves a change in the oxidation state of one element or some elements. When a molecule loses an electron, its oxidation state increases and it is therefore oxidized. When it gains an electron, it is the opposite and it is reduced. For instance, if we consider the redox reaction $H_2 + F_2 \rightarrow 2HF$, the oxidation reaction is $H_2 \rightarrow 2H^+ + 2e^-$ and the reduction reaction is $F_2 + 2e^- \rightarrow 2F^-$ so that the overall reaction can be rewritten as $H_2 + F_2 \rightarrow 2H^+ + 2F^-$. Oxidation is therefore the loss of electrons, while reduction refers to the acquisition of electrons. The chemical species that is oxidized is called the reducing agent or reductant (H_2 in the example), while the one that gets reduced is the oxidant or oxidizing agent (F_2 in the example).

Electrochemistry is a mainstay in the field of biosensing. Most routine biosensors rely on measurements based on electrical signals. Because plasmonic platforms require metallic films, the opportunity of their technological association with electrochemistry leading to the concept of electro-plasmonic devices is rather straightforward.

Electrochemistry relies on reactions that produce or consume free electrons (typically, oxidation and reduction reactions where electrons are exchanged).

Electroanalytical methods are based on electrically conductive probes named electrodes, able to generate an electrical detection of analytes in solution. These electrodes are used to bridge the electronic device to the sample and to track the presence of the target (qualitative) or to determine its concentration (quantitative). Many parameters are involved in electrochemistry, as described next. Most of these techniques rely on the flow of electrons between the electrodes and the analytes themselves. The latter can be ions, proteins, or cells and therefore need to be able to accept or give back one (or even more) electrons to the electrode, in an oxidation/reduction context [1]. To achieve this detection, different strategies can be used such as voltammetry, conductometry, amperometry, and potentiometry, among many others. All these configurations are often needed for both reference and working electrodes. Silver and its solid salt (Ag/AgCl) electrodes are commonly used as reference electrodes for both pH-meters and electroanalytical devices. They work as a reversible redox electrode so the equilibrium is between Ag and AgCl. In this way, the reaction can be written and simplified by a reversible reaction, as follows:

$$Ag_{(s)} + AgCl_{(s)} + e^- \leftrightarrows Ag_{(s)} + Ag^+ + Cl^- + e^-$$

$$AgCl_{(s)} \leftrightarrows Ag^+ + Cl^-$$

Commercial reference electrodes are based on a glass tube where a metal piece of Ag is coated with a thin layer of AgCl. A porous filter is located at the tip of the electrode and provides a contact between the solution to be measured and the electrolyte solution in equilibrium with the AgCl coated on the Ag surface, itself connected to the measuring instrument. KCl solutions are often added to stabilize the AgCl concentration. Oppositely, the working electrode is the one on which the reaction of interest occurs. When the reaction occurring on the electrode is an oxidation (reduction), it is called cathode (anode). Chemically modified (or functionalized) electrodes are used to selectively detect targets.

9.1.1 Voltammetry

Voltammetry is a well spread technique that relies on a set of electrodes (usually two or three electrodes) immersed in a liquid solution. A regular potential variation is applied to the working (or indicator) electrode relatively to the reference electrode (with a constant potential). Analytes electrochemically react at the working electrode. Sometimes, a third electrode (named counter electrode) is placed in solution to carry most of the current. In this case, the potential stays controlled between the working and the reference electrodes while the current flows between the counter and the working electrodes. The potential variation can take different ways, so many forms of voltammetry exist. Polarography, triangular wave voltammetry, and AC voltammetry are some examples. Cyclic voltammetry (CV) is largely used for sensors and is performed by cycling the potential of a working electrode and measuring the resulting current, leading to typical voltammograms (Figure 9.1). This type of graph usually obtained in CV shows on the x-axis the parameter that is imposed

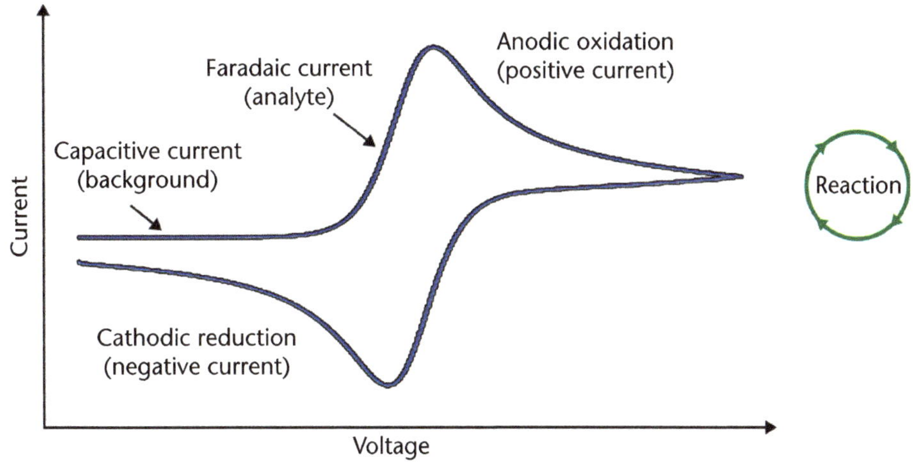

Figure 9.1 Concept of a cyclic voltammogram.

to the system, here the potential E (voltage, V), while the *y*-axis is the response, here the current. Depending on the convention used (US or IUPAC), the reading of the curve can be rotated by 180°. In this schematic example (reading following the reaction label on the right of the figure), the voltage is in excess of that predicted by the Nernst equation. The interesting part of the graph is usually caused by the electrochemical reaction, yielding a Faraday current. The capacitive current is an unwanted effect caused by the ions accumulating on the electrode and is considered background. They result in a duck-shape voltammogram, and the reaction is reversible by scanning the voltage back to higher or lower values. An interesting article with further explanations for beginners in CV is presented in [2].

9.1.2 Conductometry

Conductometry is the measurement of the capability of analytes to conduct an electrical current. This measurement is directly related to Ohm's law, where the electric current I is inversely proportional to the resistance R. The inverse of the resistance (R) is called the conductance (G); the latter increases when the ability to conduct an electric current increases. In solution, the electric current is conducted thanks to ions, so the conductance highly depends on the nature of the solution (both concentration and type of ions). It is known that small and highly charged ions present higher conductivity (they travel more rapidly with high electrostatic attraction) compared to larger and less charged ions. This method is poorly selective but is highly relevant (e.g., to verify the purity of ion-free water).

9.1.3 Amperometry

For amperometric analysis, the current between the working electrode and a second electrode is measured and related to the concentration of the target analyte. The potential of the working electrode is set to a value on the plateau phase of the voltametric wave. Usually, a calibration curve is achieved using a series of known concentrations (standard solutions). Then the target concentration is determined from that curve. Most commercially available glucose sensors rely on amperometric devices generating a current from the electrochemical reaction between glucose (target) and glucose oxidase immobilized on the working electrode [3].

9.1.4 Potentiometry

Potentiometry relies on the potential between two electrodes while the electric current between those electrodes is controlled. Usually, the potential of the working electrode varies depending on the analyte concentration while the potential of the reference electrode stays constant. Potentiometry is one of the most frequently used electroanalytical methods. The role of the working electrode is to probe the reaction of interest, so it is required to only induce the reaction of interest, without inducing any other reactions. The reference electrode provides a reference level for the potential difference between the reference and working electrode to be actually measurable. It is required to have a constant, stable, or predictable potential. A voltameter is connected in between the reference electrode and the working electrode to measure the potential difference. Finally, the counter electrode is usually additionally needed to conduct the current of interest through itself, so its surface needs to be large enough, and needs to be chemically inert (Figure 9.2). The development of electrochemical biosensors, especially miniaturized probes at large scale is a large market trusted by companies such as Zimmer and Peacock Ltd., Palmsens, Abbott, Nix Biosensors, and Masimo.

Due to the transposition of bulky electrochemical techniques to small-printed sensors, they quickly became game changers for POC, especially for food quality, environmental monitoring, and clinical diagnoses. Small electrochemical chips can indeed be easily mass-produced at extremely low cost and be directly read by portable devices. They also require small volumes such as a drop of blood or saliva to properly work. In the same philosophy, metal-coated optical fibers such as plasmonic optical fibers can be turned into small electrodes and attract targets due to an electrical current or make reactions to happen due to specific parameters, such as those involved in electrochemistry. More than passive electrodes, optical fibers still conduct light,

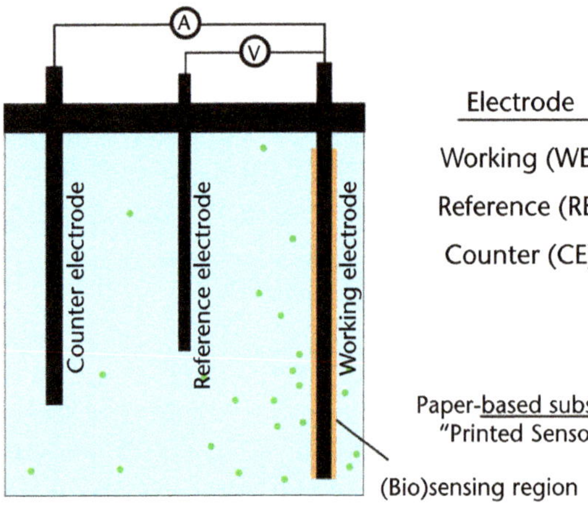

Electrode	Purpose	Requirements
Working (WE)	Reaction of interest	Only reaction of interest
Reference (RE)	Reference level	Constant and stable
Counter (CE)	Conduct the current	Large surface and inert

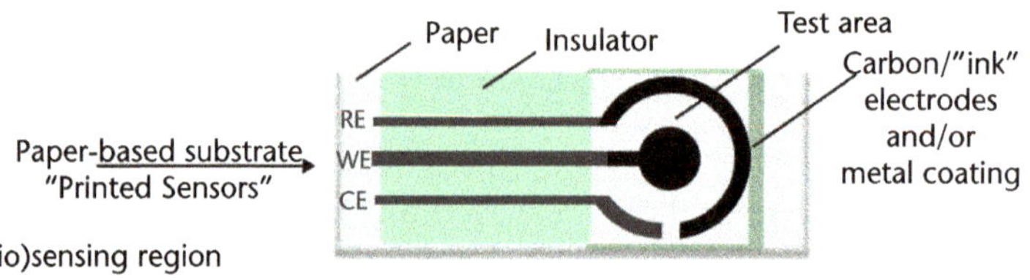

Figure 9.2 Scheme of the electrodes for electrochemical measurements. Purpose and requirements for these electrodes and scheme of a paper-based electrochemical probe, usually called a printed sensor.

an electromagnetic wave. The plasmonic effect can therefore be tuned depending on its illumination or on the current used for the electro-plasmonics generation.

9.1.5 Combination with Plasmonic Optical Fiber Sensing

Because of the presence of a thin metal film on their surface, plasmonic optical fiber configurations can be quite straightforwardly combined with electrochemistry. The first demonstration dates back to 2016 using gold-coated TFBGs for *in situ* electroactive biofilms measurements [4]. The gold-coated TFBG is used as the working electrode in a three-electrode bioelectrochemical cell. The nanometric-scale, gold coating over the fiber surface offers both the simultaneous detection of the electrochemical response (electrochemical current) and the optical signal (SPR), which allows assessing the mechanism of electroactive biofilm formation. Such devices also appear ideally suited to detect electrical processes and charging states within batteries and supercapacitors used in energy storage as their properties depend strongly on the distributions of charges along the metal coating surfaces. Placing a metal-coated fiber sensor in proximity to an electrode yields a response to the state of the electrode and to the local concentration of electrolytes. The demonstration of such a sensing capability was published in 2018 using gold-coated TFBGs. Simultaneous optical and electrochemical measurements of a supercapacitor under charge-discharge cycles were reported, as shown in Figure 9.3 [5].

More recently, it was demonstrated that an electrochemical SPR optical fiber sensor can be used for the *in situ* detection of ion kinetics at the electrolyte-electrode interface without interfering with battery operation [6]. In particular, the two-step ion intercalation process of MnO_2 as an aqueous Zn-ion battery electrode has been proved by the change of optical signal during discharge process. A semi-quantitative method based on the normalized optical rate curves has also been developed to further analyze the different ion kinetics between the two electrode materials. Using this method, it is further demonstrated that a poly(3,4-ethylenedioxythiophene) (PEDOT) layer coating on MnO_2 can significantly optimize the H^+ diffusion kinetics of the electrode, thus improving the electrochemical performance.

In [7], an experimental evidence that the limit of detection (LOD) of gold-coated TFBGs biosensors can be lowered by integrating them in a electrochemical system has been demonstrated. Employed again as a working electrode, the gold-coated TFBG served for SPR excitation and electric attraction of biological targets in the probed medium at the same time. The developed electro-plasmonic biosensor showed an increase of the sensitivity value and an LOD improvement of about two orders of magnitude with respect to the conventional method for HER2 cancer biomarker sensing. These recent works confirm the interest of the scientific community to mix plasmons with electrochemical phenomena to highlight the added value of this combination.

9.2 Magneto-Plasmonics

It is well known that surface plasmons provide light-matter interactions that enable molecular detection. However, for the past few years, many applications outside the

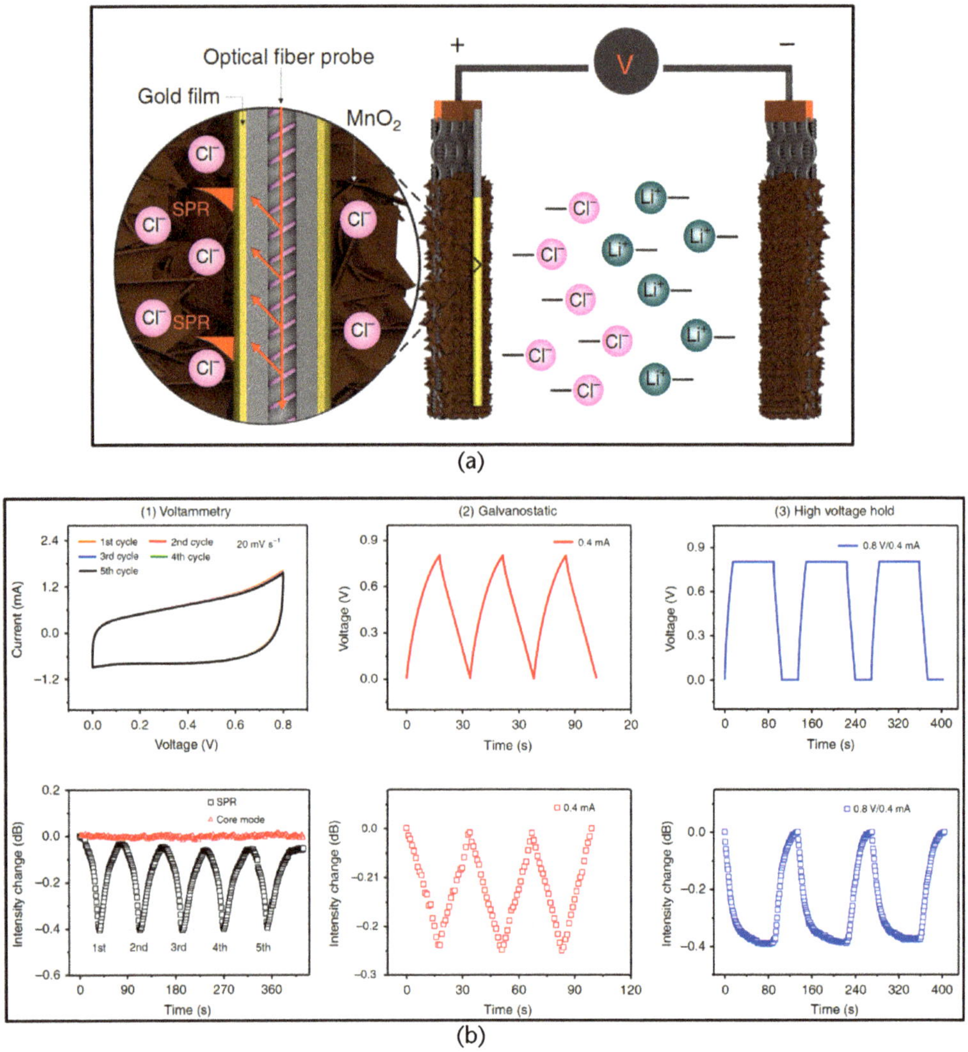

Figure 9.3 (a) TFBG-SPR measurement of the state of charge of a supercapacitor; and (b) representative electrical measurements (top) and corresponding response of the most sensitive TFBG resonance (bottom). (*From:* [5].)

biosensing area have been increasing using plasmonics, especially for temperature and pH sensing, energy harvesting, telecommunications, and quantum plasmonics [8–11]. A novel concept to achieve these goals relies on magneto-plasmonics, combining magnetism and plasmonics. It gives birth to new paradigms to study plasmon-spin interactions, to enhance the magneto-optical cavity in materials, to tune the plasmonic response under weak magnetic fields, and, finally, to use magnetic beads for biosensing [12].

Magnetism can be defined as the capacity to induce attractive and repulsive phenomena in other entities. For instance, the electric currents and the magnetic moments of particles can give rise to a magnetic field, which acts on other currents and magnetic moments. Magnetism is only one aspect of the combined phenomena called electromagnetism. Magnets are well known to induce a magnetic field.

They are made of ferromagnetic materials (also named ferrimagnetic, such as Ni, Co, Fe, and alloys) that are strongly attracted by magnetic fields and can be magnetized to allow a permanent magnetic effect. Some rare elements are also known for their magnetic properties and only magnetic dipoles have been observed so far (Figure 9.4), although predictions describe the possible existence of magnetic monopole. Dipoles imply the existence of a North (N) Pole and a South (S) Pole and magnetic field lines between them.

Ferromagnetism is responsible for most of the effects of magnetism encountered in everyday life, but there are other types of magnetism, as summarized in Table 9.1. The force of a magnet on paramagnetic, diamagnetic, and antiferromagnetic materials is usually too weak to be felt. It can only be detected with laboratory instruments, so that these substances are most often described as nonmagnetic.

As mentioned in Chapter 2, light is an electromagnetic wave. Therefore, magneto-optics refers to the influence of magnetic fields on light propagation. While polarized light is transmitted through magneto-optic material in the presence of a magnetic field, the polarization state is modified. For example, the Faraday effect explains the rotation of the polarization state (i.e., without dephasing) while the magnetic field is parallel to the propagation direction of light. Another magneto-optic effect is the Cotton-Mouton effect (also called magneto-optic Kerr effect),

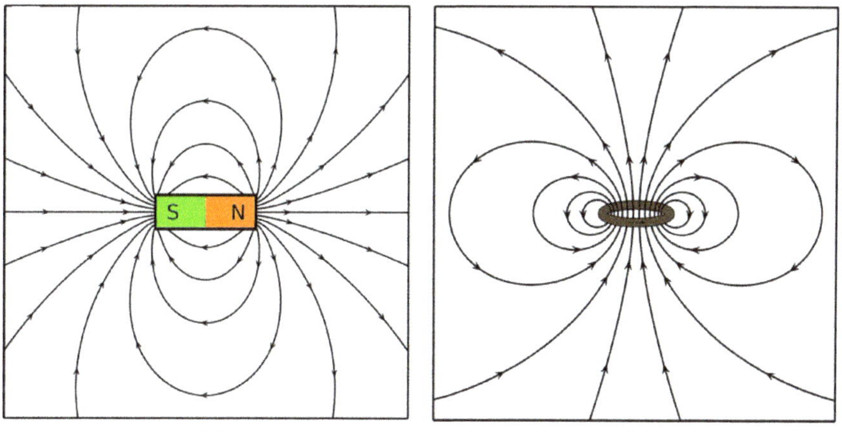

Figure 9.4 Field generated by a magnet with S/N dipole: (a) exposed horizontally and (b) exposed vertically.

Table 9.1 Other Types of Magnetisms

Effect	*Description*	*Examples*
Paramagnetism	Unpaired electrons with magnetic moments tending to align in the same direction as the field is applied	Al, O (weakly attracted)
Diamagnetism	Tendency of a material to oppose an applied magnetic field	Cu, C (weakly repelled)
Antiferromagnetism	No field is produced; magnetic moments of valence electrons to point in opposite directions (net zero magnetic moment)	Cr

which induces birefringence if the magnetic field is applied perpendicularly to the propagation direction. Moreover, the state of polarization plays a crucial role in the photonic transfer of quantum information and the integration of polarization modulation components is often needed, such as for Kerr or Faraday polarization rotation with magnetic bias. These components such as rotators or nonreciprocal isolators are often macroscopic passive devices and difficult to integrate. Playing with the magnetic properties of light coupled with plasmonic waves generation could lead to small integrated devices and overcome these limitations, especially using specific surface nanostructures [13–16].

In the context of magneto-optics, nano-antennas also play a key role [17]. Antennas are well known devices to transmit wireless information such as radio waves. These antennas are engineered to provide access to the emitted signal in a far-field distribution. Nano-antennas are specific types of antennas built at the nanoscale and optical nano-antennas are based on metal nanoparticles deposited on a dielectric substrate such that their interdistance is less than the light wavelength. They are designed to efficiently convert free-propagating optical radiation to localized energy and vice versa. They have been developed in a similar way as radio frequency (RF) antennas have been produced by electrical engineers. Today, antennas are a key enabling technology for many devices such as cellular phones and TV, using electromagnetic radiation in the radio wave or microwave regime. However, their optical counterpart is absent in today's technology but generates considerable and still growing interest as they are expected to improve lots of applications, in many different fields and in the near future. Research in this field is driven by the need for high field enhancement, strong field localization, and large absorption cross-sections. Due to the light scattering induced by nanoparticles, optical nano-antennas localize and emit propagating light waves from the electromagnetic fields trapped in the metal with subwavelength structures. Optical nano-antennas are the subject of intensive research, mainly for high-resolution microscopy, spectroscopy, photovoltaics, controlled light emission, and coherent control [18–20]. There is no universal optical nanoantenna design, the latter needing to be optimized for each specific application. Prominent examples are the Yagi-Uda and bow-tie configurations. Currently, optical nano-antennas are obtained on top of planar waveguides or on the facet of an optical fiber where an array of nanoparticles can be produced by electron-beam lithography [21], following a layer-by-layer process [22] or electrostatic self-assembly [23], to name a few techniques. Plasmonic nano-antennas are often made of silver or gold, and their operating principle is that they achieve an enhancement of the optical intensity by many orders of magnitude, due to a strong local field confinement near the metal surface. Magneto-plasmonic nano-antennas also exist and bring a novel dimension to the field. In contrast to conventional nano-antennas, these are built using ferromagnetic materials such as iron, cobalt, nickel, or alloys [24].

Looking now at the combination between magnetic materials and plasmonic optical fiber sensors, there have been several relevant works to date. Liu et al. and Weng et al. theoretically investigated magnetic-field sensors based on plasmonic side-hole fibers and D-shaped photonic-crystal-fiber, respectively [25, 26]. Rodríguez-Schwendtner et al. experimentally boosted the sensitivity up to approximately 100 pm/mT by using a tapered fiber as an SPR substrate [16]. These plasmonic-based magnetic field sensors were only able to detect magnetic field intensity, but not orientation. This issue was solved in 2016 with the achievement of a plasmonic TFBG magnetometer [27]. In this work, a gold-coated TFBG was packaged in a capillary containing a ferromagnetic fluid (Fe_3O_4), and it was shown that both the orientation (2 nm/deg) and intensity (1.8 nm/mT) of the magnetic field can be determined from the analysis of the SPR-attenuated cladding mode resonances in the amplitude spectrum of the grating. This configuration was tested in the range from 0 to 18 mT. A configuration based on plasmonic, side-polished, few-mode fibers functionalized with magnetic nanoparticles of Fe_3O_4 was studied in [28]. The noncircularly symmetric geometry of the side-polished fiber and the nonuniform distribution of the magnetic field around the fiber both allowed to sense the magnetic field's orientation with a sensitivity of −11.917 nm/deg. The sensitivity to the intensity of the magnetic field was reported to be 69 pm/mT.

In addition to the use of plasmonic optical fiber sensors for magnetic field measurement as summarized in previously, magneto-plasmonic optical fiber configurations have also been used for chemical sensing and biosensing. The detection of prostate-specific antigen (PSA), human chronic gonadotropin (hCG), and triacylglycerol has been performed recently, opening the field for novel sensing configurations. These works are summarized in Table 9.2.

9.3 Fluorescence-Based and Quantum Dot-Based Plasmonics

Biochemical assays used for rapid screening under laboratory settings often call on fluorescence. Fluorescence is a part of luminescence in which molecules emit light from electronically excited states created by a physical (e.g., the absorption of light at a specific wavelength range), mechanical (friction), or chemical reaction. The fluorescent principle occurs after a certain time gap (usually in the order of the femtosecond) named a fluorescent lifetime, which can be used to track molecules in motion (both rotational and/or translational), as they emit light in a different direction than the excitation plan. Fluorescent-labeled molecules, often called tracers, can be used on cell cultures, on slides of tissues, or in mixtures and solutions

Table 9.2 Some Examples of Magnetic-Based Optical Fiber Biochemical Sensors

Optical fiber type	*Receptor/Target*	*LOD*	*Reference*
ZnO nanowires/AuNPs, multimode optical fiber (LSPR)	PSA	2.06 and 0.51 pg/mL	[29]
Single-mode-tapered no-core-single-mode (STNCS) optical fiber using magnetic microspheres	Antibody/human chorionic gonadotropin (hCG)	0.0001 mIU/mL (international unit for hCG detection)	[30]
Multimode fiber-single-mode fiber reflector (MSR structure) using magnetic beads	Lipase/triglycerides (TG) in serum	NA	[31]
Unclad multimode fiber (SPR)	Ag-Co nanocomposite/ ammonium in water	2.9 ppm	[32]

(Figure 9.5(a)). While the tracer binds its target, it results in a higher fluorescence spot that can be further quantified or tracked as an indicator for both diagnosis and prognosis, the detection of pesticides in water, or to track metabolites in urine or serum. This process can also be a useful mapping technique to follow molecules within a cell and to understand their metabolism.

Historically, fluorescence has always been a fascinating field of research. Many biological species take part in it, in particular by their natural fluorescence. Fluorescent is defined as the absorption of electromagnetic radiation (typically from ultraviolet to visible light) from a molecule and release of a photon of lower energy (smaller frequency and longer wavelength), yielding a different color. This process is often observed for many types of corals (Figure 9.5(b)). Fluorescence can be distinguished from bioluminescence where it comes from the natural production of light by chemical reactions occurring within an organism (and not coming from the external environment). The anglerfish is a good example of a bioluminescent species (Figure 9.5(c)). This species uses a modified luminescent ray (one long filament sprouting from their heads called the illicium) as a lure. Light comes from symbiotic bioluminescent bacteria present in the bulb. Last but not least, phosphorescence is very similar to fluorescence as it also requires light as a provider of excitation energy to occur. However, the difference lies in the stability of the energized electron, which continues to glow even after the stimulating light source has been stopped (Figure 9.5(d)). In this case, phosphorescent tools can be used to glow in the dark but no truly phosphorescent organisms are known so far.

Fluorescent phenomena in the broad sense can be generated directly (by an external light excitation) but also by molecular intermediaries. Förster (or fluorescence) resonance energy transfer (FRET) is a mechanism relying on the energy transfer between two (or more) light-sensitive molecules. A donor molecule initially in its excited state (for instance, a fluorophore excited by an external light at a defined

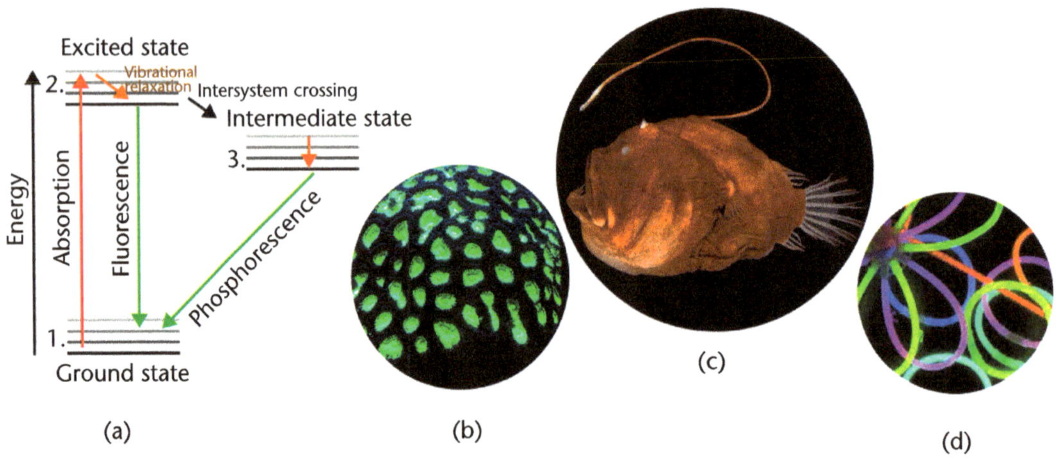

Figure 9.5 (a) Jablonski diagram representing (1) the excitation of a molecule to (2) its singlet excited state and (3) the intersystem crossing to the triplet state that relaxes to the ground level by phosphorescence. (b) Picture of a fluorescent coral (Scleractinia), exhibit in the Monterey Bay Aquarium, California. (c) Picture of an anglerfish showing bioluminescent process used as a lure for other fish. (d) Phosphorescent glow-in-the-dark wristbands. (All the pictures were available under the Creative Commons (CC) Universal Public Domain Dedication.)

wavelength) transfers its energy to an acceptor light-sensitive molecule through dipole-dipole coupling, providing a fluorescent signal from that molecule. The excited molecule emits a virtual photon that is instantly absorbed by the receptor.

These measurements are essentially exploited in biology and chemistry to determine if two molecules are within a certain distance of each other and may interact. It is considered a photonic ruler in near-field communication because the distance of interaction is much smaller than the wavelength of the emitted signal. FRET is consequently used to determine distances between two proteins but also between domains in a single protein (folding). The main issue with FRET is that it requires an external illumination to initiate the fluorescence transfer, which can interfere with the target signal and lead to high background noise levels. To avoid this issue, bioluminescence resonance energy transfer (BRET) has been developed. This strategy uses a bioluminescent molecule (e.g., the luciferase enzyme) to initiate the photon emission. Nowadays, specific luciferases have been engineered to provide bright signals and several patented reagents are associated with this technology.

Besides fluorescent dyes, quantum dots (QDs) are man-made colloidal fluorescent semiconductor crystals of few nanometers in size, which have specific optical, electronic, and photonic properties. These properties are different than those observed from larger particles usually used due to their quantum mechanics. While these QDs are illuminated with a specific wavelength range, electrons are excited and therefore able to drop back into their valence band-releasing energy, such as light. This light emission (known as photoluminescence) varies and depends on the energy difference between the valence band and the conductance band levels. QDs possess intermediate properties between bulk semiconductors and discrete atoms or molecules. Depending on their size and shape, QDs can emit in several wavelength ranges and therefore be used in different applications (solar cells, LEDs (especially in next-generation TV screens), quantum computing, single-photon sources, microscopy, medical imaging). QDs are essentially generated using colloidal synthesis, plasma synthesis, self-assembly, or electrical gating. A major concern regarding their use can be related to their nature or their solvents, which are sometimes toxic and/or not compatible with biological applications. Core type, core shell, or alloyed QDs can be easily purchased, as many providers exist. Some examples of QDs are those made of cadmium selenide (CdSe), cadmium telluride (CdTe), indium phosphide (InP), or zinc selenide (ZnSe), to name a few (Figure 9.6) [33].

More than only fluorescent enhancers, QDs possess large oscillator strengths and high photostability, which is useful to screen ultra-small events, especially single cells or single molecules. QDs can be implemented upon plasmonic transducers to couple either weakly or strongly, resulting in several properties [34]. The weak coupling regime of plasmonic structures is usually used to enhance the radiative rate of an emitter, while the strong coupling regime is used to enhance the energy level of the two systems together (QD-plasmon), forming coupled matter-light states [35]. QD nanoparticles initially studied in the early 1980s are still struggling to convince companies due to their high costs of production. However, there is a clear gain of interest as their optical properties are highly tunable. They are therefore involved in novel theranostic platforms (which means combined therapeutics and diagnostics) with applications ranging from sensing to drug delivery and biomedical imaging.

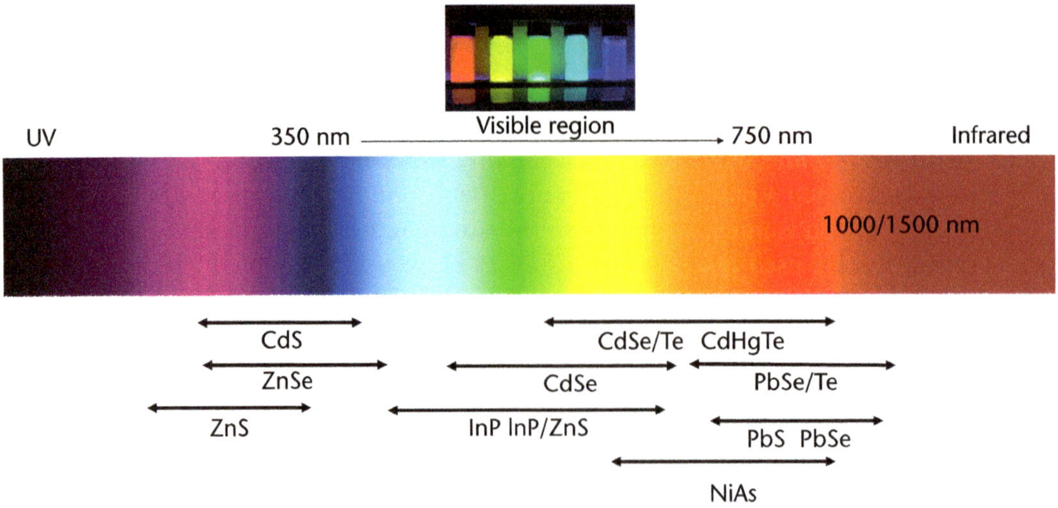

Figure 9.6 QDs working wavelength ranges.

The incorporation of fluorophores on plasmonic substrates, or, more generally, on photonic substrates, tailors the spatial distribution of light emission in the surrounding environment. The selection of adequate fluorescent molecules (or particles) therefore enables spectrally overlapping optical modes from the substrate with the ones of the fluorophore or, more generally, improving the sensing amenities by association of multiple approaches [36]. For instance, the binding of labeled molecules on plasmonic substrates can be verified under fluorescent microscopy in real time to verify their anchoring and its efficiency. In recent years, it has been realized that the traditional fluorescent assay could be significantly enriched harnessing the assets of plasmonic structures. For instance, in the field of plasmon-enhanced fluorescence readout, it has been reported that the diffraction coupling to propagating plasmon waves are able to improve the electromagnetic field intensity and the local density of optical states on the surface, yielding enhanced sensitivity [37, 38]. Some commonly used fluorophores with their associated excitation and emission range are listed in Table 9.3. These fluorophores can usually be covalently attached to specific molecules directly upon request to the provider or be further linked through specific labeling kits. Hundreds of different markers exist, covering a broad range of wavelengths and being adapted for several applications. In association with these fluorescent tags, it is important to consider the quenching effect. Indeed, the quenching is defined as a decrease of the fluorescence intensity caused by a substance. Chloride ions or acrylamide are well-known quenchers and quenching and dequenching cycles using specific molecular interactions are particularly needed for some applications such as for imaging or FRET, as described earlier. Some molecules can also carry a quencher and a fluorescent tag and release a fluorescent signal depending on their 3-D conformation, modifying the effect on the first one to the other. Quenching strategies therefore play a tremendous role in studying molecular and cellular processes.

In the continuity of these techniques of fluorescence and molecular quenching/monitoring, the optical fiber brings its advantages both by its capacity to detect the emitted signal but also to locally illuminate the sampling area. This would help to

Table 9.3 List of Some Commonly Used Fluorophores

Fluorescent Dye	*Excitation (maximum, nm)*	*Emission (maximum, nm)*	*Color*
DAPI	350	456	Blue
Alexa Fluor 405	401	421	Blue
Egfp	488	509	Green
Alexa Fluor 488	493	517	Green
Fluorescein (FITC)	494	523	Green
FAM	495	520	Green
SYBR Green I	498	522	Green
Cy3	550	570	Yellow-Orange
Rhodamin	555	580	Red
Texas red	590	615	Red
Cy5	649	670	Red
Cy7	743	767	Violet
Alexa Fluor 750	749	775	Violet

carry out tests locally and sheltered from the ambient light, which disturbs in particular the longevity of the reagents and the possibilities for outdoors manipulations. Although the principle has been widely used for molecular and cellular detection on unclad and tapered fibers (Table 9.4), doors remain open to innovation for its implementation with specific structures such as grating-based optical fibers and surface microstructures and nanostructures, which could become a new detection theme for the future of optical fiber biochemical sensors. Table 9.4 reviews some relevant configurations that make use of fluorescence, mainly in the form of localized surface plasmon coupled fluorescence [39].

A very large part of fluorescent-based biosensors rely on fluorescent proteins to visualize and/or quantify molecules or their modification within cells. This is true for *in vitro* cell cultures but also for *in vivo* imaging. FRET and BRET sensors discussed in Chapter 8 also fall withing this group of sensors, in order to locate molecular interactions. Fluorescent transducers are usually classified in two groups, depending on if they are genetically encoded and translated into fluorescent proteins, or if they are synthetically (or semi-synthetically) built from site-specific chemical modifications such as for the addition of a fluorescent dye. The most famous fluorescent protein is named GFP, as it is green fluorescent and has evolved in largely produced and engineered family of colorful mutants ranging from blue to red. These literally expanded and changed the possibilities to build new encoded fluorescent proteins for cell imaging.

9.4 Raman Scattering

Among all the optical sensing amenities, Raman scattering is a result of the inelastic light scattering process leading to the emission of scattered light with a different

Table 9.4 Some Examples of Fluorescent-Based Plasmonic Optical Fiber Biosensors

Optical Fiber Type	*Receptor/Target*	*LOD*	*Dye*	*Reference*
D-shaped microstructured optical fiber with gold film	Rhodamine B dye only (proof of concept)	NA	Rhodamine B	[40]
Multimode plastic optical fiber (localized surface plasmon coupled fluorescence (LSPCF))	Antibody/mouse IgG detection/AuNPs amplification	1 pg/mL (7 fM)	Cy5	[41]
Quartz optical fiber tip (600 μm ∅), Carbon QDs (pH-dependent)	Acetylcholine (Ach)	16.28 μM	Fluorescent carbon QDs	[42]
Quartz optical fiber tip (600 μm ∅), CQD/GOD/CA film	Glucose oxidase (GOD)/glucose	6.43 μM	Fluorescent carbon QDs	[43]
Tapered optical fiber with carbon dots	Laccase (enzyme)/dopamine	41.2 nM	Blue (405 nm), green (488 nm), red (543 nm)	[44]
Unclad PMMA fiber (LSPCF)	Antibody/Alpha-fetoprotein (AFP) in PBS and serum and AuNPs (Protein A)	0.1 ng/mL (1.4 pM) in PBS and 2.33 ng/mL in serum	DyLight 649	[45]
Unclad PMMA fiber (LSPCF)	Antibody/SARS-CoV Nucleocapsid protein (GST-N)	1 pg/mL in serum	DyLight 649	[46]
Unclad polymer optical fiber (LSPCF)	Antibody/Swine-origin influenza A (H1N1) hemagglutinin protein (HA)	13.9 pg/mL in PBS	Atto 633	[47]

frequency associated with molecular vibrations of the target. The Raman effect usually relies on complex spectra where the separation of weak inelastic scattering and intense Rayleigh scattering is difficult. These limitations led to the development of many parallel techniques such as resonance Raman spectroscopy (RRS), nonlinear Raman spectroscopy, and surface-enhanced Raman spectroscopy (SERS), among others. The discovery of SERS was made accidentally by Fleischmann and coworkers in 1974 during measurements of the Raman scattering of pyridine on rough silver electrodes [48]. Since then, thousands of articles have been published on SERS, and it has become an important research field as one of the most sensitive analytical techniques currently available.

The Raman effect can provide molecular fingerprints and present enormous assets for the identification of targets due to their vibrational states (phonons). A large part of the light's frequency is not modified while interacting with a target material due to the Rayleigh scattering. Under incident light, an inelastic scattering process may occur and result in the emission of scattered light with more (anti-Stokes) or less (Stokes) frequency, due to inner molecular vibrations (Figure 9.7). While this phenomenon happens, it yields to a Raman spectrum, where bands appear at specific positions depending on vibrational frequencies. Each functional group (e.g., –OH, –NH) can be characterized through its position and intensity. This method is nowadays one of the most common techniques used to determine the nature of substances, and large database are available for different types of molecules and compounds. The main constraint of Raman is the relatively weak signal obtained and the problem of complex media where multiple molecules are present, provoking the accumulation and overplotting of numerous peaks. They also need specific equipment such as optimized laser excitations, filtering, and lenses and highly depend on the final investigated interface. On its side, SERS is a highly sensitive technique that enhances the Raman scattering of molecules supported by nanostructured materials, enabling the detection of analytes at low concentration due to the mediation of plasmonic amplification. For this reason, there is great interest in exploited it at the end facet of optical fibers and generating surface molecular analyses.

The use of theoretical modeling has become essential to understanding SERS as a fundamental phenomenon and to interpreting and predicting experimental results obtained from various conditions and environments. There is now full evidence that the enhancement factor provided by SERS is a combination between electromagnetic enhancement associated with plasmon excitation in metal particles operating as the SERS substrate and chemical enhancement resulting from the target molecules being able to transfer electrons to or from metal particles in both ground and excited states, usually in the process of forming metal-molecule bonds.

Physically, the Raman signal involves absorption of an incident photon of frequency ω_{in}, coupling to an internal degree of freedom of the molecule, typically a

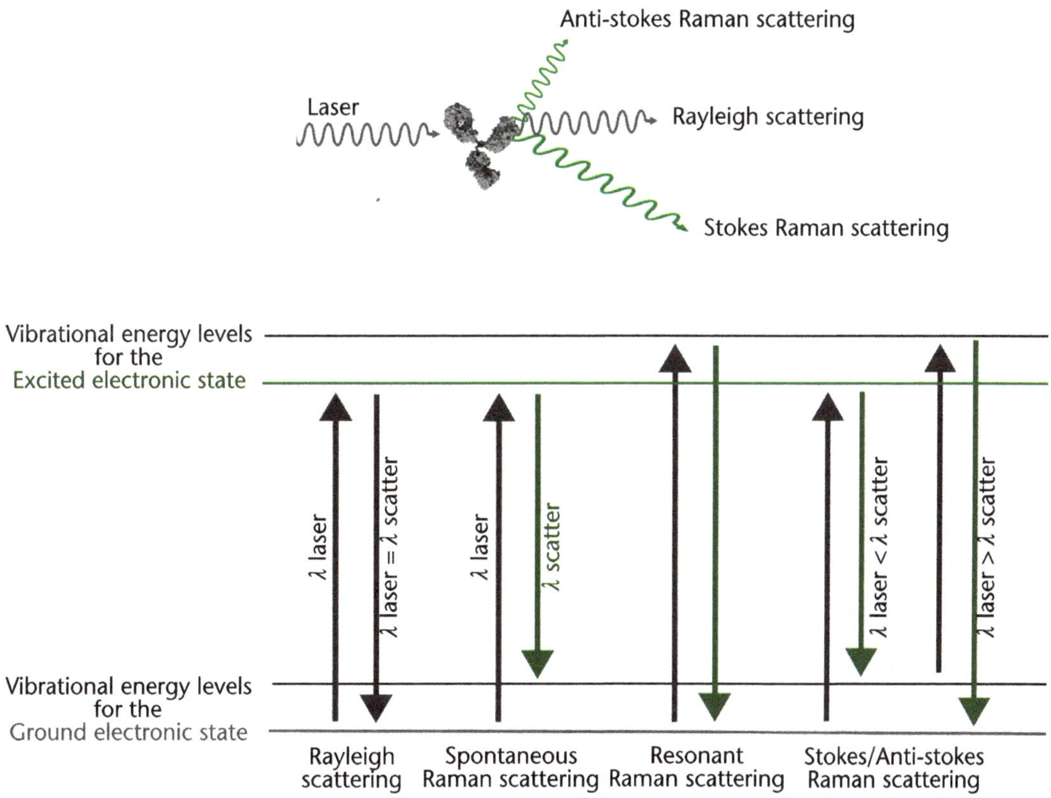

Figure 9.7 Different types of light scattering from a molecule and their corresponding energy level transitions.

molecular vibration of frequency ω_{vib}, and reemission at two different frequencies $\omega_{em} = \omega_{in} \pm \omega_{vib}$. The sum and difference result in anti-Stokes and Stokes Raman scattering, respectively. Three inelastic transitions are involved in such a process: absorption, vibrational excitation, and reemission. The vibrational excitation occurs with a probability that depends on the environment through its chemical interaction, while other two processes are controlled by the availability of photonic states at the positions of the molecules. Without a structured environment (e.g., in solution), the Raman process has a low probability of operating. The latter is quantified in terms of an optical cross-section (e.g., the area of the incident beam over which incident photons are effectively converted into emitted Raman photons). It results in values of approximately 10^{-11} to 10^{-12} nm^2, which depends on whether the process is resonant or nonresonant Raman. Resonant (nonresonant) means that the incoming light is (is not) resonant with transitions between ground and excited electronic states of the corresponding molecules. The intrinsic intensity of Raman scattering is by far too low in many practical applications, which implies finding means to enhance the Raman process. The most efficient way to do so is certainly through the large optical field enhancement produced by suitable resonant structures. As the absorption process is directly proportional to the local electric field intensity at the molecule, it can be drastically enhanced by plasmons in noble metal nanostructures. Metal nanoparticles organized with nanometer-scale gaps between them can enable the enhancement factor up to 10^5 or even 10^6 to be reached. These kinds of arrangements are usually called hotspots.

Hotspots can be produced not only from gaps between adjacent nanoparticles but also within nanoparticle junctions and flat metal surfaces supporting plasmon resonances. The resulting field strength depends strongly on the gap distance and other geometrical considerations. Typical SERS hotspots have an extension in the 2–10-nm range and are well-described within classical electromagnetism by neglecting nonlocal effects and accounting for frequency-dependent dielectric functions of the materials involved in the structure. For gap distances between nanoparticles below 1 nm, nonlocal effects come into play, requiring a more sophisticated treatment of the optical response. Moreover, the enhancement of electromagnetic fields is so strong that the corresponding optical response may become nonlinear. In this case, classical models are no longer valid and must be complemented by descriptions based on quantum-mechanical approaches. Specific and accurate modeling of SERS and competing processes in subnano-sized to nano-sized hotspots are crucial in supporting and/or interpreting experimental results and enabling the design of substrates with the desired SERS response. Readers interested in these considerations are invited to consult [48–51] to find relevant information. The modeling of the SERS process will not be further detailed here.

When implemented for biochemical sensing, SERS is usually split into two approaches called direct or indirect, depending on the use of reported molecules, also called nanotags. To perform direct SERS, nanostructures (often Ag or Au nanoparticles) are exploited as active substrate to adsorb the molecules to detect. To specifically detect a biotarget, bioreceptors can be added onto these particles and the Raman spectra are compared before and after the binding and washing. However, the addition of dyes or specific Raman reporter molecules can lead to intense bands in the response spectra, becoming an indirect measurement of the analyte. The direct approach is characterized by a sensitivity and specificity related to the properties of the substrate and its surface functionalization, while its enhancement is provided by the SPR generated through the metallic nanostructures and nanoparticles. The indirect approach calls on SERS nanotags adapted to the sensor design but also to amplification strategies using capture nanoparticles or reporters where specific coatings or protective shells need to be added. Suitable plasmonic nanostructures or SERS nanotags therefore need to be adapted according to the sample matrix and its application. To practically achieve such SERS performances, the development of plasmonic arrays is needed. The latter should be arranged periodically or regularly and needs to be stable over time and reproducible from batch to batch. The signal enhancement needs to be homogeneously distributed on the sensing area and avoid any perturbation from environmental conditions. Most sensing devices designed on optical fibers are therefore based on their end facet in order to provide a flat surface and a neat interaction between the light beam and the target sample to analyze.

Raman techniques are a growing focus for their biomedical applications, especially because they provide molecular fingerprints without need for sample preparation, work in aqueous samples, do not need specific biofunctionalization, and are suited to work *in vivo*, as the measurement is not destructive and not invasive. Important advances have been demonstrated in hyperspectral imaging where the distribution of lipids, proteins, nucleic acids, and carbohydrates could be identified and tracked. These tools are therefore key assets for the study of cellular mechanisms and their metabolic pathways. The identification of nonhealthy tissues through surface contact is also a promising field of research for early cancer diagnosis. As for many of the aforementioned techniques, miniaturization and simplification are major threads to develop these tools for real-life applications. In this way, the use of optical fiber-based Raman or SERS sensors are of great interest.

In Table 9.5, we list some relevant SERS optical fiber biosensor configurations together with their performance in terms of LOD when available. For additional considerations about the actual implementation, the reader is invited to consult the specific reference. This table is far from being exhaustive given the very important number of publications in this topic.

9.5 Ultrasound and Radio-Plasmonics

While Raman, especially SERS, is extensively used to identify the nature of surface molecules or molecules in solution, ultrasound can be exploited to image deep tissue with a high spatiotemporal resolution. Ultrasound imaging uses high-frequency sound waves (above 20 kHz, in a frequency range such that the corresponding waves cannot be heard) to capture images of tissues, cells, or bodies in real time, as sketched in Figure 9.8. Compared to X-ray, there is no ionizing source and therefore no associated radiation exposure. Classical ultrasound as found in medical practice is often based on a transducer (the probe) placed on the skin or inside an open body, usually with the addition of a specific gel to avoid the air interface. The ultrasound waves are transmitted inside the tissues and the resulting image is produced based on the reflections provided by the different types of tissues and cells and the Doppler echo they produce. Despite its many advantages, ultrasound

Table 9.5 Some Examples of SERS-Based Optical Fiber Biosensors

Optical Fiber	*Target or Sensing Strategy*	*LOD*	*Reference*
SERS/plasmonic optical fiber tip	R6G dye, IgG, DNA/nanocavity structured metal surface produced with microspheres	0.157 μM R6G	[52]
SERS fiber-PDMS cap (25 optical fibers of 100-μm core-diameter bundle)	Aβ-peptides/Anti Aβ-antibodies with gold nanostars and gold nanorods	NA	[53]
SERS hollow optical fiber	R6G and ceftriaxone/in fiber sensing using optofluidics	10^{-14} M and 10^{-6} M	[54]
SERS optical fiber tip	R6G dye/5 layers of 50-nm silver nanoparticles (AgNPs)	200 nM	[19]
SERS optical fiber tip	4-Mpy, crystal violet and maleic acid/ EDMA/GMA functionalization with Ag nanoparticles	~10^{-7} M	[55]
SERS optical fiber tip	R6G dye and *E. coli* bacteria/two-photon polymerization strategy (polymeric microstructures)	10^{-7} M	[56]
SERS optical fiber tip	Methylene blue and crystal violet, R6G/gold nanoparticles	10^{-9} and 10^{-8} M	[57]
SERS optical fiber tip	R6G/Graphene and Au nano-triangle structures	~10^{-6} M	[58]
SERS optical fiber tip (multimode) and a liquid-core photonic crystal fiber (LCPCF)	Lysozyme, cytochrome c, Shewanella oneidensis MR-1/silver nanoparticles, and nanowires	0.2 μg/mL and 10^{6} cells/mL	[59]
SERS tapered fiber	R6G dye/APTMS and gold nanoparticles	10^{-8} mol/L	[60]
SERS tapered fiber	Tetracycline hydrochloride/gold nanoparticles	NA	[61]
SERS tapered fiber	R6G/Ag nanoparticles	10^{-7} M	[62]
SERS tapered fiber	4-aminothiophenol (4-ATP)/Thiol functionalization of Ag nanoparticles	2.15×10^{-11} M	[63]
SERS U-bent plastic optical fiber	4-MBA Raman label and GaHIgG/HIgG/ electroless deposition, sputtering, and chemisorption of nanoparticles	NA	[64]

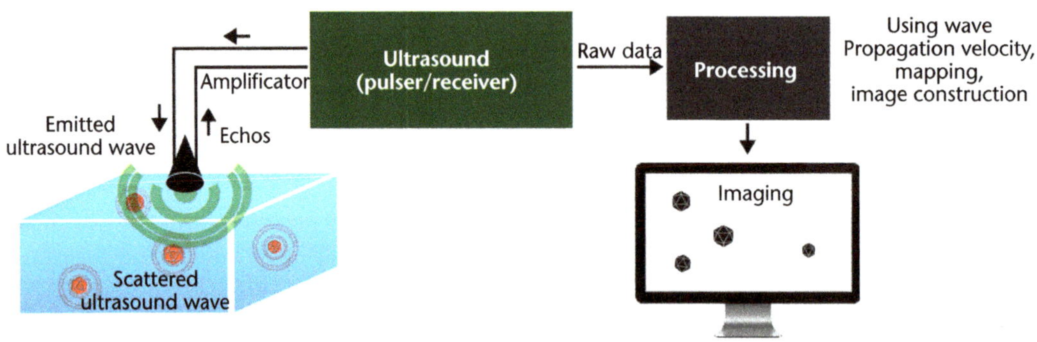

Figure 9.8 Operating principle of ultrasound imaging based on the Doppler echo.

imaging cannot be considered as sensing, as it lacks to connect its contrast to the activity of specific biomolecules. To overcome this limitation, different strategies have been implemented, especially to link certain molecules to a light-up effect in ultrasound imaging (e.g., through enzymatic reactions) [65].

Ultrasound imaging using optical fiber facets is already an important field of research for dip-and-read operations. The coupling of the ultrasound plasmonics with SPR has been studied through the integration of periodic nanostructures implemented on the end facets of optical fibers. The design of such periodic SPR nanostructures cannot straightforwardly inherit their free-space optics counterparts, so the Q-factor usually decreases significantly. This effect can be explained in two ways: first, the illumination of the periodic grating is equivalent to the grating coupling of a superposition of plane waves with different incident angles (within the numerical aperture). This therefore broadens the SPR spectrum and reduces the coupling efficiency at a single wavelength. Second, the loss of the SPR for the small core area of an SMF is so significant that it broadens the SPR spectrum and moves it far from the critical coupling conditions. Most of these nanostructured devices at a fiber-end face are therefore used for SERS instead of refractive index sensing due to the broad linewidths of the SP resonances. To overcome this limitation, different research groups have highlighted the possibility to form a cavity able to lessen the SP propagation out of modal area. Bragg reflectors or second-order distributed-feedback (DFB) gratings can be used to create a defect mode in the SPP bandgap and solve this issue, as demonstrated in [66]. In [67], it was demonstrated that SPR cavities on SMF end facets present a large bandwidth of over 125 MHz and a wide angular range of detection, which is 70° at approximately 10 MHz. For interested readers, a relevant numerical study on lab-on-fiber devices for ultrasound sensing based on four different configurations with hybrid metallo-dielectric nanostructures can be found in [68].

In order to obtain both high resolution and fast ultrasound imaging, the technique needs a low noise equivalent pressure (NEP) and a broad angular response, the ability to achieve a phase-array detection, and, finally, have a broadband frequency. The main advantage to bring OFs into the game is surely their compactness and the ability to bring the sensor intravascularly or in a small environment (e.g., inside delicate tissues). The main acousto-optical devices are based on the detection of reflection index change or morphological changes resulting in a resonance shift, changing the transmittance or reflectance of a laser beam. To achieve this using an optical fiber, its end facet is usually covered by a gold-epoxy interface to create the SPR cavity. In the simplest cases, this can be achieved by a glue-and-strip fabrication where a gold-pattern film is transferred from a planar substrate to the fiber. This method requires a precise alignment of the fiber, but is usually easier than adapting other nanopatterning strategies to small fiber end facets. Biosensing through ultrasound plasmonic optical fibers could lead to interesting sensors especially in viscous and complex media where classical molecular detection often require sample dilutions. Many steps still need to be explored in this field of research and in the implementation of a novel concept of 3-D-specific biosensing.

In conclusion, this chapter has reviewed the main technologies that can be combined with plasmonic optical fiber sensors to increase the sensing modalities and/or improve the overall metrological performance. Some are still at a very early stage like magneto-plamonics and ultrasound-plamonics, for which numerous achievements remain to be obtained. Others are much more mature. In particular, the use of electro-plasmonics and fluorescence-plasmonics with optical fiber configurations has allowed lowering the limit of detection of some relevant biosensing configurations. Besides sensing, SERS and ultrasound strongly contribute to the advent of plasmonic imaging on optical fibers. As confirmed in previous chapters, this will certainly shape the future of research activities in that growing field.

References

[1] Grieshaber, D., et al., "Electrochemical Biosensors—Sensor Principles and Architectures," *Sensors*, Vol. 8, No. 3, 2008, pp. 1400–1458.

[2] Elgrishi, N., et al., "A Practical Beginner's Guide to Cyclic Voltammetry," *J. Chem. Educ.*, Vol. 95, No. 2, 2018, pp. 197–206.

[3] Jaffari, S. A., and A. P. F. Turner, "Recent Advances in Amperometric Glucose Biosensors for In Vivo Monitoring," *Physiol. Meas.*, Vol. 16, No. 1, 1995, pp. 1–15.

[4] Yuan, Y., et al., "Electrochemical Surface Plasmon Resonance Fiber-Optic Sensor: In Situ Detection of Electroactive Biofilms," *Anal. Chem.*, Vol. 88, No. 15, 2016, pp. 7609–7616.

[5] Lao, J., et al., "In Situ Plasmonic Optical Fiber Detection of the State of Charge of Supercapacitors for Renewable Energy Storage," *Light Sci. Appl.*, Vol. 7, No. 1, 2018.

[6] Wang, R., et al., "Operando Monitoring of Ion Activities in Aqueous Batteries with Plasmonic Fiber-Optic Sensors," *Nat. Commun.*, Vol. 13, No. 1, 2022, pp. 1–11.

[7] Lobry, M., et al., "Electro-Plasmonic-Assisted Biosensing of Proteins and Cells at the Surface of Optical Fiber," *Biosensors and Biolectronics*, 2022, p. 114867.

[8] Lee, C., et al., "Quantum Plasmonic Sensors," *Chem. Rev.*, Vol. 121, No. 8, 2021, pp. 4743–4804.

[9] Jiang, N., X. Zhuo, and J. Wang, "Active Plasmonics: Principles, Structures, and Applications," *Chem. Rev.*, Vol. 118, No. 6, 2018, pp. 3054–3099.

[10] Li, Z., et al., "Operando Optical Fiber Monitoring of Nanoscale and Fast Temperature Changes During Photo-Electrocatalytic Reactions," *Light Sci. Appl.*, Vol. 11, No. 1, July 13, 2022, p. 220.

[11] Moaied, M., S. Palomba, and K. Ostrikov, "Quantum Plasmonics: Longitudinal Quantum Plasmons in Copper, Gold, and Silver," *J. Opt. (United Kingdom)*, Vol. 19, No. 10, 2017.

[12] Armelles, G., and A. Dmitriev, "Focus on Magnetoplasmonics," *New J. Phys.*, Vol. 16, 2014, pp. 14–17.

[13] Fang, Z., et al., "Gated Tunability and Hybridization of Localized Plasmons in Nanostructured Graphene," *ACS Nano*, Vol. 7, No. 3, 2013, pp. 2388–2395.

[14] Lodewijks, K., et al., "Magnetoplasmonic Design Rules for Active Magneto-Optics," *Nano Lett.*, Vol. 14, No. 12, 2014, pp. 7207–7214.

[15] Cennamo, N., et al., "Plastic Optical Fiber Sensors and Magnetic Fluids: Plasmonic Tunability and Sensing Properties for Measurements," *I2MTC 2020—Int. Instrum. Meas. Technol. Conf. Proc.*, 2020, pp. 1–6.

[16] Rodríguez-Schwendtner, E., et al., "Plasmonic Sensor Based on Tapered Optical Fibers and Magnetic Fluids for Measuring Magnetic Fields," *Sensors Actuators A Phys.*, Vol. 264, 2017, pp. 58–62.

[17] Fischer, H., and O. J. F. Martin, "Engineering the Optical Response of Plasmonic Nanoantennas," *Opt. Express*, Vol. 16, No. 12, 2008, p. 9144.

[18] Smythe, E. J., et al., "Optical Antenna Arrays on a Fiber Facet for In Situ Surface-Enhanced Raman Scattering Detection," *Nano Lett.*, Vol. 9, No. 3, 2009, pp. 1132–1138.

[19] Andrade, G. F. S., M. K. Fan, and A. G. Brolo, "Multilayer Silver Nanoparticles-Modified Optical Fiber Tip for High Performance SERS Remote Sensing," *Biosens. Bioelectron.*, Vol. 25, No. 10, 2010, pp. 2270–2275.

[20] Yap, F. L., et al., "Nanoparticle Cluster Arrays for High-Performance SERS Through Directed Self-Assembly on Flat Substrates and on Optical Fibers," *ACS Nano*, Vol. 6, No. 3, 2012, pp. 2056–2070.

[21] Lin, D., et al., "Dielectric Gradient Metasurface Optical Elements," *Science*, Vol. 345, No. 6194, 2014, pp. 298–302.

[22] Chen, H. -T., A. J. Taylor, and N. Yu, "A Review of Metasurfaces: Physics and Applications," *Reports Prog. Phys.*, Vol. 79, No. 7, 2016, p. 76401.

[23] Li, A., S. Singh, and D. Sievenpiper, "Metasurfaces and Their Applications," *Nanophotonics*, Vol. 7, No. 6, 2018, pp. 989–1011.

[24] Maksymov, I. S., "Magneto-Plasmonic Nanoantennas: Basics and Applications," *Rev. Phys.*, Vol. 1, 2016, pp. 36–51.

[25] Ying, Y., et al., "Magnetic Field Measurement Using Surface Plasmon Resonance Sensing Technology Combined with Magnetic Fluid Photonic Crystal," *IEEE Transactions on Instrumentation and Measurements*, Vol. 65, No. 1, 2016, pp. 170–176.

[26] Weng, S., et al., "High Sensitivity Side-Hole Fiber Magnetic Field Sensor Based on Surface Plasmon Resonance," *Chin. Opt. Lett.*, Vol. 14, No. 11, 2016, p. 110603.

[27] Zhang, Z., et al., "Plasmonic Fiber-Optic Vector Magnetometer," *Appl. Phys. Lett.*, Vol. 108, No. 10, 2016.

[28] Chen, Y., et al., "Magnetic Nanoparticles Functionalized Few-Mode-Fiber-Based Plasmonic Vector Magnetometer," *Nanomater. (Basel, Switzerland)*, Vol. 9, No. 5, 2019.

[29] Kim, H. M., J. H. Park, and S. K. Lee, "Fiber Optic Sensor Based on ZnO Nanowires Decorated by Au Nanoparticles for Improved Plasmonic Biosensor," *Sci. Rep.*, Vol. 9, No. 1, 2019, pp. 1–9.

[30] Kumar, R., et al., "Ultrasensitive Biosensor Based on Magnetic Microspheres Enhanced Microfiber Interferometer," *Biosens. Bioelectron.*, Vol. 145, December 1, 2019.

[31] Zhou, S., et al., "Dual-Fiber Optic Bioprobe System for Triglyceride Detection Using Surface Plasmon Resonance Sensing and Lipase-Immobilized Magnetic Bead Hydrolysis," *Biosens. Bioelectron.*, Vol. 196, January 15, 2022.

[32] Samavati, Z., et al., "Optical Fiber Sensor Based on Magneto-Plasmonic Features of Ag-Co Nanostructure for PPM Ammonium Detection in Aqueous Solutions," *Opt. Fiber Technol.*, Vol. 67, December 2021.

[33] Medintz, I. L., et al., "Quantum Dot Bioconjugates for Imaging, Labelling and Sensing," *Nat. Mater.*, Vol. 4, No. 6, 2005, pp. 435–446.

[34] Bitton, O., S. N. Gupta, and G. Haran, "Quantum Dot Plasmonics: From Weak to Strong Coupling," *Nanophotonics*, Vol. 8, No. 4, 2019, pp. 559–575.

[35] Yadav, R. K., et al., "Room Temperature Weak-to-Strong Coupling and the Emergence of Collective Emission from Quantum Dots Coupled to Plasmonic Arrays," *ACS Nano*, Vol. 14, No. 6, 2020, pp. 7347–7357.

[36] Bauch, M., et al., "Plasmon-Enhanced Fluorescence Biosensors: A Review," *Plasmonics*, Vol. 9, No. 4, 2014, pp. 781–799.

[37] Fossati, S., et al., "Multiresonant Plasmonic Nanostructure for Ultrasensitive Fluorescence Biosensing," *Nanophotonics*, Vol. 9, No. 11, 2020, pp. 3673–3685.

[38] Dutta Choudhury, S., R. Badugu, and J. R. Lakowicz, "Directing Fluorescence with Plasmonic and Photonic Structures," *Acc. Chem. Res.*, Vol. 48, No. 8, 2015, pp. 2171–2180.

[39] Chou, C., et al., "Localized Surface Plasmon Coupled Fluorescence Fiber Optic Based Biosensing," *Met. Fluoresc.*, 2010, pp. 183–259.

[40] Yong, D., et al., "Plasmonic-Enhanced Fluorescence Emission Using D-Shape Microstructured Optical Fiber," *J. Phys. Conf. Ser.*, Vol. 277, No. 1, 2011.

[41] Hsieh, B., et al., "Localized Surface Plasmon Coupled Fluorescence Fiber-Optic Biosensor with Gold Nanoparticles," *Anal. Chem.*, Vol. 79, No. 9, 2007, pp. 3487–3493.

[42] Zhang, Y., et al., "A New Optical Fiber Biosensor for Acetylcholine Detection Based on pH Sensitive Fluorescent Carbon Quantum Dots," *Sensors Actuators B Chem.*, Vol. 369, October 15, 2022.

[43] Yu, S., et al., "A Novel Optical Fiber Glucose Biosensor Based on Carbon Quantum Dots-Glucose Oxidase/Cellulose Acetate Complex Sensitive Film," *Biosens. Bioelectron.*, Vol. 146, December 15, 2019.

[44] Sangubotla, R., and J. Kim, "Fiber-Optic Biosensor Based on the Laccase Immobilization on Silica-Functionalized Fluorescent Carbon Dots for the Detection of Dopamine and Multi-Color Imaging Applications in Neuroblastoma Cells," *Mater. Sci. Eng. C*, Vol. 122, 2021, p. 111916.

[45] Chang, Y. F., et al., "Localized Surface Plasmon Coupled Fluorescence Fiber-Optic Biosensor for Alpha-Fetoprotein Detection in Human Serum," *Biosens. Bioelectron.*, Vol. 24, No. 6, 2009, pp. 1610–1614.

[46] Huang, J. C., et al., "Detection of Severe Acute Respiratory Syndrome (SARS) Coronavirus Nucleocapsid Protein in Human Serum Using a Localized Surface Plasmon Coupled Fluorescence Fiber-Optic Biosensor," *Biosens. Bioelectron.*, Vol. 25, No. 2, 2009, pp. 320–325.

[47] Chang, Y. F., et al., "Detection of Swine-Origin Influenza A (H1N1) Viruses Using a Localized Surface Plasmon Coupled Fluorescence Fiber-Optic Biosensor," *Biosens. Bioelectron.*, Vol. 26, No. 3, 2010, pp. 1068–1073.

[48] Wang, W., et al., "Raman Spectra of Pyridine Adsorbed at a Silver Electrode," *ACS Appl. Mater. Interfaces*, Vol. 8, No. 24, 2016, pp. 15591–15597.

[49] Kleinman, S. L., et al., "Structure Enhancement Factor Relationships in Single Gold Nanoantennas by Surface-Enhanced Raman Excitation Spectroscopy," *J. Am. Chem. Soc.*, Vol. 135, No. 1, 2013, pp. 301–308.

[50] Kumar, S., et al., "Surface-Enhanced Raman Scattering: Introduction and Applications," Chapter 8 in *Recent Advances in Nanophotonics: Fundamentals and Applications*, M. Kahrizi, (ed.), London, U.K.: IntechOpen, 2020.

[51] Langer, J., et al., "Present and Future of Surface-Enhanced Raman Scattering," *ACS Nano*, Vol. 14, No. 1, 2020, pp. 28–117.

[52] Liu, Y., et al., "Simple and Low-Cost Plasmonic Fiber-Optic Probe as SERS and Biosensing Platform," *Adv. Opt. Mater.*, Vol. 7, No. 19, 2019, pp. 1–11.

[53] Credi, C., et al., "Fiber-Cap Biosensors for SERS Analysis of Liquid Samples," *J. Mater. Chem. B*, Vol. 8, No. 8, 2020, pp. 1629–1639.

[54] Gao, D., et al., "Optofluidic In-Fiber Integrated Surface-Enhanced Raman Spectroscopy Detection Based on a Hollow Optical Fiber with a Suspended Core," *Opt. Lett.*, Vol. 44, No. 21, 2019, p. 5173.

[55] Liu, C., et al., "A Surface-Enhanced Raman Scattering (SERS)-Active Optical Fiber Sensor Based on a Three-Dimensional Sensing Layer," *Sens. Bio-Sensing Res.*, Vol. 1, 2014, pp. 8–14.

[56] Kim, J. A., et al., "Fiber-Optic SERS Probes Fabricated Using Two-Photon Polymerization for Rapid Detection of Bacteria," *Adv. Opt. Mater.*, Vol. 8, No. 9, 2020, pp. 1–12.

[57] Gu, C., Z. Zhao, and P. Shi, "Development of Monolayer AuNPs Decorated on an Optical Fiber Facet for SERS Analysis," *Appl. Opt.*, Vol. 60, No. 3, 2021, p. 792.

[58] Xia, M., et al., "SERS Optical Fiber Probe with Plasmonic End-Facet," *J. Raman Spectrosc.*, Vol. 48, No. 2, 2017, pp. 211–216.

[59] Yang, X., et al., "Highly Sensitive Detection of Proteins and Bacteria in Aqueous Solution Using Surface-Enhanced Raman Scattering and Optical Fibers," *Anal. Chem.*, Vol. 83, No. 15, 2011, pp. 5888–5894.

[60] Chen, Z., et al., "Gold Nanoparticles-Modified Tapered Fiber Nanoprobe for Remote SERS Detection," *IEEE Photonics Technol. Lett.*, Vol. 26, No. 8, 2014, pp. 777–780.

[61] Jin, D., et al., "SERS Detection of Expired Tetracycline Hydrochloride with an Optical Fiber Nano-Probe," *Anal. Methods*, Vol. 7, No. 4, 2015, pp. 1307–1312.

[62] Zhang, J., et al., "Tapered Fiber Probe Modified by Ag Nanoparticles for SERS Detection," *Plasmonics*, Vol. 11, No. 3, 2016, pp. 743–751.

[63] Cao, J., D. Zhao, and Y. Qin, "Novel Strategy for Fabrication of Sensing Layer on Thiol-Functionalized Fiber-Optic Tapers and Their Application as SERS Probes," *Talanta*, Vol. 194, March 1, 2019, pp. 895–902.

[64] Danny, C. G., A. Subrahmanyam, and V. V. R. Sai, "Development of Plasmonic U-Bent Plastic Optical Fiber Probes for Surface Enhanced Raman Scattering Based Biosensing," *J. Raman Spectrosc.*, Vol. 49, No. 10, 2018, pp. 1607–1616.

[65] Lakshmanan, A., et al., "Acoustic Biosensors for Ultrasound Imaging of Enzyme Activity," *Nat. Chem. Biol.*, Vol. 16, No. 9, 2020, pp. 988–996.

[66] Yang, T., et al., "Surface Plasmon Cavities on Optical Fiber End-Facets for Biomolecule and Ultrasound Detection," *Opt. Laser Technol.*, Vol. 101, 2018, pp. 468–478.

[67] Zhou, X., et al., "Ultrasound Detection at Fiber End-Facets with Surface Plasmon Resonance Cavities," *Optics Letters*, Vol. 43, 2018, pp. 775–778.

[68] Giaquinto, M., et al., "Lab-on-Fiber Plasmonic Probes for Ultrasound Detection: A Comparative Study," *Journal of Lightwave Technology*, Vol. 34, 2016, pp. 5189–5198.

CHAPTER 10

Current Developments and Future Challenges

We could have ended this book with the previous chapter focusing on the combination between plasmonic enhancement and other stimuli. Nevertheless, we consider that there is a set of important practical aspects that deserve to be summarized in a separate chapter. They compose what we believe to be the current developments and future challenges to enable the plasmonic optical fiber sensing technology to be a commercial success in the years to come. These developments come aside from optical fibers and functional materials and pertain to added values for their practical applications. For instance, with the recent integration of technologies such as 3-D printers and microfluidics, it is possible to generate portable and affordable sensing designs. The evolution and miniaturization of electronics along with new signal analysis methods are certainly also involved in the run for their industrial adoption. We also discuss about the possible areas where these types of sensors could become frontrunners.

10.1 Integrated Optical Fiber Devices

In this section, we will cover the efforts made towards the combination of plasmonic optical fiber devices with microfluidics or optofluidics as well as more integrated readout devices comprising smartphone-based interrogation and the use of integrated optics.

10.1.1 Microfluidics

Nowadays, most of plasmonic biosensors are coupled with microfluidic systems to provide a continuous and controlled supply of buffer or sample loading along the sensing area. This allows the use of small volumes (typically starting from a few microliters) provided in small size chips and capillary tubes with low energy consumption. Some devices even call on passive microfluidics where the flow is performed through capillary effect, while more advanced devices call on active microfluidics, supported by micropumps and microvalves [1, 2].

First, it is important to consider the surface tension effect. While using small optical fiber probes and metal-coated surfaces, the surface tension can lead to a high fluidic resistance, yielding poor water/surface interactions. Also, the transportation of fluids at a microscale level is different than the one usually performed with higher volumes. For instance, it is important to ensure a good wettability of the surfaces to improve the plasmonic quality and stability. Some ways to achieve this is to use low concentrations of detergents added in buffers and samples (Tween, P20, Triton X) or to pretreat the surface. A thermal annealing of the surface or a UV/O_3 treatment can also lead to significant improvements of the wettability. Moreover, the surface roughness plays a crucial role in the interaction with water molecules. Immersing the sensors in buffer for a certain time (from minutes to several hours or even days) before starting the biofunctionalization or detection experiments helps for surface loading and saves time for signal stabilization. Some studies using ultralow temperatures (−195°C using liquid nitrogen) for gold deposition also showed interesting features to improve surface morphology and enhanced optoelectronic properties, but the latter is more complex to achieve under traditional laboratory configurations [3–5]. Once the plasmonic surface is well studied and characterized, the best-suited microfluidic method can be implemented. When it comes to flat surfaces, it is easily conceivable to use channels implemented by pressure on top of the metal layer. The first material that comes to mind is undoubtedly the use of a gummy polydimethylsiloxane (PDMS)-type form. By pressure, the channels made within this polymerized material are sealed and can be connected to an external micropump system and to a liquid waste evacuation. It is not uncommon to add a degasser at the inlet of this type of microfluidics scheme to avoid the formation of air bubbles that risk being stuck in the small diameter channels and affect the monitored plasmonic signal. A flowmeter and a damper used to finely control and optimize the injection of liquid (or gas) samples are also often needed to ensure a homogeneous flow and to record any sampling perturbations [6, 7].

To properly design and build an elastomeric PDMS-based microfluidic system, the first essential step is to draw a clear map of the whole system. Both inlets, outlets, diameters of the tubing, and flow channels need to be determined and encoded in an adapted software. Then it is needed to build a wafer, playing the role of a mold for the PDMS. This wafer is usually made through photolithography, where the patterns are achieved using a photosensitive polymer that is selectively exposed to light. The surface is then immersed in successive etching solutions and thoroughly rinsed in order to clean the substrate from its upper layers, and leave the mold ready for the preparation of the PDMS (Figure 10.1) [8]. PDMS is a well-known material for research purposes and some commercial equipment. However, industrials often prefer using other polymers such as thermoplastics that are cheaper at a large scale and easier to handle using automated manufacturing tools. The absorption and hydrophobic effects provided by PDMS channels also need to be considered, especially for their long-term use inside equipment.

Nevertheless, when dealing with the cylindrical surface of an optical fiber or the extremely small end face of a fiber tip, it becomes very difficult to design an applicable flow channel applied by soft pressure. It is therefore necessary to think about new integrations for these biosensors. Most of these integrations fully embedded the optical fiber within capillaries or full fluidic chambers in order to immerse the entire plasmonic area. The fibers are also sealed on one or both ends to ensure signal stability (Figure 10.2). The integration of tapered fibers is probably one of the most challenging, as these fibers are very fragile and need to be correctly pulled before being delicately encapsulated in a protective layer of polymer. The development of

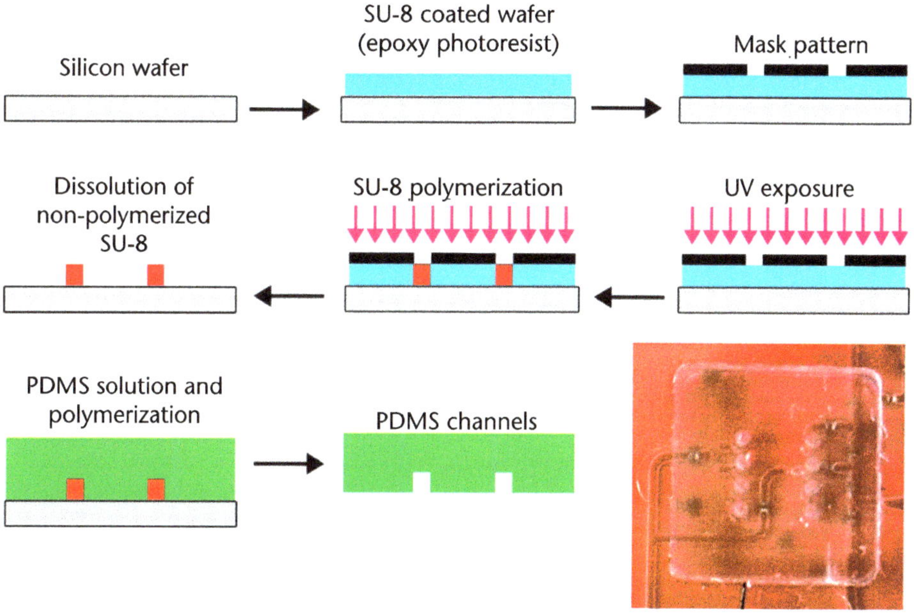

Figure 10.1 Illustration of a photolithographic process to manufacture a PDMS master wafer with a picture of a PDMS microchannels.

simple and clean techniques to integrate these probes still remains an open door in the field and will undoubtedly broaden their horizons of applications.

The encapsulation of optical fibers within standard fluidic chamber designs allows a longitudinal run along the plasmonic edifice while controlling the flow using an external micropump. The use of such chips, with a certain inclination with respect to the experimental bench, also helps avoiding bubbles formation in the chamber by causing them rise along a wall of the chamber and therefore not interfering with the sensor surface [9]. The integration perpendicularly to the optical fiber longitudinal axis has also been investigated but does not allow a good management of the flow rate and uniform distribution of the analytes along the sensing area. These single-use chips are usually commercially available and adapted on demand for a target application or can be specifically manufactured within laboratories using classical tools adapted for material works. Many research groups equipped for advanced photonics are also excited to adapt laser etching methods in order to refine the manufacture of these interfaces. Cheap 3-D printers are however not recommended in this field because they often have a too coarse resolution and do

Figure 10.2 Picture of a microfluidic chamber embedding two plasmonic optical fiber gratings. The red caps closing both two inlets and two outlets of each chamber.

not allow working with transparent materials. Also, printed polymers usually present high roughness and do not guarantee low-bind features necessary for biological applications. To overcome this issue, companies are developing new generations of 3-D printers enabling PDMS printing with the design of highly resolute microchannels [10, 11].

Finding alternatives to external microfluidic tooling is also important. Recent advances in integrated and cost-effective microfluidics have shown the relevance of simple systems generated with blotting paper. In these systems, microfluidics channels are drilled into a plastic holder containing a reservoir on one side and blotting paper on the other. By initiating the flow by means of a small blow on the surface or by an initial liquid injection, the blotting paper plays its role as an absorber and provokes a general flow, so that the biosensing area is in contact with the sample. Single-use platforms with capillary-driven fluidics could therefore be implemented with optical fibers to enable fast sensing under simple microfluidics [12–14]. It is clear that to carry out low-cost tests or POC tests, it is needed to choose between two concepts: a permanent measurement platform with disposable or reusable plasmonic chips, such as for traditional plasmonic platforms, or a fully integrated single-use cassette with this type of cost-effective microfluidics components for effective mass screening. The integration of optical fibers with blotting paper could maybe provide such a flowing effect as well, but the acquisition of the initial reference signal will probably remain an issue, as this technique only allows a one-shot solution.

Despite the efficient integration of optical fiber-based sensors on a chip (which results to the concept of lab-on-a-fiber-on-a-chip rather than lab-on-a-fiber), one can figure out that by immobilizing the sensor, two major assets of the optical fiber itself are somehow lost, namely its very thin size and its flexibility. This is the main reason why many practical applications have investigated the use of plasmonic fibers directly in solution by dipping the probes inside small volumes and even in nonliquid matrices. The use of a protective packaging or their insertion on an automated measuring setup also allows their easy handling, while allowing their use in diversified environments (see *in situ* sensing prospects in Section 10.4). However, limiting the parameters influencing the signal response such as the curvature of the fiber or temperature changes could be important assets brought from such packaging. The study of their sensitivity to molecular binding could therefore be more straightforward and reproducible using well-designed and well-controlled workstation with fixed parameters.

10.1.2 Optofluidics

Besides the aforementioned considerations that consist in dipping or integrating optical fibers with microfluidics, an emerging trend consists in directly integrating microfluidics within the optical fibers themselves. This field of research is called optofluidics and takes on its full meaning for very specific specialty optical fibers, as reviewed in Chapter 5. Obviously, these include optical fibers comprising holes in their cross-section, along the propagation axis (Figure 10.3 [15]).

The controlled injection of fluids in hollow channels within optical fibers is a technical feat. Although it brings many interesting detection features and limits the need for an external fluidic interface, it is often more restrictive in terms of

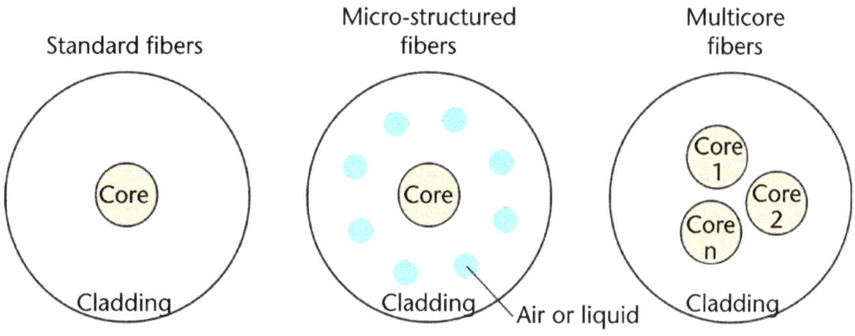

Figure 10.3 Illustration of three optical fiber configurations (cross-sections). (*Adapted from:* [15]. (RightsLink license number 5311201102204.))

connection and in terms of tunability. The use of inner channels located around the core is often limited to LSPR applications with metal nanoparticles injected into the channels because outer metal films on top of the cladding do not lead to interactions with the inner samples. The detection techniques and applications are therefore totally different, as everything must be performed internally to the fiber structure, unlike the external measurements previously mentioned. Many examples of integrated optofluidic optical fibers exist and some recent ones are listed in Table 10.1. In this direction towards microfluidics integration, it is clear that femtosecond pulse laser micromachining is very relevant. Precise apertures or channels can be made with tightly focused laser pulses. A subsequent etching process, usually with KOH, is then applied to obtain clear cuts. The concept of flat optical fibers is also an emerging trend, but it is still timidly rising in the scientific community. This can be explained by the difficulty to manage all their optical and mechanical properties. Flat fibers could represent a real scientific challenge in the near future, especially for biological monitoring processes [16–18].

10.1.3 Smartphone-Based Optical Fiber Sensors

The use of several optical fibers together can improve the detection reliability, especially if both a reference fiber and a functionalized biosensor can be tested under the same conditions. Due to their small diameters, it is possible to add multiple optical fiber sensors on one single chip, and this would probably help to improve the efficiency and reproducibility by confirming the detection by multiplying tests, against the same target. The implementation of microcircuits, electronics, and/or electrochemistry inside a 3-D-printed chamber can also drastically increase the number of final applications. Besides the fact that these sensors could be used reversibly due to advanced bioreceptors or through surface regeneration (see Chapters 6 to 8), optical fiber technologies also show a growing interest for on-site uses as POC sensors and for single-use. Lab-on-fibers can indeed be integrated into multiple interfaces such as smartphones or portable interfaces (Figure 10.4). Using a smartphone as a sensing platform is critically relevant. While thinking about an optical fiber, it is quite difficult to ignore all the equipment that connects to it: the need for a light source, a spectrometer, sometimes the addition of a polarization controller or intermediate

Table 10.1 Some Examples of Optofluidic Research Applications

Optical Fiber Configuration	*Receptor/Target*	*LOD*	*Reference*
1 × 3 multichannel optical fiber (fluorescence)	Antibody/atrazine or 2,4D (herbicides)	0.03 μg/mL atrazine and 0.04 μg/mL 2,4-D	[19]
All-in fiber Michelson interferometer in PCF with fs laser micromachining	Antibody/DNA hybridization and methylation detection	5 nM	[20]
Thin-walled hollow optical fiber coated with SiO_2 nanoparticles and Au nanorods	Streptavidin/Biotin Horseradish peroxidase enzyme (HRP)	NA	[21]
Fresnel reflection optical fiber	Antibody/SARS-CoV-2 spike protein receptor-binding domain (S-RBD)	5 pg/mL	[22]
Lng-period grating (LPG) in PCF	Antibody anti-sialoprotein (specific and nonspecific)	NA	[23]
Hollow microneedle	Acetyl-L-Lys-D-Ala-D-Ala binding site/ Vancomycin-HRP	<100 nM	[24]
Two Fabry-Perot resonators in hollow core fiber	Aptamer/Interferon-gamma	0.5 ng/mL	[25]
Miniaturized hollow core fiber	Biotin/Avidin	9.5 pM	[26]
Hollow core fiber with femtosecond laser micromachining	Antibody/Benzo[a]pyrene (PAH)	1.65 pM	[27]
All-in fiber Michelson interferometer (femtosecond laser micromachining)	DNA/DNA hybridization	1 μM	[28]
Microstructured optical fiber with Bragg grating	Peptide nucleic acid (PNA)/ DNA + Streptavidin AuNPs	NA	[29]
Microfiber in lateral contact with a silica capillary	DNA/miRNA	212 pM (1.43 ng/mL)	[30]
Tapered fiber probe inside a fluidic cell	DNA strand/Hg^{2+}	8.5 nM	[31]

elements such as a circulator or other quick connectors, and the need for fiber splicing. All these bulk parts negatively taint the small size of the biosensor itself. Even so, today, we all have a smartphone in our pocket that can provide a light source through its flashlight, and a camera or an ambient light sensor that can be exploited as a detector. Extensive studies have shown that, by fitting specific lenses to these elements, it is possible to monitor an optical fiber sensor in real time. In addition, the generation of adapted software or portable applications are plethoric and would make it possible to achieve live analysis and detection for the end user. The limitations of smartphones would come up from the achievable wavelength range or the proper characteristics of the light source, but also from the detector itself. Some optical configurations are therefore not suited to the actual features embedded in smartphones. It would then be necessary to develop a miniaturized platform finely adapted to the desired application of the sensor. This is technologically possible and

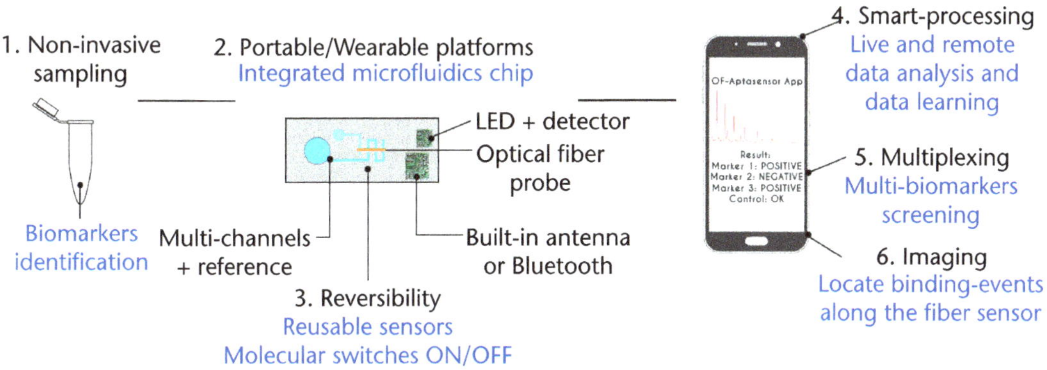

Figure 10.4 Prospects in integrated optical fiber biosensors using smartphone-based interfaces for POC. (*From:* [15]. (RightsLink license number 5311201102204.))

very encouraging for potential commercial use on the field. For instance, portable and waterproof designs and battery-operated and time-lapse stations will be needed in the coming years to be able to test the performances of such optical fiber-based plasmonic biosensors for POC or environmental monitoring [15, 32–36]. We can think that this is a long-term vision, but it is already running through the strategic lines of big companies such as Meta (to develop medical imaging and VR tools) or Apple and Microsoft (to develop live monitoring tools such as those integrated in smart watches).

The first demonstration of smartphone-based interrogation of plasmonic optical fiber sensors dates back to 2015 [37]. A 400-μm multimode optical fiber section is specifically prepared (cleaved at 45° at both ends) and connected between the flashlight and camera of a smartphone. For each measurement, an MP4-video (480×640 pixels) was recorded and processed using MATLAB. This configuration, reproduced in Figure 10.5, was used for refractometric sensing. It was demonstrated that a sensitivity of 5.96 10^{-4} RIU/pixel can be obtained in the range of

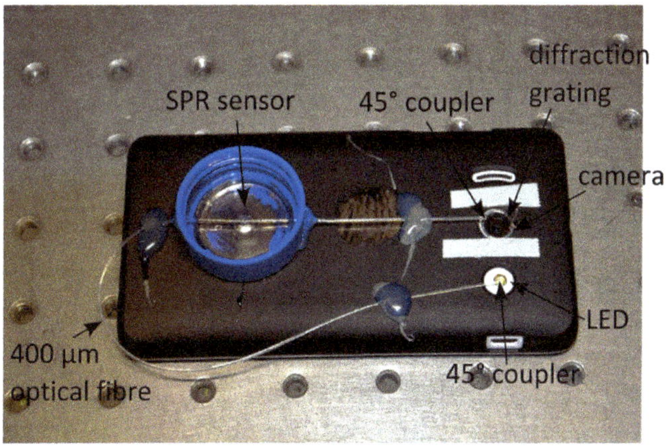

Figure 10.5 Picture of a silver-coated unclad multimode optical fiber section connected to the flashlight and camera at the back of a smartphone. (*Picture from:* [37].)

1.33 to 1.36. Smartphone-based SPR sensors have been recently reviewed in [38] where interested readers can find relevant practical information.

Many books and research articles have been published about integrated optics, but most readers are certainly not familiar with its definition. Integrated optical systems consist in most cases of an optical chip that is provided with a network of optical channel-type waveguides. Most of these integrated optical devices therefore rely on planar structures (e.g., a glass slide or a silicon wafer). However, our consideration of integrated systems for biosensing and, more importantly, for commercial purposes is not only focusing on the biosensing/optical part. The concept of integrated biosensors refers to devices where all the components are merged together in order to obtain a single reading system, running the experiment. For instance, merging a light source, a spectrometer, and a reception notch for a disposable biosensor is part of the integration process to generate new readout interfaces. Following this meaning, we can easily understand that the use of an optical fiber as a standalone probe is not sufficient to fulfill all the experimental procedure. Therefore, many companies are developing their own metrological tools under specific designs that most often simply rely on the integration of other commercial parts, available from other companies. In this way, the technological development is minimal, while the technological adaptation is enormous. This creates significant added value to the materials and increases the application possibilities.

Numerous efforts are currently being made towards such integration. The 3-D printing technology is definitely a key for such developments. The technological innovation comes from the printer itself, which allows creating whatever design is needed (dedicated optical parts, holders, covers or tooling boxes, chips), while printing an object is no more innovation but a smart exploitation that takes advantage of the 3-D printing technology. Integration is therefore a backbone in order to adapt existing technologies to a larger number of systems and the arrival of tools that can be entirely managed by the end user (such as 3-D printers and their related open-source software) set in this universality. This being said, purely integration-based research articles are therefore not innovative enough as such through their technological innovation. However, they have a large interest in adapting technologies to exploit new sensing features or to screen new types of samples. In this way, they can bridge the gap between existing elements and the need for their combination to generate new properties from where the innovation comes. Miniaturization is surely an important aspect of this integration as well. Combining advantages of several devices or technologies and step into practical reality through integration can therefore lead to novel (and sometimes unexpected) research outbreaks [34, 35, 39–42].

Fully integrated optical systems developed for research are not abundant, mainly for the reason aforementioned (integration and development are part of an industrial approach more than a technological innovation from academia). Some recent and promising examples for further biosensing implementation/optical fiber integration are highlighted in Table 10.2. Others have been successfully launched on the market and will be discussed in Section 10.2 towards commercial practices. It is important, as biosensing is part of an applied research, to think about the final integration of the developed sensor. Obviously, this is overlooked and not practically detailed in most research articles.

Table 10.2 Some Examples of Optical Fiber-Integrated Biochemical Platforms

Optical Fiber Configuration	*Reference*
Smartphone-based spectrometer for optical fiber Fabry-Perot interferometric sensor	[43]
Optofluidic fluorescent imaging cytometry on a cell phone	[44]
SPR-based optical fiber biosensor setup on a smartphone	[45]
Multimode smartphone biosensing: transmission, reflection, and intensity spectral analyzer (TRI)	[46]
Hg^{2+} detection using fluorescence optical fiber sensor handled by a smartphone	[47]
Portable smartphone-based plasmonic nanosensor readout platform for biosensing	[48]

10.1.4 Multiplexing

The interest in these optical fiber-based biosensors would also be significantly increased if the possibilities of detection were multiplexed, with the detection of several targets simultaneously. This multiplexing could be implemented through several ways: by nano-printing different bioreceptors on the optical fiber surface, using fluorescent labels absorbing or emitting in successive wavelength ranges, or using tunable bioreceptors that could bind several targets in determined conditions. These bioreceptors could, for instance, be engineered to switch on or off by progressively changing the experimental conditions (pH, temperature, ions concentration) or be activated using specific quenching molecules. Following the current knowledge and database available for tunable bioreceptor engineering (with advanced systems for DNA and RNA design such as those developed by startup compagnies like WayFinder Biosciences), this technology is on track and will slowly turn to the next level of play.

Certain configurations have been described for optical fiber multiplexing as well, starting from distributed refractive index sensors to more advanced features with the possibility of discriminating one target from the other.

The next section focuses more on the current achievements towards the industrialization of plasmonic optical fiber devices.

10.2 Towards Commercial Practices

It is clear that there is a huge gap between industrial companies research efforts by the scientific community and development efforts by industrials. As covered in previous chapters, the former tends to report new configurations and additional functionalities and ways to improve metrological performance (sensitivity, limit of detection), among other research activities. Industrial companies are much more focused on more mature technologies such as unclad fibers, which they further develop and integrate as standalone devices. Integration levels are now such that some plasmonic optical fiber sensors are commercial products.

This section gathers figures about the current state of research and the current industrial actors. First, let us mention that it is important to place the elements in their context: the total number of research projects resulting in scientific publications (any types) is increasing every year. This is globally true for all topics, so it is therefore also the trend observed for publications related to biosensors each year, with a small rebound to be expected from 2020 to 2022 following the pandemic and bolstering the growing trend (Figure 10.6). The progression of this increase is faster

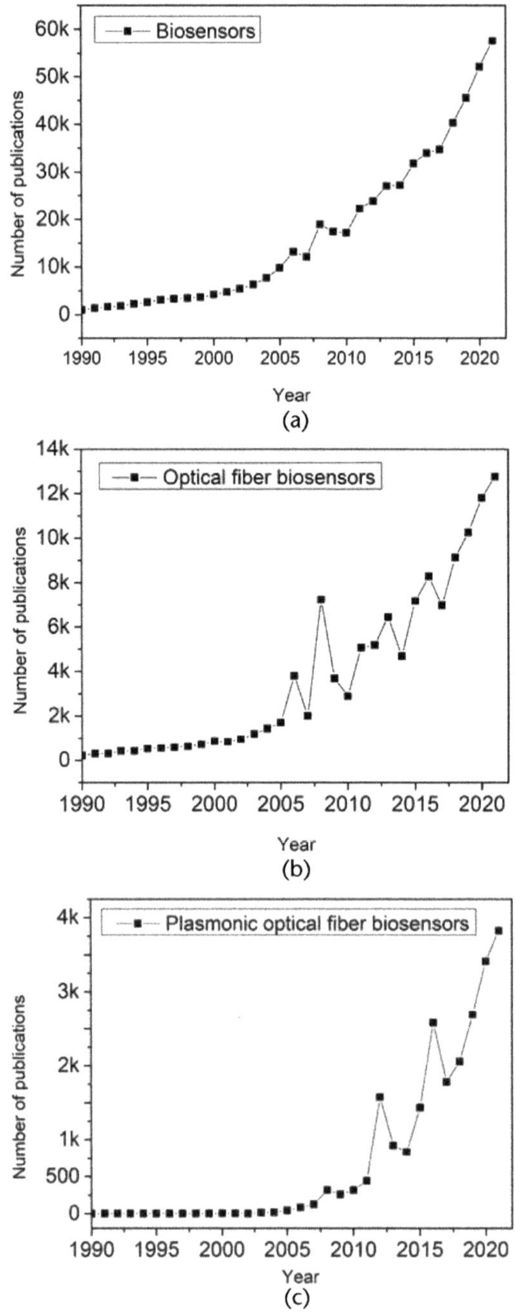

Figure 10.6 Number of publications about biosensors (a), optical fiber biosensors (b), and plasmonic optical fiber biosensors (c), generated from https://app.dimensions.ai/discover/publication in October 2022.

than in other research areas, demonstrating the growing and innovative interest in bioengineering and biotechnology.

This general trend is also depicted concerning optical fiber biosensors publications and for plasmonic optical fiber biosensors.

These trends reflect a growing interest in the field of plasmonic optical fiber biosensors and therefore demonstrate that the scientific community behind these techniques is still growing. In addition, the targeted applications of these biosensors are spread across many fields of research, which can be approximated as depicted in Figure 10.7. Biochemistry and engineering together are the main areas involved in most optical fiber-based research works, which is also highlighted throughout this book, with two possible readings depending on the readers' experience and affinity for these.

Besides these considerations that pertain to academic research activities, one can also note a significantly growing number of companies active on the development and commercialization of components useful for plasmonic biosensors or their integration. Most of these companies are not yet involved in optical fiber-based sensing. There are listed for their relevance with respect to plasmonics in Table 10.3 and to make interested readers aware of their existence. While a published article is sustainable as a function of time, the existence of a commercial depends strongly on its demand on the market and/or investments. We therefore realize how Table 10.3 may evolve in the years to come. Hence, it is only perfectly valid at the time of writing and the nonexhaustive list of important actors in the field is provided in order to help readers wishing to implement elements of microfluidics, optofluidics, platform designs, or optimization.

Overall, there is a growing demand for biosensors and devices that can be used outside of laboratory settings. In the following section, we give some considerations about the possibility to implement plasmonic optical fiber sensors in that context.

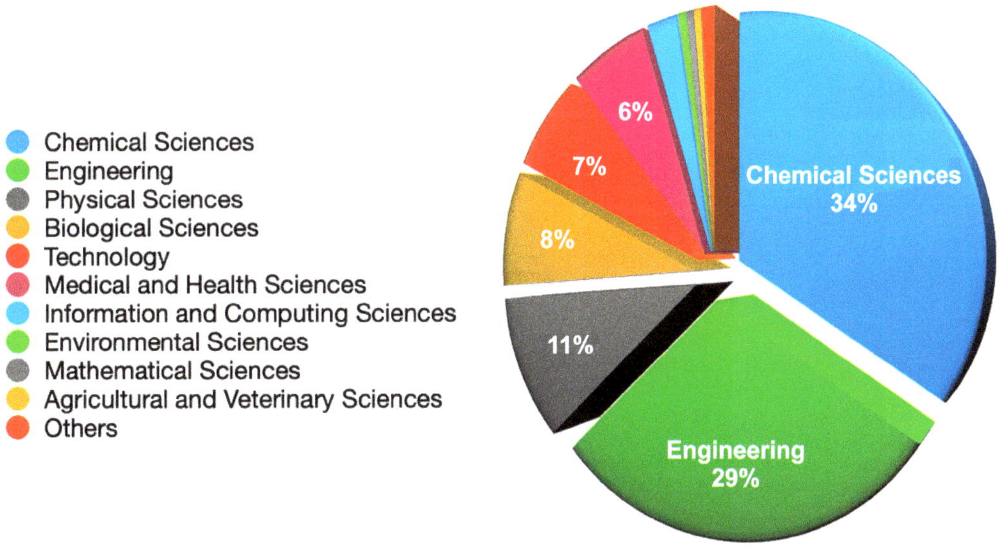

Figure 10.7 Proportion of publications related to biosensors classified per field of expertise.

Table 10.3 Nonexhaustive List of Industrial Actors for the Development of Biosensors and Their Integration, Microfluidic Interfaces, or Optical Fiber-Based Biosensing Devices

Major Industrial Actors and Startup Companies			
Company	*System*	*Website*	*Country*
Affinité Instruments	Miniaturized SPR sensing interfaces	https://affiniteinstruments.com	Canada
Bartels mikrotechnik	Microfluidics facilities and miniaturized devices	https://www.bartels-mikrotechnik.de	Germany
Bio-rad	ProteOn and associated plasmonic interfaces	https://bio-rad.com	International
BioNavis	SPR Navi-instruments	https://www.bionavis.com	Finland
Bioptix analytical	SPR instruments	http://www.bioptixanalytical.com	United States
Biosensing Instruments	Bulk SPR instruments (BI series); SPR microscopy (SPRM series) and SPR electrochemistry (EC SPR series)	https://biosensingusa.com	United States
Biosuplar	Open-configuration SPR sensing interfaces	http://www.biosuplar.de/	Germany
Bruker	Sierra SPR devices and SPR^{+} combining imaging techniques	https://bruker.com	International
Cadworks 3D	Microfluidic interfaces, PDMS master molds, and 3-D printers	https://cadworks3d.com/	Canada
Carterra	High throughput SPR devices	https://carterra-bio.com	United States
Creative Biostructure	SPR services (MagHelix devices)	https://creative-biostructure.com	United States
Cytiva	Biacore instruments including the Kretschmann prism (previously GE Healthcare)	https://cytivalifesciences.com	International
Darwin Microfluidics	Microfluidics facilities/reseller	https://darwin-microfluidics.com	France
Dolomite-Microfluidics	Specialized in droplets microfluidics and 3-D printed chips	https://www.dolomite-microfluidics.com	United Kingdom, international
Elveflow	Microfluidics facilities and 3-D printed chips	https://www.elveflow.com	France
Fluigent	Microfluidics facilities	https://store.fluigent.com	Germany
Formlabs	Dedicated 3-D printers and materials	https://formlabs.com	International
Fox Biosystems	Plasmonic unclad fibers inside an automated lab-based interface	https://foxbiosystems.com	Belgium
Horiba Scientific	Bulk SPR and SPR imaging instruments	https://horiba.com	International
Ibis Technologies	Ibis MX SPR equipment/SPR imaging	https://www.ibis-spr.nl	Netherlands
iclueBio	SPR instruments	https://www.icluebio.com	South Korea
imec	Chips design and microfluidics facilities	https://www.imec-int.com	Belgium, international

Table 10.3 *(Continued)*

Major Industrial Actors and Startup Companies			
Company	*System*	*Website*	*Country*
LSPR	LSPR-based devices for DNA sequencing (prototype stage)	https://lspr.swiss	Switzerland
Microfluidic Chip-shop	Design services, chips, and accessories	https://microfluidic-chipshop.com	Germany
NanoSPR	NanoSPR suite, microfluidics and chips	https://nanospr.com	United States
Nicoya	Alto and OpenSPR systems	https://nicoyalife.com	Canada, United States
Novalix	Drug discovery services with SPR facilities	https://novalix.com/	France, Belgium
Ocean Insight	Customized fiber probe services and miniaturized light sources/ spectrometers	http://oceaninsight.com	International
Phaselab Instruments	Compact phase-sensitive SPR devices	https://phaselabinstrument.com	France
Plasmetrix	Compact SPR devices (Corgi suite)	https://plasmetrix.com	Canada
Plexera	SPR imaging device (Plexarray)	http://www.plexera.com	United States, Japan, China
Presens	Optical interfaces (industrial and research monitoring and imaging)	https://presens/de	Germany
Reichert Technologies (Xantec bioanalytics)	4SPR series Bulk SPR instruments	https://reichertspr.com	International
Research International	Integrated devices and optical fiber-based devices, specialized in defense/security	https://www.resrchintl.com	United States
Sartorius	Octet series (SPR instruments)	https://sartorius.com	International
Sofchip	SPR sensors chips and nanoparticles	https://sofchip.com	United States
uFluidics	Microfluidics chips and dedicated manufacturing services	https://www.ufluidix.com	Canada
Ward's Science	Open SPR device	https://www.wardsci.com	United States
Xona Microfluidics	Chips and microfluidics facilities	https://xonamicrofluidics.com	United States

10.3 POC Sensing and Related Innovation

In recent years, the world has become aware of its fragility in several areas: medicine (viruses, cancers), climate, energy, and military defense, to name but a few. The need for fast field measurements to monitor more parameters including biological and chemical targets is clearly evidenced. This is precisely the role of POC devices. POC sensing is a general trend that consists in bringing medical devices including biosensors at the patient home (ambulatory practices). Recent technologies in mobile-based diagnoses constitute the next wave of innovations in POC sensing. Beyond the optical fiber field, some current and emerging biosensor designs enable a transition between laboratory settings and home-based applications. Blood glucose testing and ecotoxicology screening certainly remain the most prominent examples nowadays. Some important developments are therefore focused on the end user as the core element of a multipurpose screening and information acquisition platform. This trend is facilitated both by progress in data processing with the aid of artificial intelligence (as outlined in Section 10.5), as well as the integration of miniaturized sensors with smartphones (and also with tablets and wearables). The majority of these novel designs and devices are usually compatible with an Internet-of-Things setup, allowing for their implementation in a global dimension.

The coronavirus pandemic (SARS-CoV-2, Covid-19) of 2020 has revealed the need to find POC alternatives to the ELISA and the PCR techniques, which remain the gold standard for diagnosis in clinical settings. Dedicated paper-based biosensors can be found in supermarkets and many other POC sensors are still emerging, fulfilling more tasks to deliver a clear result directly to the patient. As a result, both scientists and industrials rushed to find alternatives to existing diagnostic techniques in order to find new products to develop in a short time and to sell them at a large scale. Despite the many promises that can be found in the literature and the long duration of the health crisis, it is clear that very few viable alternatives have actually been brought to the market but all these efforts contribute to a global progress. Plasmonic optical fiber configurations have been proposed to target SARS-CoV-2 antigens, which somehow confirms that the technology can be quite rapidly transposed to hot topics, when required.

This being said, let's concentrate now on what we believe are key aspects to possibly turn plasmonic optical fiber sensors into actual POC devices. According to our background, the following points of attention must be considered:

1. The sensor tip itself, which needs to be robust enough for manipulation by the end user. We have already discussed integration aspects in the previous section. Here, the integration should be more considered in the form of a housing or protection to enable the use of the sensors by nontrained people. This housing should encompass the way that the analytes are brought into contact with the sensing region. Preparing the analytes solution in a small vial or Eppendorf is probably the easiest way to go in the context of POC devices, but there are many more options.
2. The connection of the fiber tip with the readout device should be optimized. One can imagine the use of LC connectors for relatively easy plug-in in the interrogator.
3. The readout device that should certainly be miniaturized, cost-effective, and reliable. Integrated optics is one way to go. The use of a smartphone is certainly another important way. It then means that a small plug-in device has to be prepared to enable the easy connection between the back of the smartphone and the sensor tip. The latter can certainly be prototyped using additive manufacturing or 3-D printing.

4. Data processing is another important consideration. Obviously, this can be made in the form of an application whenever a smartphone or tablet is involved. A script should be optimized to analyze the sensor signal in real time and produce a response that is easy to interpret. This important aspect of the device comes along with an exhaustive training process, to account for numerous data sets and a form of artificial intelligence for their interpretation (see Section 10.5).
5. The last aspect is the communication of the processed data. This one requires less effort because it is common to many other developments.

Besides the use as POC devices, another important aspect of sensing that the optical fiber technology can enable is the possibility of *in situ* sensing, the subject of the next section.

10.4 Towards *In Situ* Sensing

Optical fiber tip-sensors are well suited for *in situ* sensing. *In situ* literally means "in its original place," in complementarity to *in vitro*, which refers to an experiment performed in a test tube or other laboratory conditions (petri dish, Eppendorf tube, microtiter plate). *In vivo* is even more restrictive because it means "inside a living organism," while *ex vivo* refers to measurement outside the organism (e.g., inside an organ after its resection). *In situ* sensing therefore encompasses *in vitro* and *in vivo* measurements.

It is well established that very few sensors are able to perform *in situ* measurements, due to their geometrical configuration or their need to be handled in confined environments (control of external influential parameters). Even if they could be, they are usually not sufficiently certified by the demanding quality and sterility protocols that allow this equipment to be used in practice. Because optical fibers can deal with these mechanical constraints, it is important to exploit their full integration capabilities. To reach that goal, many adaptations have to be considered:

1. *The nature of the matrix (composition and density):* For example, performing measurements in a filtered sample diluted in buffer is very different from working in whole blood, which will be more viscous and contain cells. Its variation over time can be problematic, especially during coagulation. Finally, when it comes to working inside tissues or organs, the density variability must be compensated and the optical fiber must be sufficiently protected to avoid any perturbation. This means that adapted packaging/housing should be considered *in vivo*, while it is usually useless *in vitro*. This makes an important difference of practical implementation and requires the involvement of biomechanical considerations.
2. *The temperature of the sample:* The temperature of the matrix is also important and must not lead to large signal deviations during measurement. A system may be needed to ensure a constant temperature or compensation of the signal may be required. In this case, the best is to get an intrinsic

compensation of temperature variations like the one provided by TFBGs, for instance (see Chapter 4).

3. *The conditions for molecular interactions:* The molecular interaction can be affected by environmental factors or the nature of the sample. For example, some molecular interactions may not take place in urine or saliva because of the pH or the presence of enzymes or cells. The contact with the sample must also be optimized to ensure sufficient flux in the probed sample.
4. *The sterilization and quality control:* Some environments such as in the biomedical context would require compliance with certain standards, including sterilization. Most biosensors are sterilized with steam in an autoclave or under ethylene oxide gas (sometimes ozone) to maintain the properties of the bioreceptors. Quality checks are needed to ensure the effectiveness of the process. The sensing edifice (optical fibers, metal coatings, and biofunctionalization) should survive this sterilization process. Besides works published from Stolov et al. in 2015 [49, 50] on the sterilization effects in optical fibers for clinical use, everything remains to be done in this field.
5. *The washing and regeneration:* Finally, nonspecific interactions should be avoided, especially when dealing with complex matrices. This is the major obstacle for any *in situ* measurements (in both biomedical and environmental fields). For the biosensor to work enough times, it must be protected from the elements of the matrix. This can be done through blocking strategies or filtration systems placed upstream of the sensitive area. Regeneration and rinsing steps must be implemented, but very often these conditions are unthinkable for medical uses where only a single insertion or measurement is allowed. The constraint therefore strongly depends on the integration of the sensor.

Examples of integrated optical fibers for *ex vivo* and *in vivo* measurements have been quite extensively developed in our research group, with the insertion of biofunctionalized gold-coated TFBGs in an endoscope, in order to detect lung cancer biomarkers (Figure 10.8). The first encouraging results obtained with plasmonic optical fibers *in vivo* will require further characterization and sterilization to consider their development and long-term use. The study highlighted in this case is, to our knowledge, the first *ex vivo* detection and *in vivo* plasmonic excitation using TFBGs. As depicted in Figure 10.8, a very specific catheter in polyoxymethylene (POM) has been developed to embed two plasmonic optical fiber sensors. The latter are exposed to a lateral aperture made in the catheter so that they can sense the surrounding medium. The end of the catheter takes the form of a needle for easy insertion into tissues. The dimensions of the packaging are such that it can be inserted without friction in the operating channel of an endoscope (a bronchoscope on the picture). The insertion of this packaged sensor in a living pig can be seen in Figure 10.9 together with the corresponding SPR signal.

Besides *in vivo* sensing, an integrated strategy that could perhaps lead to contact biosensors inside textiles was first highlighted by the group with Professor Y. Fink at the Massachusetts Institute of Technology (MIT), only with physical measurements so far, but with promising future for innovating sensing. The implementation of wearable sensors for the detection of molecules on the skin or through sweat could lead to breakthrough applications for healthcare. For instance, monitoring

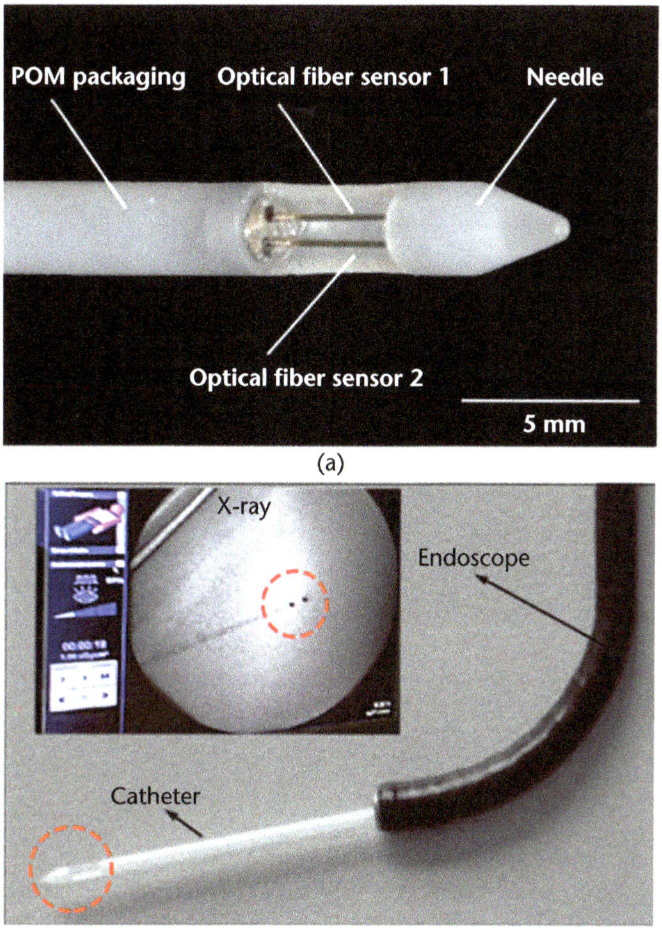

Figure 10.8 (a) Picture of two embedded plasmonic TFBGs inside a bioengineered packaging with an opened window and a conic tip to penetrate inside tissues. (b) Picture of the packaging inserted inside an endoscope with two crimping collars in stainless steel to maintain the fiber in position. The latter can be used by the surgeon through X-ray to position the sensors inside the lungs.

the presence of bacteria or toxins under a bandage or near a wound could be of great interest. To reach the objective, we need to solve important problematics concerning the miniaturization and the remote control of the optical probe. Their conditioning and lifetime are also something to push forward to achieve fabrics as a service, as described in recent works [51, 52]. Nowadays, wearable biosensors are increasingly popular and raised over tens of billions of dollars in market parts in the United States in 2021. Moving forward the plasmonic-based technology to high-level implementation on skin or for biofluids analysis is an open breach to find new concrete biomedical applications.

As stated in the beginning of the section, *in situ* means much more than *in vivo*. For the sake of completeness, it is therefore relevant to end this section by recalling that different *in situ* measurements have been achieved so far with the plasmonic optical fiber technology. One can think about their use for: (1) bulk plasmonic

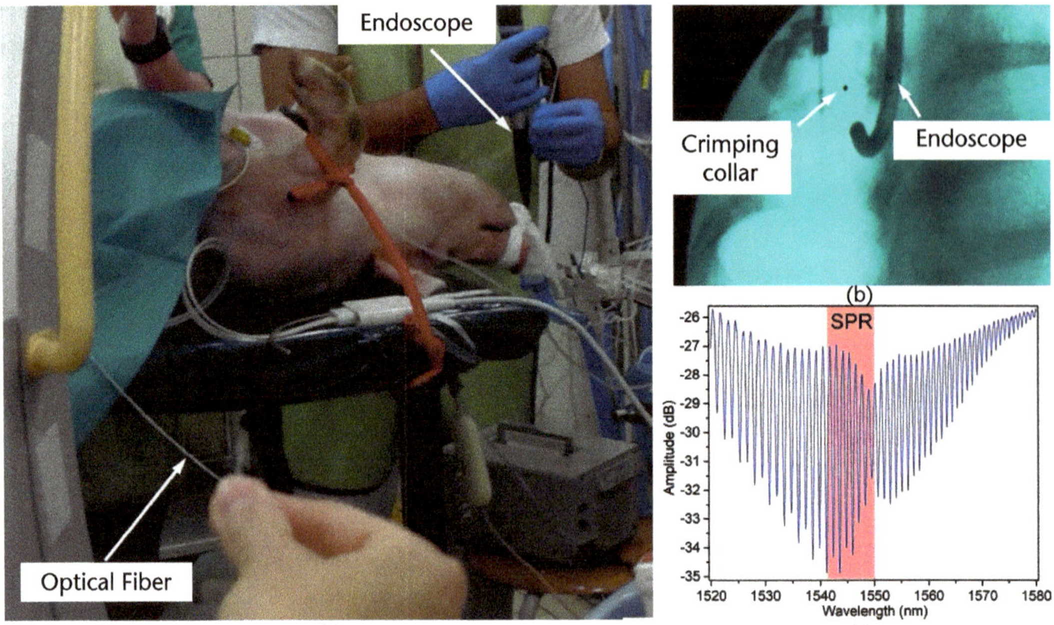

Figure 10.9 (a) Picture of an anesthetized pig with the live endoscope navigation performed at the Erasmus Hospital in Brussels, ULB. (b) Chest radiograph showing pig lungs and the endoscope, with metallic crimping collar indicating the location of the needle. (c) First plasmonic signature obtained in pig lungs after the insertion of a catheter-protected gold-TFBG tip sensor in the upper lobe region.

environmental sensing [53, 54], (2) thin film growth monitoring [55], (3) for the detection of the state of charge of supercapacitors [56] and more recently of their temperature and capacitance [57], (4) for battery monitoring [58], or even (5) for gas (H_2S) sensing [59]. These are prominent examples among a much longer list that confirm the relevance of optical fibers to enable sensing directly in the environment of interest. The miniaturized size, the easy light injection, and the remote operation are obviously the keys enabling this sensing modality.

The next section will focus on another important contemporary aspect of sensing (i.e., the use of artificial intelligence in the demodulation process and signal analysis).

10.5 Artificial Intelligence-Assisted Sensing

The latest digital revolution has enabled the development of smart devices encompassing sensor hardware and artificial intelligence software to perform intelligent tasks. Smart sensors are now present everywhere in applications ranging from voice-enabled home devices to the industrial Internet of Things (IIoT). This important and recent progress has been made possible by the miniaturization of sensing hardware, the development of big data storage, the access to high-performance computing, and the latest achievements in machine learning and artificial intelligence technologies. Artificial intelligence results from research and developments in deep learning, a subfield of machine learning that makes use of bio-inspired neural networks to perform learning tasks [60]. Artificial intelligence technologies have now matured to such a degree that their performance for practical tasks such image and speech recognition or machine translation is comparable with, if not better than, human skills. A review about the recent advances in machine learning for optical fiber sensing applications can be found in [61].

Since the advent of artificial intelligence a decade ago or so, this technology has been obviously applied to plasmonic optical fiber sensing. Relevant implementations will be reviewed in this section. In biosensing, machine learning can evaluate large data, produce appropriate results, even from noisy and low-resolution sensing data, and find interrelationships between signals and bio-events, as sketched in Figure 10.10. Looking at the typical spectra that need to be demodulated, machine learning algorithms are able to analyze all changes in the spectrum and automatically highlight the most significant features. Machine learning algorithms are therefore efficient to increase sensitivity, solve problems of weak selectivity, or automatically calibrate sensors. They also lead to an increase in resolution and a decrease in noise level. The most prominent machine learning algorithms are artificial neural networks [60], k-nearest neighbors [62], decision tree algorithms such as gradient boosting [63], and principal component analysis [64].

Given the richness of their spectrum, gold-coated TFBGs are very well suited to be used with machine learning algorithms for data processing. In [65], machine learning algorithms have been applied to obtain the value of the surrounding refractive index from the experimentally measured transmitted amplitude spectrum of a gold-coated TFBG sensor. Data recorded with two nonidentical plasmonic TFBG sensors were processed with machine learning algorithms. To determine the surrounding refractive index value, dimensionality reduction with principal component analysis and polynomial regression were both used. A model trained on data from one sensor was successfully applied to process optical spectra from the other TFBG. An accuracy of $9\ 10^{-6}$ in the determination of the surrounding refractive index changes has been achieved with this method. In [66], bare TFBGs were demodulated with convolutional neural networks. The model was trained on a few thousand spectra recorded in calibrated refractive index solutions. A mean square error of $2.8\ 10^{-7}$ RIU was reported. These figures clearly demonstrate the usefulness of machine learning algorithms in the enhancement of the metrological performance of optical fiber-based refractometric sensors. Hence, there is no doubt that many more implementations with plasmonic optical fiber devices will be reported in the future [65].

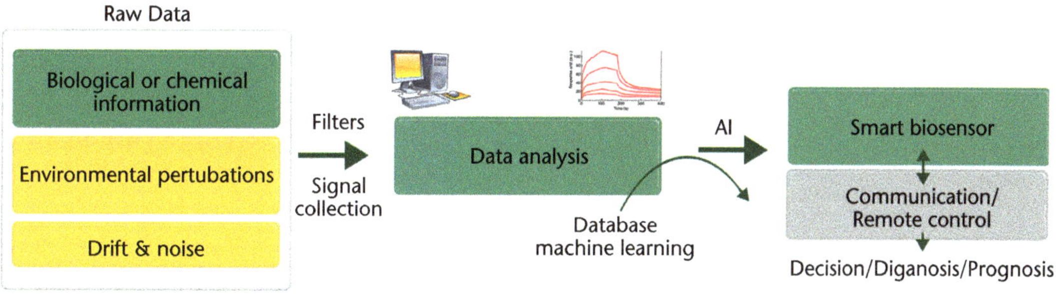

Figure 10.10 Scheme of the artificial intelligence principle based on raw data collection, analysis, and exploitation.

In the last section of this chapter, we provide some brief considerations about another emerging trend in plasmonic optical fiber devices (i.e., enabling their capability for imaging instead of sensing).

10.6 From Sensing to Imaging

Imaging is definitely a very important area of research and development and numerous books are entirely devoted to this topic. There are many competing technologies in imaging. Some of them were approached in Chapter 9, notably when combining plasmonics with fluorescence, ultrasound, or Raman scattering. In this section, we will provide some considerations about SPR imaging (SPRi). We will stay very concise as imaging remains aside from sensing considerations. Also known as SPR microscopy, SPRi follows the same general principles of SPR. The measured information and the method of detection are different. This technique was inspired by the need for a higher-throughput method of studying binding interactions based on the SPR phenomenon.

There are various configurations of SPRi [67]. Most commonly, a monochromatic polarized beam of light is launched through a prism onto a 2-D array of binding sites immobilized on a thin film of gold. A CCD camera is used to capture high-resolution images of the binding area, as sketched in Figure 10.11. In this technique, measurements are therefore taken for a constant wavelength and angle. Light is captured by each spot of the CCD camera simultaneously and, because of continuous monitoring, it provides real-time kinetic data in a high-throughput manner. The change in reflected light intensity (which is proportional to the change of refractive index produced by compounds bound to the functional sites on the metal film) is measured using images from the CCD camera. Intensity changes of the reflected light result from the shift of the reflected angle due to the surface refractive index change. In theory, such a setup can be used to study hundreds of binding sites simultaneously. In practice, the inherent limitations (such as picture resolution) resulting from the use of camera images usually provide lower sensitivity measurements compared to other SPR techniques. The main advantage of this techniques is therefore its high throughput. The most prominent application of SPRi

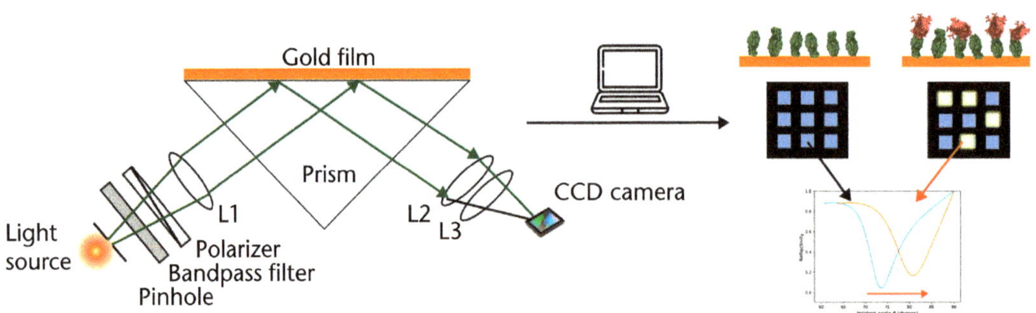

Figure 10.11 (Left) General principle of SPRi and the analyte-ligand interaction shifts the SPR curve towards a higher angle (red to orange). (Right) Due to the measurement confinements (fixed wavelength and angle of incidence θ), changes in the reflectivity (Δ%R) at a single spot of the array can be simultaneously detected. (*Adapted from:* [69].)

so far is certainly the demonstration of direct multiplexed detection of unlabeled low molecular weight protein biomarkers [68].

SPRi has been successfully applied to gold-coated unclad multimode optical fiber configurations in [70]. In this work, light from an LED source was launched into the fiber and the transmitted light was detected by a camera. The overall setup was made very compact, in contrast to otherwise much bulky configurations in the absence of optical fibers. This pioneering work will certainly pave the way to much more developments of SPRi on optical fibers in the years to come, as the advent of highly resolved and high-speed cameras can further improve the performance of the device.

In conclusion, this chapter has highlighted what we believe are the current hot topics and next steps to take in research activities related to plasmonic optical fiber sensing. Some will contribute to the development of even more robust industrial sensors that can meet requirements in smart sensing or POC.

References

[1] Soler, M., C. S. Huertas, and L. M. Lechuga, "Label-Free Plasmonic Biosensors for Point-of-Care Diagnostics: A Review," *Expert Rev. Mol. Diagn.*, Vol. 19, 2019, pp. 71–81.

[2] Brolo, A. G., "Plasmonics for Future Biosensors," *Nat. Photonics*, Vol. 6, No. 11, 2012, pp. 709–713.

[3] Uelzen, T., and J. Müller, "Wettability Enhancement by Rough Surfaces Generated by Thin Film Technology," *Thin Solid Films*, Vol. 434, No. 1–2, 2003, pp. 311–315.

[4] Lemasters, R., et al., "Ultrathin Wetting Layer-Free Plasmonic Gold Films," *ACS Photonics*, Vol. 6, No. 11, 2019, pp. 2600–2606.

[5] Antohe, I., et al., "Chemical Thermal Annealing of Gold Coated Fiber Optic Surfaces for Improved Plasmonic Biosensing," *Sensors Actuators B. Chem.*, Vol. 229, 2016, pp. 678–685.

[6] Sackmann, E. K., A. L. Fulton, and D. J. Beebe, "The Present and Future Role of Microfluidics in Biomedical Research," *Nature*, Vol. 507, No. 7491, 2014, pp. 181–189.

[7] Kim, J., "Joining Plasmonics with Microfluidics: From Convenience to Inevitability," *Lab Chip*, Vol. 12, No. 19, 2012, pp. 3611–3623.

[8] Raj M, K., and S. Chakraborty, "PDMS Microfluidics: A Mini Review," *J. Appl. Polym. Sci.*, Vol. 137, No. 27, 2020, pp. 1–14.

[9] Lobry, M., et al., "Plasmonic Fiber Grating Biosensors Demodulated Through Spectral Envelopes Intersection," *J. Light. Technol.*, Vol. 39, No. 22, 2021, pp. 7288–7295.

[10] Nielsen, A. V., et al., "3D Printed Microfluidics," *Annu. Rev. Anal. Chem.*, Vol. 13, 2020, pp. 45–65.

[11] Ho, C. M. B., et al., "3D Printed Microfluidics for Biological Applications," *Lab Chip*, Vol. 15, No. 18, 2015, pp. 3627–3637.

[12] Achille, C.,et al., "3D Printing of Monolithic Capillarity-Driven Microfluidic Devices for Diagnostics," *Adv. Mater.*, Vol. 33, No. 25, 2021, pp. 1–7.

[13] Dal Dosso, F., et al., "Creasensor: SIMPLE Technology for Creatinine Detection in Plasma," *Anal. Chim. Acta*, Vol. 1000, 2018, pp. 191–198.

[14] Gervais, L., and E. Delamarche, "Toward One-Step Point-of-Care Immunodiagnostics Using Capillary-Driven Microfluidics and PDMS Substrates," *Lab Chip*, Vol. 9, No. 23, 2009, pp. 3330–3337.

[15] Loyez, M., et al., "Overview and Emerging Trends in Optical Fiber Aptasensing," *Biosens. Bioelectron.*, Vol. 196, January 15, 2022.

[16] Kalli, K., et al., "Flat Fibre and Femtosecond Laser Technology as a Novel Photonic Integration Platform for Optofluidic Based Biosensing Devices and Lab-on-Chip Applications: Current Results and Future Perspectives," *Sensors Actuators B Chem.*, Vol. 209, 2015, pp. 1030–1040.

[17] Geng, Z., et al., "A Route to Low-Cost Nanoplasmonic Biosensor Integrated with Optofluidic-Portable Platform," *Sensors Actuators B Chem.*, Vol. 195, 2014, pp. 682–691.

[18] Wu, J., et al., "Emerging Optofluidic Technologies for Biodiagnostic Applications," *View*, Vol. 2, No. 4, August 2021, p. 20200035.

[19] Li, W., et al., "Integrated Multichannel All-Fiber Optofluidic Biosensing Platform for Sensitive and Simultaneous Detection of Trace Analytes," *Anal. Chim. Acta*, Vol. 1040, 2018, pp. 112–119.

[20] Gao, R., et al., "Fiber Optofluidic Biosensor for the Label-Free Detection of DNA Hybridization and Methylation Based on an In-Line Tunable Mode Coupler," *Biosens. Bioelectron.*, Vol. 86, 2016, pp. 321–329.

[21] Mao, J., et al., "Nanomaterial-Enhanced Fiber Optofluidic Laser Biosensor for Sensitive Enzyme Detection," *J. Light. Technol.*, Vol. 38, No. 18, 2020, pp. 5205–5211.

[22] Xu, W., et al., "Development of a Novel Label-Free All-Fiber Optofluidic Biosensor Based on Fresnel Reflection and Its Applications," *Anal. Chim. Acta*, Vol. 1181, 2021.

[23] He, Z., et al., "Long-Period Gratings in Photonic Crystal Fiber as an Optofluidic Label-Free Biosensor," *Biosens. Bioelectron.*, Vol. 26, No. 12, 2011, pp. 4774–4778.

[24] Ranamukhaarachchi, S. A., et al., "Integrated Hollow Microneedle-Optofluidic Biosensor for Therapeutic Drug Monitoring in Sub-Nanoliter Volumes," *Sci. Rep.*, Vol. 6, April 2016, pp. 1–10.

[25] Gao, R., et al., "Dual-Optofluidic Waveguide In-Line Fiber Biosensor for Real-Time Label-Free Detection of Interferon-Gamma with Temperature Compensation," *Opt. Express*, Vol. 28, No. 7, 2020, p. 10491.

[26] Yang, X., et al., "A Sequentially Bioconjugated Optofluidic Laser for Wash-Out-Free and Rapid Biomolecular Detection," *Lab Chip*, Vol. 21, No. 9, 2021, pp. 1686–1693.

[27] Gao, R., et al., "Optofluidic Immunosensor Based on Resonant Wavelength Shift of a Hollow Core Fiber for Ultratrace Detection of Carcinogenic Benzo[a]pyrene," *ACS Photonics*, Vol. 5, No. 4, 2018, pp. 1273–1280.

[28] Hu, X., et al., "Chemical In-Fiber Optofluidic Michelson Interferometer for Detecting Small Volume and Low Concentration Chemicals with a Fiber Ring Cavity Laser," *Sensors Actuators B Chem.*, Vol. 370, November 1, 2022, p. 132467.

[29] Bertucci, A., et al., "Detection of Unamplified Genomic DNA by a PNA-Based Microstructured Optical Fiber (MOF) Bragg-Grating Optofluidic System," *Biosens. Bioelectron.*, Vol. 63, 2015, pp. 248–254.

[30] Liang, L., et al., "Interferometric Detection of MicroRNAs Using a Capillary Optofluidic Sensor," *Sensors Actuators B Chem.*, Vol. 242, 2017, pp. 999–1006.

[31] Zhou, Y., et al., "Simple, Rapid, and Sensitive On-Site Detection of Hg2+ in Water Samples Through Combining Portable Evanescent Wave Optofluidic Biosensor and Fluorescence Resonance Energy Transfer Principle," *Anal. Chim. Acta*, Vol. 1155, 2021, p. 338351.

[32] Cheng, Y., et al., "Reusable Smartphone-Facilitated Mobile Fluorescence Biosensor for Rapid and Sensitive On-Site Quantitative Detection of Trace Pollutants," *Biosens. Bioelectron.*, Vol. 199, March 1, 2022, p. 113863.

[33] Dutta, S., "Point of Care Sensing and Biosensing Using Ambient Light Sensor of Smartphone: Critical Review," *TrAC—Trends Anal. Chem.*, Vol. 110, 2019, pp. 393–400.

[34] Pereira, S. O., et al., "Cost-Effective Fiber Optic Solutions for Biosensing," *Biosensors*, Vol. 12, 2022, p. 575.

[35] Geng, Z., et al., "Recent Progress in Optical Biosensors Based on Smartphone Platforms," *Sensors*, Vol. 17, 2017, p. 2449.

[36] Jain, S., et al., "Smartphone Integrated Handheld Long Range Surface Plasmon Resonance Based Fiber-Optic Biosensor with Tunable SiO_2 Sensing Matrix," *Biosens. Bioelectron.*, Vol. 201, April 1, 2022, p. 113919.

[37] Bremer, K., and B. Roth, "Fibre Optic Surface Plasmon Resonance Sensor System Designed for Smartphones," *Opt. Express*, Vol. 23, No. 13, 2015, p. 17179.

[38] Singh, G. P., and N. Sardana, "Smartphone-Based Surface Plasmon Resonance Sensors: A Review," *Plasmonics*, Vol. 17, No. 5, 2022, pp. 1869–1888.

[39] Chen, Y. T., et al., "Review of Integrated Optical Biosensors for Point-of-Care Applications," *Biosensors*, Vol. 10, No. 12, 2020, pp. 1–22.

[40] Gupta, B. D., A. Pathak, and A. M. Shrivastav, "Optical Biomedical Diagnostics Using Lab-on-Fiber Technology: A Review," *Photonics*, Vol. 9, No. 2, 2022.

[41] Estevez, M. C., M. Alvarez, and L. M. Lechuga, "Integrated Optical Devices for Lab-on-a-Chip Biosensing Applications," *Laser Photonics Rev.*, Vol. 6, No. 4, 2012, pp. 463–487.

[42] Ricciardi, A., et al., "Lab-on-Fiber Technology: A New Vision for Chemical and Biological Sensing," *Analyst*, Vol. 140, No. 24, 2015, pp. 8068–8079.

[43] Markvart, A., et al., "Continuous Hue-Based Self-Calibration of a Smartphone Spectrometer Applied to Optical Fiber Fabry-Perot Sensor Interrogation," *Sensors (Switzerland)*, Vol. 20, No. 21, 2020, pp. 1–16.

[44] Zhu, H., et al., "Optofluidic Fluorescent Imaging Cytometry on a Cell Phone," *Anal. Chem.*, Vol. 83, No. 17, 2011, pp. 6641–6647.

[45] Liu, Y., et al., "Surface Plasmon Resonance Biosensor Based on Smart Phone Platforms," *Sci. Rep.*, Vol. 5, 2015, pp. 1–9.

[46] Long, K. D., et al., "Multimode Smartphone Biosensing: The Transmission, Reflection, and Intensity Spectral (TRI)-Analyzer," *Lab Chip*, Vol. 17, No. 19, 2017, pp. 3246–3257.

[47] Liu, T., et al., "Quantitative Remote and On-Site Hg^{2+} Detection Using the Handheld Smartphone Based Optical Fiber Fluorescence Sensor (SOFFS)," *Sensors Actuators B Chem.*, Vol. 301, December 12, 2019.

[48] Fu, Q., et al., "A Portable Smart Phone-Based Plasmonic Nanosensor Readout Platform That Measures Transmitted Light Intensities of Nanosubstrates Using an Ambient Light Sensor," *Lab Chip*, Vol. 16, No. 10, 2016, pp. 1927–1933.

[49] Stolov, A. A., et al., "Effects of Sterilization on Optical and Mechanical Reliability of Specialty Optical Fibers and Terminations," *Proc. SPIE, Optical Fibers and Sensors for Medical Diagnostics and Treatment Applications XIV*, February 20, 2014.

[50] Stolov, A. A., et al., "Effects of E-Beam and Gamma Sterilization on Attenuation of Selected Single-Mode and Mutimode Optical Fibers," *Proc. SPIE*, Vol. 10872, February 2019.

[51] Loke, G., et al., "Digital Electronics in Fibres Enable Fabric-Based Machine-Learning Inference," *Nat. Commun.*, Vol. 12, No. 1, 2021, pp. 1–9.

[52] Yan, W., et al., "Single Fibre Enables Acoustic Fabrics Via Nanometre-Scale Vibrations," *Nature*, Vol. 603, No. 7902, 2022, pp. 616–623.

[53] Ibrahim, M. K., S. A. Kadhim, and N. I. Fawaz, "Water Pollution Fiber Sensor Based on Surface Plasmon Resonance Technique; Implementation and Characterization," *J. Phys. Conf. Ser.*, Vol. 1963, No. 1, 2021.

[54] Samavati, Z., et al., "Optical Fiber Sensor Based on Magneto-Plasmonic Features of Ag-Co Nanostructure for ppm Ammonium Detection in Aqueous Solutions," *Opt. Fiber Technol.*, Vol. 67, December 2021.

[55] Tsao, Y. -C., et al., "An In-Situ Real-Time Optical Fiber Sensor Based on Surface Plasmon Resonance for Monitoring the Growth of TiO2 Thin Films," *Sensors*, Vol. 13, No. 7, 2013, pp. 9513–9521.

[56] Lao, J., et al., "In Situ Plasmonic Optical Fiber Detection of the State of Charge of Supercapacitors for Renewable Energy Storage," *Light Sci. Appl.*, Vol. 7, No. 1, 2018.

[57] Qian, S., et al., "Plasmonic Fiber-Optic Sensing System for In Situ Monitoring the Capacitance and Temperature of Supercapacitors," *Opt. Express*, Vol. 30, No. 15, 2022, pp. 27322–27332.

[58] Wang, R., et al., "Operando Monitoring of Ion Activities in Aqueous Batteries with Plasmonic Fiber-Optic Sensors," *Nat. Commun.*, Vol. 13, No. 1, 2022, pp. 1–11.

[59] Prado, A. R., et al., "Surface Plasmon Resonance-Based Optical Fiber Sensors for H2S In Situ Detection," *Plasmonics*, Vol. 16, No. 3, 2021, pp. 787–797.

[60] LeCun, Y., Y. Bengio, and G. Hinton, "Deep Learning," *Nature*, Vol. 521, No. 7553, 2015, pp. 436–444.

[61] Venketeswaran, A., et al., "Recent Advances in Machine Learning for Fiber Optic Sensor Applications," *Adv. Intell. Syst.*, Vol. 4, No. 1, 2022, p. 2100067.

[62] Altman, N. S., "An Introduction to Kernel and Nearest-Neighbor Nonparametric Regression," *Am. Stat.*, Vol. 46, No. 3, 1992, pp. 175–185.

[63] Friedman, J. H., "Greedy Function Approximation: A Gradient Boosting Machine," *Ann. Stat.*, Vol. 29, No. 5, 2001, pp. 1189–1232.

[64] Gorban, A. N., et al., *Principal Manifolds for Data Visualization and Dimension Reduction*, New York: Springer, 2008.

[65] Chubchev, E. D., et al., "Machine Learning Approach to Data Processing of TFBG-Assisted SPR Sensors," *J. Light. Technol.*, Vol. 40, No. 9, 2022, pp. 3046–3054.

[66] Cao, Z., et al., "Improved Spectral Interrogation of Tilted Fiber Bragg Grating Refractometer Using Residual Convolutional Neural Networks," *J. Light. Technol.*, 2022, pp. 1–9.

[67] Wang, D., et al., "Recent Advances in Surface Plasmon Resonance Imaging Sensors," *Sensors*, Vol. 19, No. 6, 2019.

[68] Lee, H. J., D. Nedelkov, and R. M. Corn, "Surface Plasmon Resonance Imaging Measurements of Antibody Arrays for the Multiplexed Detection of Low Molecular Weight Protein Biomarkers," *Anal. Chem.*, Vol. 78, No. 18, 2006, pp. 6504–6510.

[69] Nguyen, H. H., et al., "Surface Plasmon Resonance: A Versatile Technique for Biosensor Applications," *Sensors (Basel)*, Vol. 15, No. 5, 2015, pp. 10481–10510.

[70] Peng, W., et al., "Compact Surface Plasmon Resonance Imaging Sensing System Based on General Optoelectronic Components," *Opt. Express*, Vol. 22, No. 5, 2014, pp. 6174–6185.

CHAPTER 11

Conclusion

As highlighted in previous chapters, plasmonic sensing with fiber optics is a transposition of physical concepts developed in bulk optics, essentially with the Kretschmann prism configuration. Optical fibers bring numerous practical assets, including straightforward light injection with low loss, remote operation, and *in situ* operation.

As of this writing, the vast majority of commercial products exploiting the surface plasmon resonance phenomena encompasses the Kretschmann prism, which is considered the true cornerstone of such devices. This genericity strongly contrasts with the plurality of configurations enabled by optical fibers, depending not only on the fiber type (multimode, single-mode, specialty) but also on the chosen architecture (unclad/etched, side-polished, U-bent, fiber grating). Chapters 3 to 5 attest to that important richness. Hence, the scientific community active on plasmonic optical fiber sensing may be seen as divided into as many expert groups as existing configurations. This coexistence of competing configurations is particularly intriguing as each one comes with its own practical advantages and limitations. Some are easier to produce, others can be developed without important investment, and others are more sensitive or temperature-independent. It is trivial to write that none of the developed optical fiber platforms outperforms all others. However, some are much more advanced than others in terms of the technology readiness level. Unclad multimode optical fibers have already succeeded the transition between academic research efforts and industrial developments. Other configurations such as tilted fiber Bragg gratings have rapidly progressed over the past decade. Considering that physical sensors based on the fiber Bragg grating technology are the core business of many companies worldwide, plasmonic optical fiber grating sensors are certainly promised to a bright future, considering all research efforts that have been accumulated to date. If not already in place, a transition between academicians and industrials should take place to catalyze this integration process. It is therefore hoped that the quite exhaustive progress review made in this book will be useful to foster this transition.

In the race to develop and create ambitious and novel biosensors, we believe that there is a need to look back to what has been done over the last years. The development of plasmonic sensors with different materials, patterns, and particles; the integration of microfluidics, imaging, and electrochemistry; the detection of single cells or single molecules; their use for detection in the environment, in human samples or in organs; the miniaturization and their transition to point-of-care using smartphones or portable interfaces; and the use of antibodies, DNA or RNA, and synthetic molecules. All these aspects were therefore previously covered.

Most technologies highlighted in this book are often at an early research stage and need more experimental runs. Their improvement could lead to large-scale and breakthrough commercial biosensors such as for those slowly arriving on the market. The combinations of new paradigms and detection strategies that fulfill real expectations are inspiring and will help to move the field forward in this direction. From a fundamental point of view, the achievement of new surface immobilizations and molecular blocking techniques would allow the design of new testing configurations. Moreover, each innovation in these areas (both optical and biological) will not be limited to plasmonic sensors or optical fiber-based sensors but they will open up new research amenities in several domains. We also believe that it is required to bring a higher critical mind to the technology, which deserves more open discussion and transparency. For all these reasons, we also hope that this book will give the keys to open new research paths. Biosensors, the many hidden satellite topics, and applicative ambitions beyond this technology are our driving forces.

Going back to the ever-growing research community active in plasmonic optical fiber sensing, it is obvious that it gathers the skills of various scientific profiles. Scientists with a background or expertise in photonics, materials science, engineering or bioengineering, or chemistry or biochemistry largely contribute to the continuous progress. During their formation or before starting their activities, these people need to be trained and, depending on their background, the training will be focused on one discipline or another. We have made editorial choices in our book content with this in mind, hoping to build a book with content as attractive and pedagogical as possible, taking into account the pluridisciplinarity of plasmonic optical fiber sensors. As we cannot pretend to be experts in all these aspects, we preferred to detail certain sections to a greater or lesser extent to provide the most relevant information based on our own expertise and knowledge as objectively as possible. To keep the originality and distance regarding what has already been achieved in previous works and books, we decided to describe a series of topics only briefly, for which relevant literature references are cited. We hope that this strategy to focus on certain hot spots will help to find inspiration and to build new and innovative research avenues in this field, without cluttering the reading with too many details.

Last but not least, we wanted to conclude on a more personal note. The writing of this book has been a long and intense journey, which allowed us to further improve our basic knowledge and broaden our vision of the field. If you read these lines, you have probably gone through the entire book or the most relevant parts depending on your actual needs. We sincerely hope that the information provided in this book has met your expectations. Be ensured that we remain open to all comments, criticism, suggestions, and ideas, and we therefore invite you to provide some feedback. Let us all contribute to the expected progress of industrialized plasmonic optical fiber sensors.

Acronyms and Abbreviations

AFM Atomic force microscopy

AFP Alpha fetoprotein

Ag Silver

AI Artificial intelligence

AIBN Azobis (isobutyronitrile)

Al Aluminum

Al_2O_3 Aluminum oxide

AmpC β-lactamase

APC Antigen-presenting cells

APTES (3-Aminopropyl)triethoxysilane

APTMS (3-Aminopropyl)trimethoxysilane

Ara h1 Peanut allergen

Au Gold

AuNPs/GNPs Gold nanoparticles

B_2O_3 Boron trioxide/diboron trioxide

BRET Bioluminescence resonance energy transfer

BSA Bovine serum albumin

CCD Charge-coupled device

CD44 Cell surface glycoprotein

CdSe Cadmium selenide

CdTe Cadmium telluride

CE Counter electrode

CEA Carcinoembryonic antigen

CFU Colony forming unit

CGM Continuous glucose monitoring

CK Cytokeratin

CoA Coenzyme A

CRP C-reactive protein

CS Chitosan

Ctn-I Cardiac Troponin I

Cu Copper
CV Cyclic voltammetry
CW Continuous wave
CymMVV Cymbidium mosaic virus
CYTOP Amorphous fluorinated polymer
DBDS Dibenzyl disulfide
DCA Dichloroethane
DGD Differential group delay
DMSO Dimethyl sulfoxide
DNA Deoxyribonucleic acid
EBL Electron beam lithography
EDC 1-Ethyl-3-(3-dimethylaminopropyl)carbodiimide
EDFA Erbium-doped fiber amplifier
EGDMA Ethylene glycol dimethacrylate
EH Hybrid modes in which the TM component dominates
ELISA Enzyme-linked immunosorbent assay
EM Electromagnetic
EP Entire part
Ex-TFG Extremely tilted fiber grating
FAD Flavin adenine dinucleotide
FBG Fiber Bragg grating
Fc Fragment crystallizable region (antibodies)
FDA U.S. Food & Drug Administration
FEM Finite element method
FOM Figure of merit
FRET Förster (or fluorescence) resonance energy transfer
FTHH Fiber-to-the-home concept
FTIR Fourier transform infrared spectroscopy
FWHM Full width at half-maximum
GeO_2 Germanium dioxide
GFP Green fluorescent protein
GO Glucose oxidase
H_2O_2 Hydrogen peroxide
H_2S Hydrogen sulfide/sulfane
H_2SO_4 Sulfuric acid
HBsAg Hepatitis B antigen
HCAb Heavy chain antibody
hCG Human chorionic gonadotrophin hormone

HE Hybrid modes in which the TE component dominates
HeCd Helium cadmium
HER2/ErbB2 Human epidermal growth factor receptor 2
Hf Hafnium
HF Hydrofluoric acid
hIg Human immunoglobulin
HTR Human serum transferrin
IA Immunoassay
ICP-MS Plasma atomic mass spectroscopy
IFX Infliximab
Ig Immunoglobulin
IIoT Industrial Internet of Things
InP Indium phosphide
IoT Internet of Things
ISF Interstitial fluid
ISPR Inverted surface plasmon resonance
ITO Indium tin oxide
IUPAC International Union of Pure and Applied Chemistry
Ka Association constant
Kd Dissociation constant
Km Michaelis constant
KOH Potassium hydroxide base
KrF Krypton fluoride
LbL Layer-by-layer
LED Light-emitting diode
LFA Lateral flow assay
LiCl Lithium chloride
LLOQ Lower limit of quantification
LNA Locked nucleic acid
LOD Limit of detection
LOQ Limit of quantification
LP Linearly polarized
LPFG (or LPG) Long period fiber grating
LRSPP Long-range surface plasmon polariton
LSPR Localized surface plasmon resonance
MDC Metal dichalcogenide
MET Mediated electron transfer
MHC Major histocompatibility complex

MIP Molecularly imprinted polymers
miRNA Micro ribonucleic acid
MMF Multimode optical fiber
MOF Microstructured optical fiber
mPOF Microstructured polymer optical fiber
mRNA Messenger ribonucleic acid
MS Mass spectrometry
MUA Mercapoundecanoic acid
N-BK7 Name of a pure silica made by Schott
NA Numerical aperture
NAD+ Nicotinamide adenine dinucleotide
NaOH Sodium hydroxide base
Nb Nanobody
ncRNA Noncoding ribonucleic acid
NDV Newcastle Disease Virus
NEP Noise equivalent pressure
NHS N-Hydroxysuccinimide
NPs Nanoparticles
OF Optical fiber
OI-RD Oblique-incidence reflectivity-difference
P_2O_5 Phosphorus pentoxide
p53 Tumor suppressor protein
PAE Phthalate ester
PANI Polyaniline
PBS Phosphate buffer saline
PC Polycarbonate
PCF Photonic crystal fiber
PCR Polymerase chain reaction
Pd Palladium
PDMS Polydimethylsiloxane
PEDOT Poly(3,4-ethylenedioxythiophene)
PFA Perfluorinated alkylated substances
***Pf*GDH** Plasmodium falciparum glutamate dehydrogenase
***Pf*HRP2** Plasmodium falciparum histidine rich protein 2
PMD Polarization mode dispersion
PMF Polarization maintaining fiber
PMMA Poly(methylmetacrylate)
PMT Photomultiplier tube

PNA Peptide nucleic acid
POC Point-of-care
POF Polymer optical fiber
POFBG Polymer optical fiber Bragg grating
PS Polystyrene
PSA Prostate-specific antigen
Pt Platinum
PTO Phosphorothioate oligonucleotide
Q-factor Quality factor
QD Quantum dot
RE Reference electrode
RF Radio frequency
RI Refractive index
RIA Radioimmunoassay
RIU Refractive index unit
RNA Ribonucleic acid
RRS Resonance Raman spectroscopy
RT-PCR Reverse transcription polymerase chain reaction
SAM Self-assembled monolayer
scaRNA Small Cajal body-specific ribonucleic acid
SELEX Systematic Evolution of Ligands by Exponential Enrichment
SERS Surface-enhanced Raman spectroscopy
SEW Surface electromagnetic waves
SiO_2 Silicon dioxide (silica)
SMF Single-mode fiber
SMS Single-multi-single optical fiber
SNPs Silver nanoparticles
SNR Signal-to-noise ratio
SOP State of polarization
SP Surface plasmon
SPO Surface plasma oscillation
SPP Surface plasmons polariton
SPR Surface plasmon resonance
SPRi Surface plasmon resonance imaging
SPW Surface plasma waves
SRI Surrounding refractive index
SRSPP Short-range surface plasmon resonance
Sulfo-EMCS N-maleimidocaproyl-oxysulfosuccinimide ester

TAP Turn-around point

TE Transverse electric field

TECS Technology-enhanced clad silica

TEOS Tetrathoxysilanes

TFBG Tilted fiber Bragg grating

TH Transverse magnetic field

TiN Titanium nitride

TiO_2 Titanium dioxide

TM Transverse magnetic

TMOS Tetramethoxysilanes

TNF-α Tumor necrosis factor alpha (inflammatory cytokine)

TOPAS Cyclic Olefin Copolymer

TTF1 Thyroid transcription factor 1

UV/O_3 Ultraviolet/ozone

VCSEL Vertical cavity surface emitting laser

VHH Variable region of the heavy chain (antibody-antigen binding site)

VLP Virus-like particle

W Tungsten

WE Working electrode

WO_3 Tungsten trioxide

WS_2 Tungsten disulfide

XNA Synthetic nucleic acid

ZEONEX Cyclic Olefin Copolymer

ZnSe Zinc selenide

Zr Zirconium

ZrO_2 Zirconium dioxide

About the Authors

Christophe Caucheteur has received his master's degree in electrical engineering from the Faculté Polytechnique de Mons, Belgium, in 2003. In 2007, he was awarded a PhD in applied sciences from the same institution. He is now a senior research associate of the F.R.S.-FNRS and the head of the Advanced Photonic Sensors Unit and has a background in photonics. At the time of this writing, Professor Caucheteur has 20 years of expertise in optical fiber sensing, mainly based on the fiber Bragg grating technology. In 2016, he cofounded the spin-off company B-SENS, which develops and sells smart sensing solutions based on the fiber Bragg grating technology.

Médéric Loyez is a postdoctoral researcher who studied and developed plasmonic optical fiber biosensors while pursuing his PhD thesis. He has a background in biochemistry. After learning new techniques in biological, chemical, and photonics teams in Canada and the United States during a year of postdoctoral work, he is currently working as a research fellow of the F.R.S-FNRS in the laboratory of Professor Ruddy Wattiez at the Faculty of Science and in the Advanced Photonic Sensors Unit of Professor Caucheteur.

As these authors were two colleagues from the University of Mons in Belgium, they believe that their backgrounds, profiles, and expertise are highly complementary and essential in the multidisciplinary context of plasmonic biosensors.

Index

Artech House Applied Photonics Library

Brian Culshaw and Marc Wuilpart, Series Editors

The ABCs of Fiber Optic Communication, Sudhir Warier

Acousto-Optic Signal Processing: Fundamentals and Applications, Pankaj K. Das and Casimer De Cusatis

Advanced Optical Communication Systems and Networks, Milorad Cvijetic and Ivan B. Djordjevic

Applications of Modern RF Photonics, Preetpaul Singh Devgan

Bistabilities and Nonlinearities in Laser Diodes, Hitoshi Kawaguchi

Chemical and Biochemical Sensing with Optical Fibers and Waveguides, Gilbert Boisdé and Alan Harmer

Coherent and Nonlinear Lightwave Communications, Milorad Cvijetic

Coherent Doppler Wind Lidars in a Turbulent Atmosphere, Viktor Banakh and Igor Smalikho

Coherent Lightwave Communication Systems, Shiro Ryu

Digital Optical Measurement Techniques and Applications, Pramond Rastogi, Editor

DWDM Fundamentals, Components, and Applications, Jean-Pierre Laude

Electromagnetic and Photonic Simulation for the Beginner: Finite-Difference Frequency-Domain in MATLAB®, Raymond C. Rumpf

Electro-Optical Systems Performance Modeling, Gary Waldman

Elliptical Fiber Waveguides, R. B. Dyott

Engineering Optical Networks, Sudhir Warier

Fiber Bragg Gratings: Fundamentals and Applications in Telecommunications and Sensing, Andrea Othonos and Kyriacos Kalli

Fiber Optics Communications, Henry F. Taylor

The Fiber-Optic Gyroscope, Third Edition, Hervé C. Lefèvre

Fiber-Optic Sensors for Biomedical Applications, Daniele Tosi and Guido Perrone

Field Theory of Acousto-Optic Signal Processing Devices, Craig R. Scott

Finite Element Modeling Methods for Photonics, B. M. Azizur Rahman and Arti Agrawal

Frequency Stabilization of Semiconductor Laser Diodes, Tetsuhiko Ikegami, Shoichi Sudo, and Yoshihisa Sakai

Fundamentals of Multiaccess Optical Fiber Networks, Dennis Mestdagh

Germanate Glasses: Structure Spectroscopy and Properties, Alfred Margaryan and Michael Piliavin

Handbook of Distributed Feedback Laser Diodes, Second Edition, Geert Morthier and Patrick Vankwikelberge

Highly Coherent Semiconductor Lasers, Motoichi Ohtsu

High-Power Optically Activated Solid-State Switches, Arye Rosen, Fred Zutavern

Integrated Optics: Design and Modeling, Reinhard Marz

Introduction to Electro-Optical Imaging and Tracking Systems, Shaheen A. Hovanessian and Khalil Seyrafi

Introduction to Glass Integrated Optics, S. Iraj Najafi

Introduction to Infrared and Electro-Optical Systems, Third Edition, Ronald G. Driggers, Melvin H. Friedman, John W. Devitt, Orges Furxhi and Anjali Singh

Introduction to Radiometry and Photometry, Second Edition, William Ross McCluney

Introduction to Semiconductor Integrated Optics, Hans P. Zappe

Laser Communications in Space, Stephen Lambert

Liquid Crystal Devices: Physics and Applications, Vladimir G. Chigrinov

Lithium Niobate Photonics, James E. Toney

Military Laser Technology and Systems, David H. Titterton

Natural Photonics and Bioinspiration, Sébastien R. Mouchet and Olivier Deparis

New Photonics Technologies for the Information Age: The Dream of Ubiquitous Services, Shoichi Sudo and Katsunari Okamoto, editors

Optical Document Security, Third Edition, Rudolf L. van Renesse

Optical FDM Network Technologies, Kiyoshi Nosu

Optical Fiber Amplifiers: Materials, Devices, and Applications, Shoichi Sudo, editor

Optical Fiber Communication Systems, Leonid Kazovsky, Sergio Benedetto and Alan Willner

Optical Fiber Sensors, Volume One: Principles and Components, Brian Culshaw and John P. Dakin

Optical Fiber Sensors, Volume Two: Systems and Applications, Brian Culshaw and John P. Dakin

Optical Fiber Sensors, Volume Three: Components and Subsystems, John Dakin and Brian Culshaw, editors

Optical Fiber Sensors, Volume Four: Applications, Analysis, and Future Trends, John Dakin and Brian Culshaw, editors

Optical Interconnection: Foundations and Applications, John Caulfied and Christopher S. Tocci

Optical Measurement Techniques and Applications, Pramod Rastogi

Optical Network Theory, Yitzhak Weissman

Optical Transmission for the Subscriber Loop, Norio Kashima

Optical Transmission Systems Engineering, Milorad Cvijetic

Optoelectronics for Low-Intensity Conflicts and Homeland Security, Anil K. Maini

Optoelectronic Techniques for Microwave and Millimeter-Wave Engineering, William M. Robertson

Passive Optical Components for Optical Fiber Transmission, Norio Kashima

Photonic Applications for Radio Systems and Networks, Fabio Cavaliere and Antonio D'Errico

Plasmonic Optical Fiber Biosensors, Christophe Caucheteur and Médéric Loyez

Polarization in Optical Fibers, Alan Rogers

Principles of Modern Optical Systems, Volume 1, Deepak Uttamchandani and Ivan Andonovic, editors

Principles of Modern Optical Systems, Volume 2, Deepak Uttamchandani and Ivan Andonovic, editors

Reliability and Degradation of III-V Optical Devices, Osamu Ueda

Securing Information and Communications Systems, Stephen Furnell, Sokratis Katsikas, Javier Lopez, and Ahmed Patel

Semiconductor Raman Lasers, Ken Suto and Jun-Ichi Nishizawa

Semiconductors for Solar Cells, Hans Joachim Moller

Semiconductor Quantum Wells and Superlattices for Long-Wavelength Infrared Detectors, M. Omar Manasreh

Signal Processing and Performance Analysis for Imaging Systems, S. Susan Young, Ronald G. Driggers, and Eddie L. Jacobs

Single-Mode Optical Fiber Measurement: Characterization and Sensing, Giovanni Cancellieri and Franco Chiaraluce

Smart Structures and Materials, Brian Culshaw

Substrate Surface Preparation Handbook, Max Robertson

Surveillance and Reconnaissance Imaging Systems: Modeling and Performance Prediction, Jon C. Leachtenauer and Ronald G. Driggers

Tunable Laser Diodes, Markus-Christian Amann and Jens Buus

Ultrafast Diode Lasers: Fundamentals and Applications, Peter Vasilev

Understanding Optical Fiber Communications, Alan Rogers

Wavelength Division Multiple Access Optical Networks, Andrea Borella, Giovanni Cancellieri and Franco Chiaraluce

*Wired and Wireless Seamless Access Systems for Public Infrastructur*e, Tetsuya Kawanishi

For further information on these and other Artech House titles, including previously considered out-of-print books now available through our In-Print-Forever® (IPF®) program, contact:

Artech House	Artech House
685 Canton Street	16 Sussex Street
Norwood, MA 02062	London SW1V 4RW UK
Phone: 781-769-9750	Phone: +44 (0)20 7596-8750
Fax: 781-769-6334	Fax: +44 (0)20 7630-0166
e-mail: artech@artechhouse.com	e-mail: artech-uk@artechhouse.com

Find us on the World Wide Web at: www.artechhouse.com